Hydrogen Recycling at Plasma Facing Materials

NATO Science Series

A Series presenting the results of scientific meetings supported under the NATO Science Programme.

The Series is published by IOS Press, Amsterdam, and Kluwer Academic Publishers in conjunction with the NATO Scientific Affairs Division

Sub-Series

I. Life and Behavioural Sciences	IOS Press
II. Mathematics, Physics and Chemistry	Kluwer Academic Publishers
III. Computer and Systems Science	IOS Press
IV. Earth and Environmental Sciences	Kluwer Academic Publishers

The NATO Science Series continues the series of books published formerly as the NATO ASI Series.

The NATO Science Programme offers support for collaboration in civil science between scientists of countries of the Euro-Atlantic Partnership Council. The types of scientific meeting generally supported are "Advanced Study Institutes" and "Advanced Research Workshops", and the NATO Science Series collects together the results of these meetings. The meetings are co-organized bij scientists from NATO countries and scientists from NATO's Partner countries – countries of the CIS and Central and Eastern Europe.

Advanced Study Institutes are high-level tutorial courses offering in-depth study of latest advances in a field.
Advanced Research Workshops are expert meetings aimed at critical assessment of a field, and identification of directions for future action.

As a consequence of the restructuring of the NATO Science Programme in 1999, the NATO Science Series was re-organized to the four sub-series noted above. Please consult the following web sites for information on previous volumes published in the Series.

http://www.nato.int/science
http://www.wkap.nl
http://www.iospress.nl
http://www.wtv-books.de/nato-pco.htm

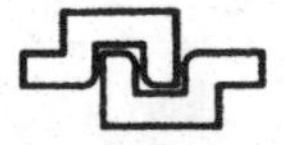

Series II: Mathematical and Physical Chemistry – Vol. 1

Hydrogen Recycling at Plasma Facing Materials

edited by

C.H. Wu

EFDA, Max-Planck-Institut für Plasmaphysik,
Garching, Germany

Kluwer Academic Publishers

Dordrecht / Boston / London

Published in cooperation with NATO Scientific Affairs Division

Proceedings of the NATO Advanced Research Workshop on
Hydrogen Recycling at Plasma Facing Materials
St. Petersburg, Russia
September 15-19 1999

A C.I.P. Catalogue record for this book is available from the Library of Congress.

ISBN 0-7923-6629-8 (HB)
ISBN 0-7923-6630-1 (PB)

Published by Kluwer Academic Publishers,
P.O. Box 17, 3300 AA Dordrecht, The Netherlands.

Sold and distributed in North, Central and South America
by Kluwer Academic Publishers,
101 Philip Drive, Norwell, MA 02061, U.S.A.

In all other countries, sold and distributed
by Kluwer Academic Publishers,
P.O. Box 322, 3300 AH Dordrecht, The Netherlands.

Printed on acid-free paper

Printed in the Netherlands.

CONTENTS

Hydrogen Transport in Materials

Hydrogen Retention and Release

Effects of Hydrogen, Helium and Plasma Impurities on PFM Behavior

Experimental Methods and Technique

PREFACE

HYDROGEN RECYCLE AT PLASMA FACING MATERIALS

Further R&D in the field of nuclear fusion aimed to give the humanity a new inexhaustible and ecologically satisfactory source of energy is supposed to be one of the main directions in applied physics and engineering in the beginning of next century.

The issue of adequate materials is one of the most important for the construction of future fusion machines with the magnetic plasma confinement. Recycle of hydrogen isotopes (fuel) at the plasma facing materials is, in turn, one of the key points at the proper material choice. In the case of actual machines operating with tritium, the points of safety and economy gain a primary importance, which demands a lowest possible tritium inventory and permeation in plasma facing materials. Besides, one should know the recycle behavior of the in-vessel components in order to make the plasma operation more predictable and controllable. Finally, special materials and methods may be developed to actively control the fuel recycle and the plasma operation regimes.

Hydrogen recycle problem becomes particularly complicated because of the materials interacting with the plasma. In particular, the chemical state of the surface becomes crucially important for the re-emission, absorption and permeation of hydrogen isotope particles under such conditions; therefore, these phenomena are strongly affected by the physico-chemical environment inside fusion machine, with the rates and parameters of the above processes hardly predictable in advance. The radiation damage and radiation induced processes (re-emission, diffusion, etc.) result in additional peculiarities in recycling at the plasma facing materials. The issue is further complicated by a number of limitations on the choice of materials specific for fusion machines, such as the necessity to use materials with a low induced radiation. Wall conditioning and using of the low Z material coatings for plasma facing components produces additional problems.

The Workshop was aimed at the discussion of both the laboratory experiments directed on investigations into the basic properties of non-equilibrium hydrogen – solid systems (diffusion, absorption, boundary processes) and of the experimental results obtained at existing fusion devices under conditions simulating to some extent the situation in the future fusion machines.

With account for an interdisciplinary character of the Workshop topicality, experts in the fields of nuclear fusion, material science, surface science, vacuum science and technology, and solid state physics, participated.

The workshop was founded by the scientists of Tokyo University, Japan and Bonch-Bruyevich State University, Russia in 1992 and was held every year.

Previous meeting in this series were :

Tokyo 1992, 1994, 1996, 1998,
St. Petersburg 1993,
Moscow 1995,1997.

Since that more then 300 experts in plasma physics, solid state physics and material engineering had an excellent opportunity to meet each other, to exchange views and to discuss the latest research results in the field of the plasma surface interaction relevant to hydrogen recycling in fusion devices.

A number of most important ideas and results in the field of controlled nuclear fusion were originated from former Soviet Union, and corresponding scientific communities and schools continue to exist, to generate new ideas and to carry out the original experimental investigations (though because of economical difficulties that are mainly laboratory experiments without using the big fusion devices). So Western colleagues are rather interested in the getting of these newest results. In a long term outlook NATO countries are interested in the involvement of a great intellectual potential of the partner countries in such having common to all mankind Projects as Thermonuclear Energy. One of very important point is the conversion of military organizations of former Soviet Union and an involvement of their specialists in the international scientific projects. In particular, International Scientific Technical Center (ISTC) is arranging such international projects. The results of such Projects (including operation with Tritium) are long term benefits for the NATO Advanced Research Workshop (ARW). The aim of this NATO Advanced Workshop was to give an opportunity for Scientist and engineers from NATO countries and NATO partner countries to come together and to discuss with Western (and Japanese) colleagues the newest results on the one of important engineering aspects of actual fusion devices and future thermonuclear power stations: recycling of Deuterium/Tritium particles.

This NATO Advanced Workshop has been held successfully from September 15-17, 1999 in Oak Room, St. Petersburg Scientists Club, St. Petersburg, Russia. We sincerely appreciated the strong support of NATO Scientific Program to this workshop .

C. H. Wu

Co-Director of the
NATO Advanced Research Workshop

A. Livshits

Co-Director of the
NATO Advanced Research Workshop

Density control and plasma wall interaction in Tore Supra

C Grisolia

Association EURATOM-CEA sur la fusion contrôlée
CEA CADARACHE, 13108 Saint Paul Lez Durance, FRance

I Introduction

In an actual tokamak, the discharge duration is much longer than the average particle confinement time, τ_p. Fuel particle must then leave and re-enter the plasma many times over the course of one discharge; this process is termed recycling. Fuelling and recycling determine the boundary conditions imposed on the hot plasma in the discharge centre (for deeper insight on recycling and plasma wall interaction see [1]).

Plasma wall interaction is a complex system. Plasma and walls are connected through the plasma edge and scrape-off layer where all the particle ionisation processes take place. The density and temperature in the boundary plasma determine the flux and energy of incident particles impinging the wall.

Recycling process from surfaces surrounding the plasma control, to some extent, the plasma conditions and the key of this control is the surface properties. For example, strong pumping by the walls (weak recycling) is thought to be linked to a higher confinement mode as H mode operation in axisymetric divertor machine [2] or supershot mode in TFTR [3], while strong local recycling affects the structure of the boundary plasma. Consequently, the fuel particle cycle is extremely complicated with plasma processes dependent on the wall processes and vice versa.

Control of recycling fluxes is of major importance especially during long discharge operation where density control is needed. In Tore Supra, long discharge plasmas are obtained with the help of Lower Hybrid Current Drive (LHCD) whose current efficiency decreases with density increase.

Recycling fluxes have to be controlled for many reasons in addition to density control. The influx of cold neutrals cools the plasma. If the recycling fluxes are too high, the energy used to ionise the incoming neutrals exceeds the plasma stored energy leading to plasma thermal collapse. High confinement plasmas require good control of particle sources around the plasma. For example in TFTR, the best plasma performances was achieved using lithium pellet injection which leads to a decreased wall outflux as seen on Hα. Recycling processes

C.H. Wu (ed.), Hydrogen Recycling at Plasma Facing Materials, 1–8.

are also important for tritium inventory, a crucial issue for existing and next step machine DT machines. Lastly, control of recycling fluxes is the key to control helium pumping which is the ash of future tritium reactor.

In the following, we will focus on the effect of wall properties on recycling during plasma operation. Different walls are in interaction with plasma leading to different consequences on density control.

II) density control during long discharge operation

In Tore Supra, 15 discharges longer than 60s have been performed[4]. They are realised at relatively low density ($<n_e>=10^{19}$ m^{-3}) with the help of a lower hybrid current drive. No external active pumping was used during these experiments. The maximum time duration of a plasma discharge was 120 s. An increase of plasma density (which accelerates in time) is often observed during long pulse operation (see fig. 1) .

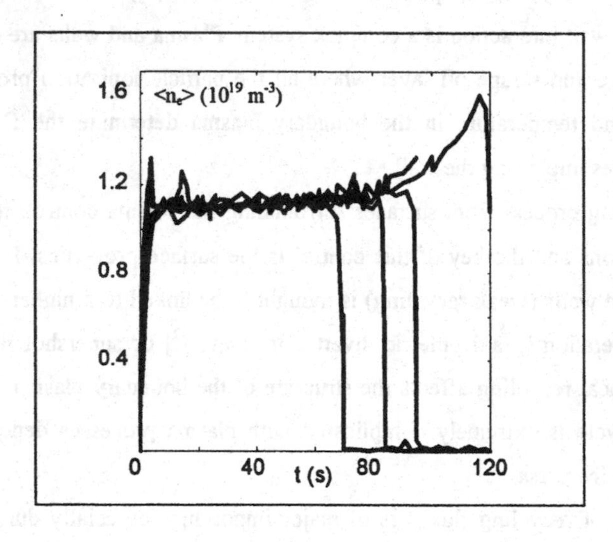

Fig. 1 : Volume average density evolution for very long discharges in Tore Supra.

The higher the injected and radiated power, the sooner the increase of density appears. This behaviour is correlated with an increase of the oxygen and hydrogen plasma density and it is attributed to water desorption induced by wall heating due to the plasma high radiated energy. The water originates from surfaces that are far from the plasma and are not baked during the conditioning procedures usually used at Tore Supra. When a series of long duration high energy discharges is realised, the density rise decreases from pulse to pulse showing evidence for a conditioning effect. However, this improvement is nullified if a disruption occurs after which the density increase regains the level observed at the beginning of the series.

To avoid such a density increase in long pulse operation, water outgassing must be controlled by an uniformly high baking temperature all around the vessel. The baking temperature should correspond to the highest temperature reached during plasma operation. It has to be pointed that walls are acting as high pumping capability devices for water. External pumping can be used to decrease in-vessel water inventory only if it is coupled with a high and homogeneous baking temperature. It has to be pointed that wall properties can play an important role on water trapping and additional work must be done to choose the best material as possible for these plasma remote surfaces.

Apart from the limitation on the heat exhaust capability on Tore Supra, the observed increase of the electron density during long pulse discharges is the main limitation preventing steady state operation. This underlines the importance of having an effective steady state particle exhaust scheme for long pulse operation. The Tore Supra Upgrade project (called CIEL) is designed to address this issue. A toroidal pump limiter will be installed the year 2000. It is designed to sustain a heat conducted flux up to $10MW/m^2$ and remove a total convective power of 15MW for a pulse duration of 1000 s. The calculated pumping efficiency of this limiter is of the order of 12% which corresponds to an extracted flux of $4Pam^3/s$. This capability is designed to be sufficient to ensure a good density control and to maintain the density profile (through pellet injection) needed to generate a significant bootstrap current. Around the vessel, stainless steel actively cooled wafer panels will remove a large amount of radiated power (up to 15 MW) enabling input powers in the range to 20-25MW. In spite of this high radiated energy, wall temperature will be constant during all the discharge to control water detrapping and therefore avoid any density excursion due to impurity release.

III) density control and saturation of walls in direct interaction with plasma.

Carbon materials exposed to plasma are covered by an amorphous carbon layer and they are characterised by their high capability to trap deuterium[[5],[6],[7]]. The total number of particle trapped is limited by a saturation value nominally 0.4 D/C but can be significantly higher or lower than this value [[8], [9], [10]]. This value is several orders of magnitude higher than that in a pure crystal (10^{-3}). Actually, D/C is not directly measured during a plasma discharge. But, it can be deduced from plasma behaviour and the wall saturation status is interpreted in term of D/C. When D/C is equal to its maximum value, no more particle can be trapped in the wall. This value depends strongly to the wall temperature and that at high temperatures can be much lower than 0.4. D/C can be adjust using conditioning procedures.

If D/C is less than the saturated value, carbon walls act as a particle sink for an impinging deuterium flux. The amount of gas recovered after a discharge is very small, around 20% of that input at carbon wall temperature of 170°C. This value depends on wall temperature and can be much smaller at lower temperature. For instance, particle recovered is 10 times greater for 170°C wall temperature than for 100°C (see figure 2).

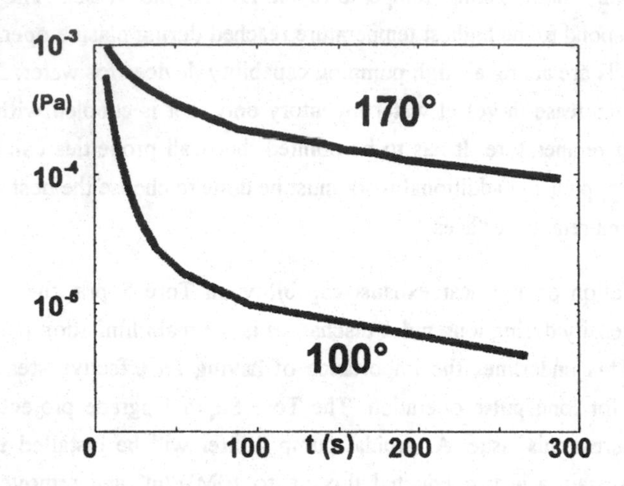

Figure 2 : time evolution of D2 pressure after a plasma D2 discharge for 2 Tore Supra wall temperatures.

D/C ratio increases toward saturation as gas is added during discharges and the wall pumping speed decreases due to particle trapping. Therefore, D/C reaches the saturation value and the resulting high recycling induce plasma thermal collapse[[11]].

The study of wall saturation is based on particle balance analysis where the wall and the plasma are considered as reservoirs exchanging particles[[12],[13],]. Particle balance consists of a careful estimation of the change in wall inventory due to particle trapping. At Tore Supra, a dedicated series of D2 ohmic shots with same plasma history were used to follow carbon wall saturation. However, as pointed previously in the introduction, particle wall trapping depends strongly on plasma properties. Therefore, it is not possible to extrapolated these results to long shot operation done with ICRH (Ion Cyclotron Resonance Heating) or LHCD heating.

The complexity of wall saturation properties appears in the following (see fig 3) where wall saturation is rapidly obtained at low edge temperature (Te~30eV) but wall desaturation is gained again just by using ICRH plasma where Te~300eV.

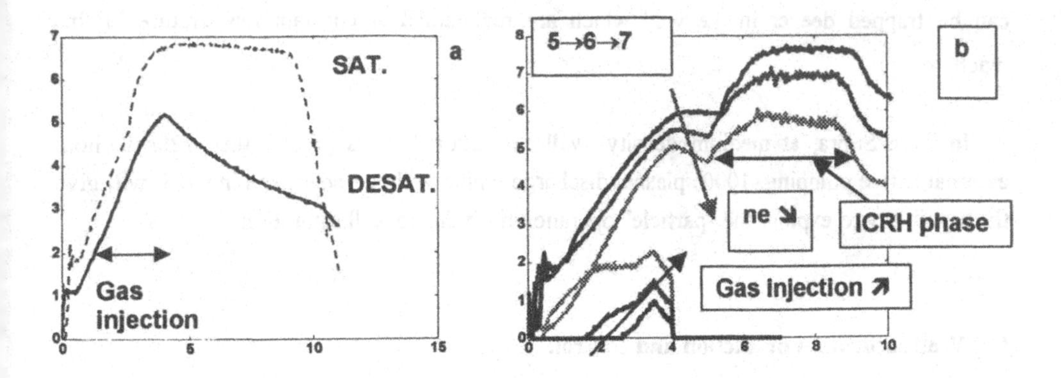

Figure 3 :

a) central linear density during 2 ohmic shots. First shot is preceding the ED shots, wall are desaturated (DESAT.) as shown by the rapid density decrease after gas injection. Second shot is following the ED shots, density stays constant after gas injection : wall is saturated (SAT.) .

b) central linear density and gas injection evolution during 3 ICRH discharges.

In this series of shots, wall saturation is rapidly obtained with ergodic divertor (ED) experiments where plasma edge temperature is low (30eV). For the same target density, gas injection is much larger in the ED configuration than for the limiter one and neutral pressure around the plasma is also higher (~ two orders of magnitude greater). This leads to much more rapid wall saturation during ED experiments. However, wall saturation is reached only close to the wall surface. This is illustrated in figure 3-a where 2 comparable ohmic plasmas are presented. The first one (DESAT.) is preceding the ED shots. The second one (SAT.) is following them. In the latter, a constant density behaviour after gas injection is got which is a clear indication of wall saturation.

Therefore, an effective wall "desaturation" can be easily obtained by raising the edge electron temperature (for example here with additional ICRH heating). During ICRH plasma, a density decrease is observed with an increase of gas injection (see figure 3-b).

The simplest picture to explain such a behaviour is that edge particles recycled from saturated surface impinge the walls at higher energy (due to higher edge temperature). They can be trapped deeper in the wall which are maintained at constant temperature in this machine.

In Tore Supra, at medium density, wall saturation time is greater than 120s without external active pumping. 1000s plasma discharge which will be undertaken in CIEL will give the possibility to explore the "particle" operation limit due to wall saturation.

IV) Wall saturation prediction and control.

Empirical measurements of the wall saturation status during tokamak experiments are used to control and predict the recycling flux which governs the plasma equilibrium density. Moreover, low recycling fluxes are needed to get a reproducible start-up conditions and obtain a good plasma current ramp up. It has to be pointed that during current ramp up, active external plasma particle pumping are not efficient. Therefore, control of recycling fluxes depends only on wall pumping efficiency.

At JET [[14]], wall saturation status is estimated following from pulse to pulse the total number of injected particles normalised to the density obtained. A desaturated wall correspond to a low value of this ratio.

Wall saturation status is also correlated with internal plasma inductance (li) during current ramp up[[15]]. As the wall saturation status increases, particle outflux and Hα intensity increases. The plasma edge becomes more and more resistive so that the current profile peaks giving an increase of li during ramp up. All of these techniques give an estimation of the wall saturation status during the plasma and allow adjustments to conditioning before the subsequent discharge.

A pre-discharge estimation of wall saturation status may be obtained by a technique based on deuterium plasmas during the initial phase of the discharge when the poloidal coils are energised[[16]]. The difference in the shape of the small plasma currents obtained come from distinct wall saturation status and gives real time estimation of D/C. This is then used to feedback control prefill gas injection and plasma breakdown voltage in order to sustain current ramp up even if the walls are close to saturation.

However, if walls are too close to saturation, a conditioning procedure can be turned on. One of the most efficient technique to lower the wall saturation status is helium glow discharge (HeGD) at high wall temperature (T>200°C). Helium ions fluxes impinging the wall surface induce a deuterium detrapping which leads to a decrease of D/C ratio[[17]]. A typical HeGD at Tore Supra with 200°C wall temperature, Iglow=6A, P~0.3Pa releases between $3x10^{21}$ to $6x10^{21}$ atoms per hour. This value has to be compared with the total number of atoms injected in a discharge, between $4x10^{21}$ and $8x10^{22}$ atoms during high density ergodic divertor operation. This simple technique is used in most tokamaks.

Experimental observations [18] show that HeGD does not release hydrogenic atom in presence of a permanent toroidal field. Therefore, new conditioning techniques have been developed for superconducting devices. In the past 4 years, ICRH heating used with a helium injection (vessel pressure~10^{-2}Pa) generated plasma which has proved to efficiently release hydrogen [19, 20]. With ICRH heating conditioning, an average deuterium release production of $1.3x10^{22}$ per hour is found in Tore Supra. In TEXTOR, 2.7 10^{20} atoms are removed within 10 sec ICRH pulses [21].

Recently, a new conditioning technique which uses only poloidal breakdown has been developed[22]. It is used routinely to lower the wall saturation status after hard plasma disruptions and to allow a normal plasma current ramp up. The average deuterium production is comparable with that of HeGD and a 2 minutes procedure is sufficient to recover good wall pumping efficiency. This new procedure operating with constant magnetic field gives equivalent results to the HeGD done between shots in TFTR for instance without toroidal field[23].

And lastly, pumped limiter or divertor can be used to reduce to the wall inventory by pumping the recycling particle fluxes. It has been shown [24] that during steady state phase of actively cooled discharge, the number of particles pumped exceeds the number of particle injected : walls are depleted. This is confirm by D2 pressure measurements done by a quadrupole mass spectrometer which are decreasing from shot to shot.

V) Conclusions

Control of recycling fluxes is needed for density control. The walls are the major source of particles around the plasma and they are the key of density control. Different walls are involved during plasma operation.

Walls far from plasma desorb water if heated during operation. In future tokamak as in Tore Supra Upgrade (CIEL), temperature around the vessel will be controlled to avoid such a behaviour. One can also modify walls physical and chemical properties to reduce water inventory and/or prevent water desorption.

Walls in direct interaction with plasma trap plasma particles and can be saturated. Some experiments exist to study such saturation behaviour but extrapolation to reactor is not possible. In Tore Supra, no evidence of wall saturation is observed during long discharge operation which are done at low plasma density.

Conditioning plays an important role to reduce wall particle trapping and wall saturation status. New conditioning procedure are actually under development for next step devices which are operating with permanent toroidal field. At Tore Supra, ICRH conditioning and repetitive plasma breakdown have proven to be efficient. At the same time, new measurements are used to estimated in real time what is the wall saturation status. They are needed to get a reproducible start-up conditions for tokamak operation.

Ad last, walls are remainig which are not treated by helium glow discharge for instance. They are playing an important role to particle trapping and tritium inventory. Even if some new procedure as oxygen flushing are used presently, new methods must be developed in the future to control this wall particle trapping.

---

1 Federici et al, to be published, Nuclear Fusion (2000).
2 Madhavi M.A., et al, J. Nucl. Mater. 176-177 (1991) 32.
3 Strachan J.D. et al., Phys. Rev. Lett., 58 (1987) 1004.
4 Grisolia C. et al., J. Nucl. Mater. 266-269 (1999) 146.
5 Winter, J., J. Vac. Sci. Technol. (1987) 2286.
6 Robertson, J., Advances in Physics, (1986), Vol. 35 (4) 317.
7 Causey R.A., J. Nucl. Mater 162-164 (1989) 151.
8 Pillath J and Winter J, J. Nucl. Mater. 176-177 (1990) 319.
9 Maruyama K., et al, J. Nucl. Mater. 264 (1999) 56]
10 Peacock A.T. et al, J. Nucl. Mater 266-269 (1999) 423.
11 Grisolia C. Ghendrih P., et al, J. Nucl. Mater 196-198 (1992) 281.
12 Cohen S., NATO ASI Series, Vol. 131, Plenum Press, NY (1986) 773.
13 TFR Group, J. Nucl. Mater., 111-112 (1982) 199.
14 Saibene G. et al., J. Nucl. Mater. 241-243 (1997) 476
15 Grisolia C. et al, 22nd Annual Meeting of the European Physical Society (Bornemouth, 1995), Vol. 19C-IV, 349. *(proceedings may be purchased from "the publication officer, JET, Abingdon, Oxon, OX14 3EA, UK)*
16 Grisolia C. et al., J. Nucl. Mater. 241-243 (1997) 538.
17 Dylla H.F. et al, J. Nucl. Mater. 162-164 (1989) 128.
18 Grisolia C. et al., J. Nucl. Mater. 241-243 (1997) 538.
19 Lyssoivan A. et al, 22nd Annual Meeting of the European Physical Society (Bornemouth, 1995), Vol. 19C-III, 341. *(proceedings may be purchased from "the publication officer, JET, Abingdon, Oxon, OX14 3EA, UK)*
20 Gauthier E. et al, J. Nucl. Mater. 241-243 (1997) 553.
21 Esser H. G., Lyssoivan A., Freisinger M., R. Koch R., Weschenfelder F., Winter J;. Nucl Mat. 241-243 (1997) 861
22 Grisolia C. et al., J. Nucl. Mater. 266-269 (1999) 146.
23 Dylla H.F. et al, J. Nucl. Mater. 162-164 (1989) 128.
24 Loarer T. at al, J. Nucl. Mater. 220-222 (1995) 183.

Wall Pumping and Hydrogen Recycling in TEXTOR 94

M. Mayer, V. Philipps, H.G. Esser and P. Wienhold
Forschungszentrum Jülich GmbH, Institut für Plasmaphysik, EURATOM Association, Trilateral Euregio Cluster, D-52425 Jülich, Germany

M. Rubel
Alfvén Laboratory, KTH Stockholm, Association EURATOM-NFR, 10405 Stockholm, Sweden

Abstract. The pumping of hydrogen by the TEXTOR vessel walls during ohmic discharges and the subsequent hydrogen release between discharges have been studied at wall temperatures between 150°C and 350°C. At the beginning of a discharge the vessel walls show a high, but transient wall pumping speed which decreases as the discharge proceeds. At the end of a discharge the wall pumping speed tends towards a steady state value of about 1×10^{20} hydrogen atoms/s. Hydrogen outgasing after discharges follows a power law $\propto t^{-0.7}$.

1. Introduction

The recycling of hydrogen from the vessel walls of thermonuclear fusion experiments and the retention of hydrogen isotopes in the vessel wall materials are of crucial importance for the particle control and the in-vessel particle inventory.

Hydrogen is retained in the plasma facing first wall during a plasma discharge. This phenomenon has been called wall pumping. It is observed both for metal [1, 2, 3, 4, 5] and carbon walls [6, 7]. Some fraction of the retained hydrogen is released from the walls after the discharge [5, 8], resulting in a dynamic retention. Another fraction is retained permanently, leading to the accumulation of hydrogen in the vessel walls. The present paper describes wall pumping in TEXTOR 94 during ohmic discharges and subsequent hydrogen release between and after discharges.

2. Experimental

2.1. THE TEXTOR VESSEL WALL

The total plasma exposed wall area of TEXTOR 94 is about 38 m^2, of which about 9.5 m^2 consist of carbon (toroidal ALT-II belt limiter, inner bumper limiter, poloidal limiters, antenna limiters). The rest of the wall area (the so called liner) is made of stainless steel (Inconel), covered

C.H. Wu (ed.), *Hydrogen Recycling at Plasma Facing Materials,* 9–15.

with 50–100 nm amorphous hydrogenated boron (a-B:H) or silicon (a-Si:H) layers. These layers are created regularly during boronizations or siliconizations for wall conditioning.

The TEXTOR liner temperature could be changed between 150°C and 350°C, the limiter temperature was identical to the liner temperature.

2.2. Plasma operation

Ohmic discharges in deuterium with a plasma current of 350 kA and plasma densities between 2×10^{19} m^{-3} and 4×10^{19} m^{-3} were performed. Series of about 20 similar discharges were made, subsequent plasma discharges are separated by breaks of at least 300 s.

All 8 ALT-II limiter pumps were closed for the whole day of operation. In the breaks between plasma discharges pumping of the torus was provided by the cryo pump in neutral injector box 2. The pumping speed of the cryo pump during plasma discharges is negligible.

2.3. Neutral gas pressure measurements

The amount of hydrogen gas supplied to the torus is determined from the pressure drop in a supply volume. The total accuracy of this measurement is about 1%. The neutral gas pressure in the torus is measured between discharges with 2 ionization gauges, which are regularly calibrated for different gases with a spinning rotor gauge (accuracy 2%). The overall accuracy of the pressure measurement is about 5% for a known gas composition. The composition of the neutral gas is determined with 2 calibrated quadrupole mass spectrometers.

3. Results and Discussion

3.1. Hydrogen release after discharges

After a discharge hydrogen is released from the vessel walls. The release rate of deuterium atoms from the wall $\Gamma(t)$ can be calculated from

$$\Gamma(t) = 2\frac{V\dot{p}(t) + Sp(t)}{k\,T_{Liner}}. \tag{1}$$

$p(t)$ is the partial pressure of D_2, $\dot{p}(t)$ the derivative of the partial pressure, V the TEXTOR volume, S the pumping speed for D_2, k the Boltzmann constant and T_{Liner} the liner temperature. A small error is introduced in Eq. 1 due to the neglect of HD molecules. The deuterium

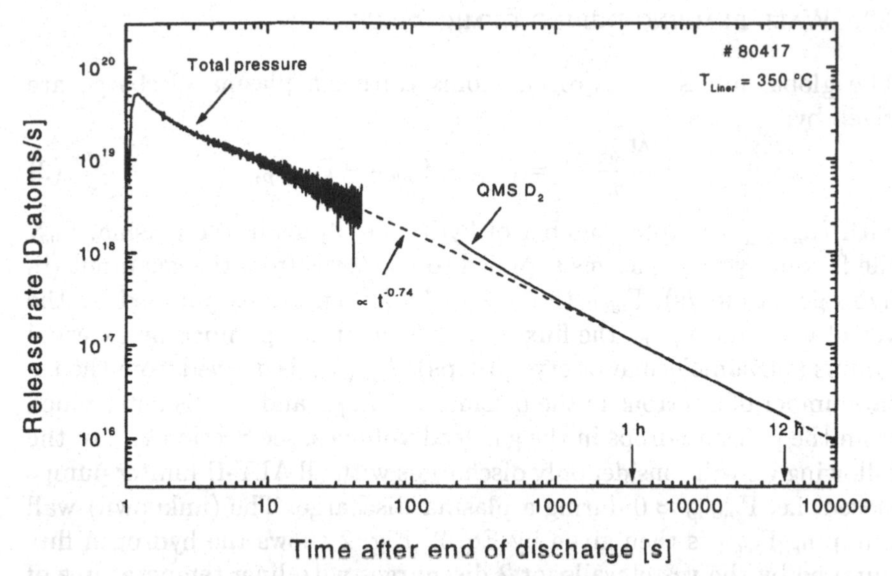

Figure 1. **Deuterium wall release rate in TEXTOR 94 after an ohmic plasma discharge in D. The release rate was derived from the total pressure and from the partial pressure of mass 4 (D_2), measured with a quadrupole mass spectrometer (QMS). Liner temperature 350° C, boronized walls.**

release rate after an ohmic discharge is shown in Fig. 1 for a liner temperature of 350° C and boronized walls. The release rate $\Gamma(t)$ follows a uniform power law $\propto t^{-\alpha}$, with $\alpha = 0.70 \pm 0.05$. A similar power law with $\alpha = 0.73 \pm 0.2$ was already observed at JET [5]. Andrew and Pick [8] have shown that for hydrogen implanted in carbon $\alpha = 2/3$ is expected due to thermal detrapping of bound hydrogen atoms and volume recombination to hydrogen molecules. The calculation of Andrew and Pick is based on the well acknowledged model of Möller and Scherzer about hydrogen trapping in carbon [9, 10]. A diffusion controlled outgassing process, which is characteristic for hydrogen implanted in many metals, yields a $t^{-1.5}$ behaviour [5], in contrast to the experimental observation at TEXTOR.

Siliconized walls show a similar behaviour as boronized walls with identical time dependence of the wall release rate. However, at identical wall temperatures the release from siliconized walls is somewhat larger than from boronized ones.

In the 300 seconds time interval between plasma discharges about 2×10^{20} – 2×10^{21} hydrogen atoms are released from the vessel walls, depending on the wall temperature. This is less than 1% of the total hydrogen inventory of TEXTOR 94.

3.2. WALL PUMPING DURING OHMIC DISCHARGES

The global fluxes of hydrogen atoms during a plasma discharge are given by:

$$\frac{dN_{plasma}}{dt} = \Gamma_{gas} - \Gamma_{wall} - \Gamma_{pump}, \qquad (2)$$

with N_{plasma} the total number of hydrogen atoms in the plasma, Γ_{gas} the flux of hydrogen atoms supplied to the torus from the gas input (in hydrogen atoms/s), Γ_{wall} the flux of hydrogen atoms pumped by the vessel wall, and Γ_{pump} the flux of hydrogen atoms pumped by external pumps (turbomolecular or cryo pumps). N_{plasma} is derived from the total number of electrons in the plasma and Z_{eff}, and Γ_{gas} is determined from the pressure drops in the gas feed volumes, see Section 2.3. In the following we will consider only discharges with all ALT-II limiter pumps closed, i.e. $\Gamma_{pump} \approx 0$ during a plasma discharge. The (unknown) wall pumping Γ_{wall} is then given by Eq. 2. Fig. 2 shows the hydrogen flux pumped by the vessel walls for 2 discharges with liner temperatures of 150°C (left) and 350°C (right), together with the plasma current and the line averaged electron density.

During the start-up phase of the discharge, while the plasma density is ramped up, the wall pumping shows a complicated behaviour. We will not try to treat this phase of the discharge. When the flat-top phase of the discharge is reached (at about 1.5 s in #81765 and 2 s in #80352, see Fig. 2), the walls show an initial wall pumping which is high, but which decreases as the discharge proceeds. It is noteworthy that the wall pumping does not decrease towards zero, but towards a steady state value. This steady state wall pumping of about 1×10^{20} hydrogen atoms/s is reached after about 2.5 s in discharge #81765. At the higher liner temperature (#80352), a steady state wall pumping is not fully obtained. The wall pumping is still decreasing, with a lowest value of about 1.3×10^{20} at the end of the discharge. The external gas feed is switched off at the start of the ramp-down phase. During the ramp-down phase, again a more complicated behaviour of the wall pumping is observed. In a series of 20 identical discharges separated by breaks of 360 s, the wall pumping is more or less identical in each discharge — this means that the high wall pumping capability at the beginning of the discharge is restored between the end of the preceding discharge and the start of the next one.

As will be shown in a different paper, the rate of hydrogen trapping by codeposition is about 6×10^{19} hydrogen atoms/s [11]. The accuracy of this number is a factor of about 2. Codeposition is mainly observed at deposition dominated areas of the ALT-II limiter [11]. The steady-state hydrogen wall pumping rate of about 1×10^{20} hydrogen atoms/s at the end of the discharges agrees within the experimental errors with the hydrogen codeposition rate. Codeposition is the only known trapping mechanism, which is able to provide a steady state wall pumping. Therefore we assume that the observed steady-state pumping of about 1×10^{20} hydrogen atoms/s is due to codeposition of hydrogen with eroded carbon.

However, the high, but transient wall pumping rate at the beginning of the flat-top phase of a discharge cannot be explained by codeposition. This transient wall pumping is due to a different mechanism: Trapping by implantation of energetic neutral hydrogen atoms at trap sites which were depleted by thermal desorption between the discharges. These energetic neutral hydrogen atoms are created in charge-exchange collisions [12]. It was already shown at Tore-Supra that implantation of charge-exchange neutrals is an important contribution to wall pumping (see i.e. [13] and the references therein). However, the crucial point is that implanted hydrogen can only be trapped if free trapping sites are available. Already occupied trapping sites are liberated by thermal release of hydrogen between the discharges, thus restoring the wall pumping capability.

The transient wall pumping due to implantation of charge-exchange neutrals explains qualitatively the following observations:

- At elevated liner temperatures the transient wall pumping increases, see Fig. 2. This is due to the enhanced release of trapped hydrogen from the wall between discharges by thermal desorption at elevated temperatures.
- The transient wall pumping decreases as the discharge proceeds. This is due to saturation of hydrogen in carbon, a-B:H and a-Si:H layers.

4. Conclusions

The pumping of hydrogen by the TEXTOR vessel walls during ohmic discharges can be separated into 2 different contributions:

1. At the beginning of each discharge the walls show a high, but transient wall pumping, which decreases as the discharge proceeds. This transient pumping is due to implantation of CX-neutrals into a-B:H or a-Si:H layers covering the vessel walls. The transient wall pumping capability is restored between discharges by thermal outgasing of implanted hydrogen.

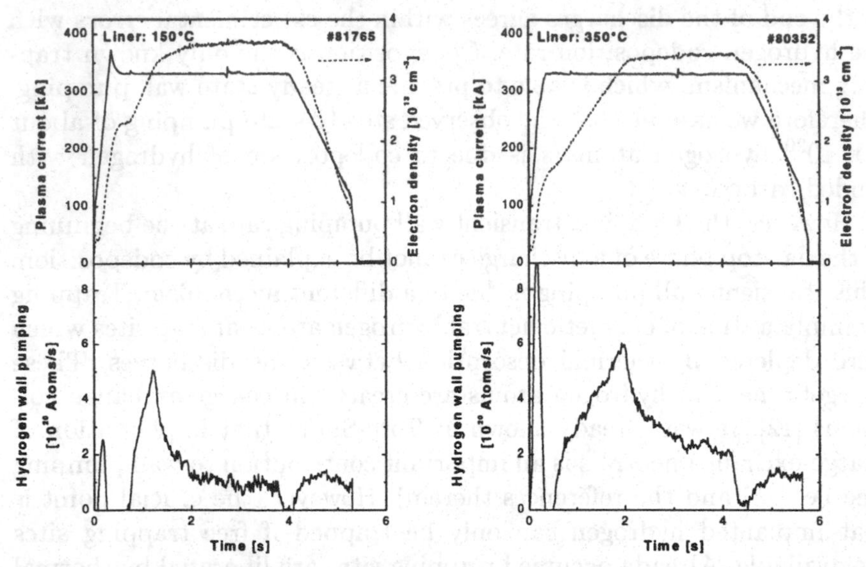

Figure 2. Plasma current (top, solid line), line averaged electron density (top, dashed line) and hydrogen flux pumped by the vessel walls, as derived from Eq. 2 (bottom) for 2 discharges. Left: Liner temperature 150°C, siliconized walls; Right: Liner temperature 350°C, boronized walls.

2. The walls show a steady state wall pumping of about 1×10^{20} hydrogen atoms/s. The steady state wall pumping is due to codeposition of hydrogen with carbon mainly on the ALT-II limiter. The observed steady state wall pumping speed agrees within the error bars with the codeposition rate derived from wall analysis [11].

Codeposition is the only known wall pumping mechanism, which is able to provide a steady state wall pumping. Wall pumping by codeposition is not limited in time, as codeposited layers grow about linearly. The wall pumping speed provided by codeposition is comparable, if not even larger, than active pumping with limiter pumps.

From the viewpoint of plasma density control elevated wall temperatures are advantageous: They result in a larger wall pumping capability due to larger hydrogen depletion of the walls between discharges. However, this beneficial effect is only transient: The walls saturate during the discharge, resulting in a decrease of the wall pumping speed.

References

1. F. Waelbroeck, J. Winter, and P. Wienhold. J. Nucl. Mater. 103-104 (1981) 471.

2. F. Waelbroeck, P. Wienhold, and J. Winter. J. Nucl. Mater. 111-112 (1982) 185.
3. J. Winter, F.G. Waelbroeck, P. Wienhold, E. Rota, and T. Banno. J. Vac. Sci. Technol. A2, 2 (1984) 679.
4. J. Ehrenberg, V. Philipps, L. de Kock, R.A. Causey, and W.L. Hsu. J. Nucl. Mater. 176-177 (1990) 226.
5. V. Philipps and J. Ehrenberg. J. Vac. Sci. Technol. A11, 2 (1993) 437.
6. J. Winter. J. Vac. Sci. Technol. A5, 4 (1987) 2286.
7. J. Ehrenberg. Hydrogen and helium recycling in tokamaks with carbon walls. Tech. Rep. JET-P(88)57, JET Joint Undertaking, 1988.
8. P. Andrew and M. Pick. J. Nucl. Mater. 220-222 (1995) 601.
9. W. Möller and B.M.U. Scherzer. J. Appl. Phys. 64, 10 (1988) 4860.
10. W. Möller. J. Nucl. Mater. 162-164 (1989) 138.
11. M. Mayer, V. Philipps, H.G. Esser, P. Wienhold, and M. Rubel. Codeposition and hydrogen wall pumping in textor 94. To be published in J. Nucl. Mater.
12. H. Verbeek, J. Stober, D.P. Coster, W. Eckstein, and R. Schneider. Nucl. Fusion 38 (1998) 1789.
13. M. Sugihara, G. Federici, C. Grisolia, P. Ghendrih, J.T. Hogan G. Janeschitz, G. Pacher, and D.E. Post. J. Nucl. Mater. 266-269 (1999) 691.

Hydrogen Recycling in the RFX Reversed Field Pinch

D. BETTELLA, L. CARRARO, M.E. PUIATTI,
F. SATTIN, P. SCARIN, P. SONATO, M. VALISA
Consorzio RFX
Corso Stati Uniti, 4 – 35127 Padova, Italy

1. Introduction

RFX (Reversed Field eXperiment) is a large toroidal device (a=0.46 m, R= 2 m) designed to explore the properties of the Reversed Field Pinch configuration up to 2 MA of plasma current [1]. So far it has been operated in hydrogen, with plasma currents up to 1.2 MA, electron densities between 1 10^{19} and 1 10^{20} m^{-3} and electron temperatures between 200 and 400 eV. The first wall is made of graphite tiles covering all over the inconel vessel and has a total surface of approximately 36 m^2. Neither protruding limiters nor divertors are present (see Fig.1). The average particle confinement time is of the order of few milliseconds and in order to sustain the electron density the fluxes of atomic hydrogen are of the order of 10^{23} s^{-1}. In most of the circumstances the hydrogen flux required to refuel the plasma originates from the recycling processes at the wall. An amount of pre-filling gas adequate to allow the breakdown may in fact be sufficient to drive a whole discharge,

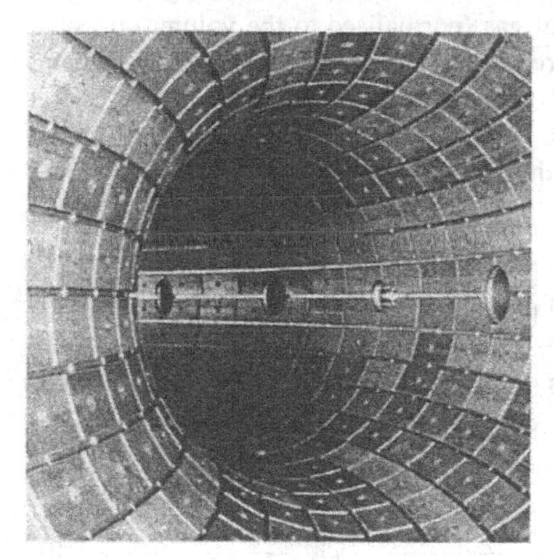

Fig.1 - The RFX first wall is made of carbon tiles that cover almost completely the vacuum vessel.

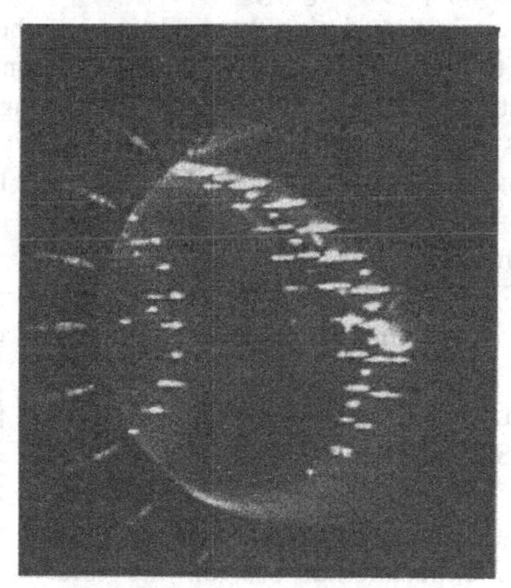

Fig.2 –Helical footprints of the plasma column deformed by the modes locking. The image was taken with a C II filter. Overheating of the tile edges is clearly observable.

whose typical duration is about 150 ms, much shorter than the characteristic pumping time (≈1 s). In absence of any wall conditioning procedures too large filling pressures may easily saturate the wall, leading to a complete loss of density control. Thus, the

C.H. Wu (ed.), Hydrogen Recycling at Plasma Facing Materials, 17–23.

capability to control the plasma density strictly reflects the capability to control the hydrogen concentration on the tile surfaces facing the plasma. Various techniques have been experimented to condition the wall: hot wall operations, (up to 280°C), Glow Discharge Cleaning (GDC) in He, boronisation with GDC in Diborane. In the following the degree of density control gained with such methods is described in detail.

It has to be emphasised that in RFX the ohmic input power required to drive the plasma current is relatively high and ranges between 20 and 80 MW. A significant fraction of the injected power is dissipated onto the relatively narrow region (few square meters) where the locking in phase of several MHD modes (m=1, n=6-15) deforms the plasma column (see Fig.2). Locally the power density may exceed 100 MWm^{-2} and the surface of the tiles can reach temperatures for which Radiation Enhanced Sublimation and strong hydrogen outgassing (blooming) are likely to occur [2]. In general the region of enhanced interaction produces ≈ 50% of the total hydrogen and main impurity influxes (carbon and oxygen). However, in high power discharges, the density build-up that accompanies blooming events may easily lead to wall saturation with loss of density control, up to represent an operational limit. Similar situation is found in the EXTRAP experiment [3]. Actively dragging of the locked modes around the torus smearing the power dissipation onto a larger area has represented a breakthrough [4]. This procedure has also lead to a general improvement in the density control.

The study of the wall response presented in this contribution is mainly based on two parameters. The first is the measurement of the desorbed gas, performed by means of absolute pressure gauges .The second is the ratio between the electron density reached at a certain time during the discharge and the total gas (normalised to the volume) injected up to that time. The latter ratio can be considered as an "effective" recycling coefficient. With this definition the effective recycling is 1 when the plasma density equals the total gas injected. The main difference between the two approaches, pressure gauge measurements and effective recycling, is that the former one includes the effects of the processes occurring during the termination of the pulse.

2. Experimental observations

2.1. EFFECTS OF GLOW DISCHARGE CLEANING

Fig. 3 shows for several discharges of one experimental session the total desorbed gas, the total gas puffed into the vessel and the effective recycling as a function of the shot number. The experimental session was preceded by 40' of He GDC (400 V and 2.4 A). The plasma current was ~ 650 *kA*, the first wall was at room temperature and the effects of the last boronisation negligible. The "effective" recycling

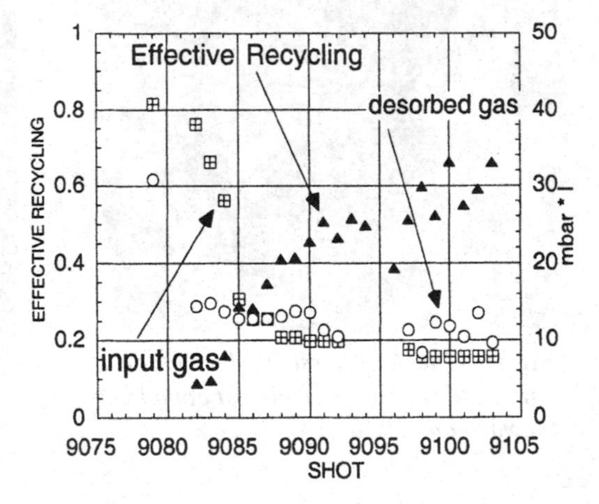

Fig.3 - Total gas injected and gas extracted after the pulse together with the effective recycling for one experimental sessions preceded by 40' of GDC in He

coefficient in approximately 20 shots moves from 0.1 to 0.7. At the beginning of the session relatively large quantities of filling gas have to be used and only 40 to 60 % of it is desorbed after the pulse. One can estimate the amount of gas required to reach wall saturation by summing up the differences between the amount of injected gas and the one desorbed after the pulse for the various discharges from the GDC till the desorbed fraction reaches 100%. It turns out that in this case the wall reaches saturation after absorbing approximately 1 10^{20} atoms/m^2. This value is comparable within a factor of two with the amount of hydrogen pumped out during GDC in He. For instance, 40' of He GDC that are typically performed to empty the surface reservoir and restore a reference wall condition, extract from the wall the equivalent of approximately 2.4 10^{20} atoms/ m^2.

2.2 EFFECTS OF BORONISATION

Soon after a boronisation the first wall has a strong pumping capability and larger amounts of gas (~70 *mbar l*) have to be used to sustain the discharge. For approximately 25 shots only 20% of the injected fuelling gas is desorbed after the discharge as shown in Fig. 4. Correspondingly the effective recycling coefficient may be very low as also shown in Fig 4. This ratio slowly increases and stabilises below 100%. He GDC sessions of 40'in this phase can bring the gas desorbed fraction to 20%. The amount of gas that the wall can absorb before it reaches saturation turns out to be of the order 1 10^{21} atoms/m^2, that is ten times higher than in a non-boronised wall conditioned with He GDC (40'). The extended reservoir capacity associated to the boronised wall makes the density control relatively simple: it is for instance possible to perform a complete density scan in both directions during an experimental session while with a non-boronized wall it is very difficult to recover from high density regimes. It should be emphasised that soon after boronisation a fairly large amount of He is trapped into the wall. The spectra of the SPRED spectrometer survey are indeed dominated by He, estimated to be around 10% of the electron density, increasing during the discharge. The presence of He may affect the overall recycling and certainly compensates for the efficient hydrogen pumping capability of the wall sustaining the density. Too an empty surface soon after a wall conditioning procedure may result in an irreproducible start up for a number of discharges as well as in an excessive variation of the density value during the discharge. One efficient way to reliably control the density in such conditions is to add to the standard 40' of GDC in He few minutes of GDC in H in order to provide an adequate and reproducible gas reservoir.

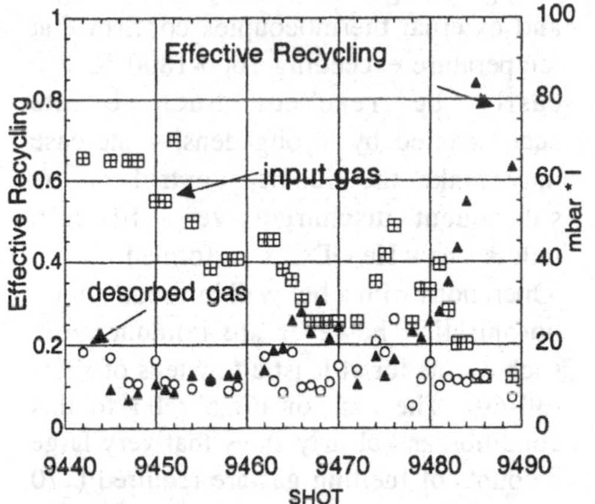

Fig.4 - Total gas injected and gas extracted after the pulse together with the effective recycling for two experimental sessions preceded by boronisation

2.3 HOT WALL OPERATION

When baked up to 280°C the behaviour of the wall response is by far less homogeneous and the fraction of desorbed gas after the shot is always greater than 50%, with an average value close to 100%. A very large amount of filling gas is to be injected into the vessel to sustain the plasma current ramp up phase, in order to compensate for the very low recycling immediately after the breakdown. However, especially at plasma currents approaching 1MA, typically characterised by large ohmic input power, the density often builds up rapidly during the discharge. An excess of density may cool the plasma and lead to a premature quenches of the plasma current. In such circumstances the "effective " recycling varies during the discharge and from relatively low values may reach and even exceed one. A possible explanation is that the strong power loading (of the order 100 MW/m^2) associated to the locking of the modes leads to a local overheating of the tiles, at temperatures that imply in particular strong hydrogen outgassing. CCD images and external thermocouples confirm that temperature exceeding 1600-1800 °C can easily be reached. Such events accompanied by strong density increase may make the density control in the subsequent discharges very difficult, unless a new He GDC is performed.

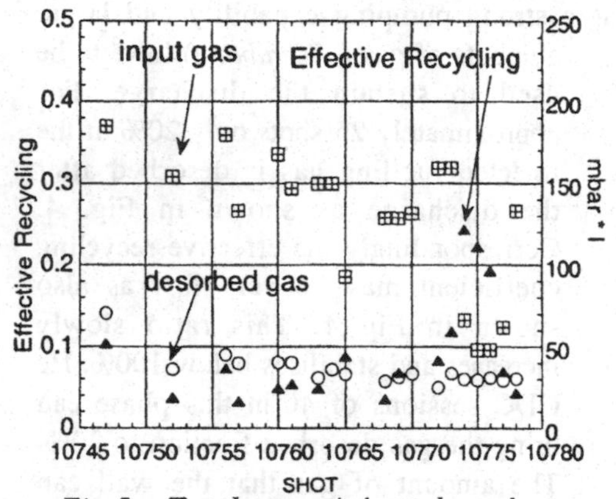

Fig.5 - Total gas injected and gas extracted after the pulse together with the effective recycling for three experimental sessions with a baked wall (280°C) after bonisation

Operations with a hot wall but soon after boronisation however are immune from such events for at least a few tens of shots (30-60). The data of Fig.5 refer to this condition and clearly show that very large amounts of fuelling gas are required (170 *mbar l*) because of the very low effective recycling values.

2.4. LOCKED MODES INDUCED ROTATION

At high currents, around the MA range, the beneficial effects of boronisation are limited to a few tens of shots. Wall-mode locking interaction, exacerbated by the higher power involved (50 – 80 MW), causes a rapid carbon redeposition over the boron layer [5]. Few blooming events are then sufficient to saturate the wall surface and re-establish critical experimental conditions.

The introduction of a toroidally rotating magnetic perturbation with adequate amplitude and frequency has been successful in dragging the modes locking around the torus [4]. Smearing the power dissipation onto a larger area prevents the occurrence of blooming and has beneficial effects on the density behaviour, extending the possibility of a good density control to several hundred discharges following boronisation. In particular it is possible to recover from high-density discharges by simply reducing the amount of pre-filling gas.

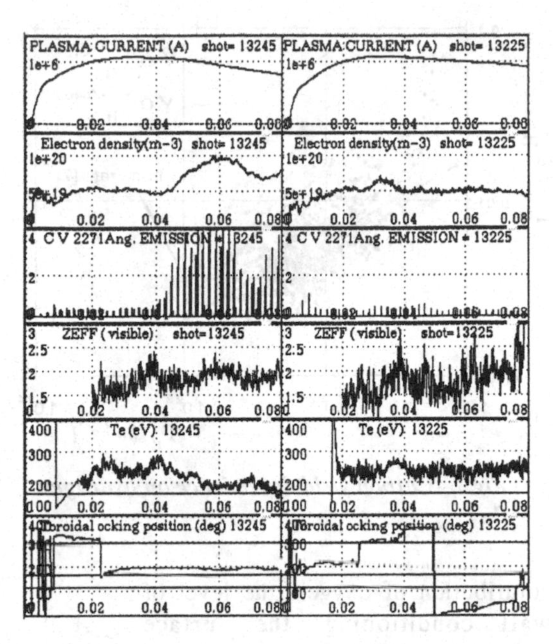

Fig.6 - Comparison among some time waveforms of two shots: one with wall locked modes (13245) and the other with induced mode-locking rotation (13225). From top to bottom: plasma current, electron density, C V line emission, the effective charge, electron temperatue and the toroidal position of the locked modes.

Unlocking of the modes avoids the blooming events and the associated build-up of the hydrogen reservoir within the wall but has not lead to significant changes in terms of plasma dilution. Impurity and hydrogen influxes as well as Zeff do not show statistical differences in the two operational regimes, with and without induced rotation. Fig. 6 shows some relevant waveforms for two high current shots with (right) and without (left) locked modes rotation. In the case of modes locked to the wall, at about 45 ms the density suddenly builds up causing a reduction of the temperature and a more pronounced decrease of the plasma current. The density increases together with the C V line emission. However Zeff does not show a pronounced change. In other cases it even decreases: this may occur either because of recombination of the highest ionisation states after the temperature drop or just because the hydrogen outgassing is higher than the carbon influx.
When the modes rotate around the torus, peak values of particle influxes are reduced, while the average influx does not change significantly, apart for the cases with large blooming events.

2.5 HYDROGEN RECYCLING AND IMPURITIES

Previous paragraphs have already shown that strong recycling does not seem to negatively affect the impurity content. In fact the effective charge *Zeff* decreases with the electron densities [6]. Moreover the high-density regimes associated to the strong recycling increase the impurity screening by reducing the ionisation length [6].
It is found that the carbon yield (the ratio between carbon and hydrogen influxes) in RFX decreases at higher hydrogen fluxes. Fig. 7 shows this fact for a line of sight lying on the poloidal pane and intercepting the outboard side of the torus. Carbon and hydrogen influxes are evaluated taking into account the temperature and density dependence of the related ionisation event per photon [9]. Local temperature and density are assumed equal

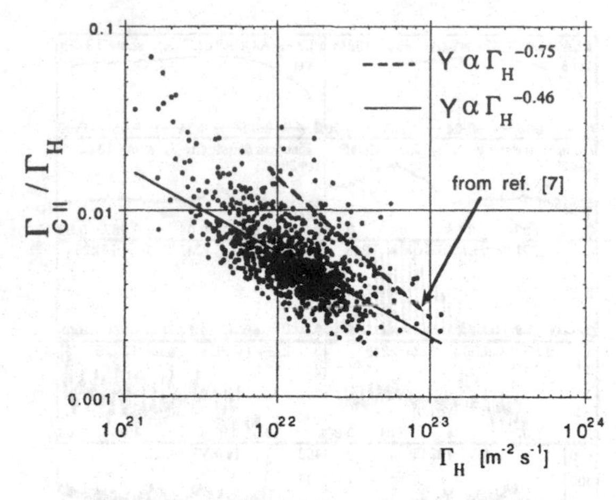

Fig.7 –Carbon yield dependence on hydrogen flux.

to those measured on the equatorial plane from the line ratio of spectroscopic lines emitted by a thermal He beam [10]. The decreasing slope is less pronounced than what is experimentally found in the divertor regions of several Tokamaks [7,8]. The weaker dependence might be associated to a contribution of the physical sputtering in RFX. Fluctuations in the carbon yield, even within the same experiment, may be ascribed to various reasons; these comprise contribution of oxygen, the level of wall conditioning, the surface temperatures of the tiles, changes in the edge parameters as well as to the uncertainties in the determinations of both carbon and hydrogen influx [6].

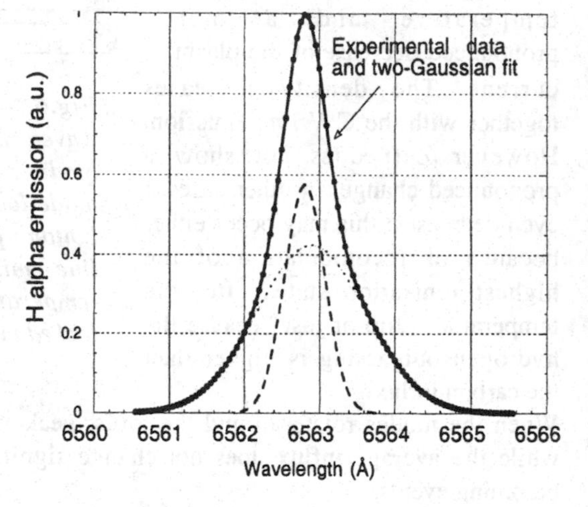

Fig.8 - Shape of the Ha emission line and the two gaussian fit

On the other hand the energies involved in the Hydrogen recycling process may be estimated from the observation of the line shape of the Hα emission, a typical example of which is reported in Fig. 8. A two-gaussian fit seems to reasonably reproduce the experimental line as the sum of a low energy component with approximately 1.5 eV and a more energetic one of 15 eV. The line is typically well symmetric, apart for a slight shift of the order of 0.1Å between the two components. Zeeman effects are negligible. The low energy component may be associated to the Frank-Condon energies with which the hydrogen atoms are produced in a molecular dissociation process [11]. The second component may be due to charge exchange processes. The shift between the two components might be the signature of a collective motion of the edge ions. However, the absence of an asymmetric feature might also be due to the line of sight not being perpendicular to the tile surface.

3. Conclusions

In most of the circumstances the density behaviour in RFX strongly depends on recycling at the wall, entirely made of graphite tiles. Boronisation increases by approximately ten times the wall reservoir with respect to the unconditioned tiles and allows a better density control in the entire range of plasma current so far experimented. Wall modes locking concentrates the dissipation of the relatively high power (20-80 MW) on to a narrow region of the wall and strongly affects recycling. Localised wall surface heating in high power discharges entails blooming processes with strong density build-up, wall saturation and also fast annihilation of boronisation through strong carbon redeposition. Dragging around toroidally the locked modes by introducing rotating perturbation modifies the topology of the plasma wall interaction, smears the power dissipation onto a larger surface, prevents the occurrence of blooming and improves the density control capability.
Considering that the energy confinement time in RFX scales positively with density, one may deduce that since high densities are associated to high recycling at the wall, the latter is not detrimental per se as to the plasma performance. Impurity concentration in general is not a severe issue especially in high density regimes, apparently due to the strong impurity screening at the edge.
It is however believed that a reduction of the wall recycling combined with other means of plasma refuelling such as pellet injection would beneficially affect the overall confinement. Pellet injection experiments have produced plasmas with the best dilution ever reached in RFX, with values close to one. In addition, the reduction of wall recycling and therefore of the neutral particle density at the edge should mitigate the effects of the related viscosity . In particular it should beneficially influence both topology and magnitude of the (**EXB**)/B^2 flow shear at the edge, which appears to be important for improving transport through turbulence stabilisation. In this perspective, an experiment to actively control the wall recycling has been envisaged: a vented pump limiter will be soon experimented on RFX [12] to verify its pumping efficiency in the Reversed Field Pinch configuration.

References

[1] Rostagni G. et al , *Fusion Eng. and Des.*, **25** (1995) 301-496
[2] Valisa M., Bolzonella T., Carraro L. et al. , *Journ. Nucl.Mat.*, **241** (1997) 988
[3] Larsson D, Bersaker H and Hedqvist A. *Jour.Nucl.Mat.* , **266** (1999) 266
[4] Bartiromo R.et al., *Phys. Rev. Lett.* **83** (1999) 1779
[5] L.Tramontin, et al., *Jour. Nucl. Mat* .**266-269** (1999) 709
[6] Carraro L., Puiatti M.E., Sattin F. et al , *Jour.Nucl.Mat.* , **266** (1999) 446
[7] Kallembach A.et al, *Proc 17th IAEA Conf., Oct. 1998, Yokohama,* CN-69/EXP/02
[8] Tobin S.J., DeMichelis C. et al, *Plasma Phys and Contr. Fusion,* **40** (1999) 1335
[9] Summers H.P., unpublished, *JET report* JET-IR (94) 06
[10] L.Carraro et al., unpublished, *Proc. of EPS Conference on Plasma Phys. And Contr. Fus., Maastricht,* (June-1999) paper P3.054
[11] Reiter D., Bogen, Samm U., *Jour. Nucl. Mat.,* **196** (1992) 1059
[12] Sonato P., *Jour. Nucl. Mat.,* **39-40** (1998) 333

ACTIVE CONTROL OF HYDROGEN RECYCLING BY THE PERMEATION AND ABSORPTION TECHNIQUES

N. OHYABU, Y. NAKAMURA AND Y. NAKAHARA
National Institute for Fusion Science, 322–6 Oroshi-cho, Toki, Gifu-ken, Japan
A. LIVSHITS, V. ALIMOV, A. BUSNYUK, M. NOTKIN, A. SAMARTSEV AND A. DOROSHIN
Bonch-Bruyevich University, 61 Moika, St. Petersburg 191186, Russia

1. Introduction

Thy potential barrier of hydrogen dissociative absorption plays an outstanding role in the interaction of energetic hydrogen with solid [1–3]. The surface barrier does not impede the implantation of suprathermal hydrogen particles of an energy higher than ~ 1 eV, but it drastically impedes the thermal reemission of absorbed atoms. As a result, the presence of a surface barrier results in a dramatic increase of the absorption and permeation of suprathermal hydrogen. At definite conditions, the permeation reaches its conceivable limit when virtually the whole incident flux passes through the solid membrane irrespective of its temperature and thickness (*the superpermeation*).

Thin (monolayer and even submonolayer) films of nonmetallic impurities are responsible for the barrier on the surface of transient metals, and the surface covered by a monolayer nonmetallic film is the usual state of a metal surface in vacuum (the "real" surface). Two implications important for fusion follow from that.

First, the properties of PFM that govern reemission, absorption and permeation depend on physico-chemical environment. Thus the PFM behavior in fuel recycling and tritium inventory/permeation may be dramatically changed during the operation, and such an evolution is not easily predictable.

Second, the superpermeation and enhanced absorption can be specially employed for a short-way separation of D/T from He [4] and for an active particle control in fusion devices [5].

2. Examples of the possible employment of permeation and absorption technique for active particle control

2.1. Short-way separation of D/T and He

He is supposed to be pumped by cryopumps in the existing ITER Project. We were suggesting [4] to install a superpermeable membrane along the pumping duct walls. The membrane can isolate a major part of D/T mixture before the cryogenic panels, automatically compress D/T by a few orders of magnitude, purge it of any impurities including He, and return back into the fueling system. If the membrane is installed far enough from the divertor, D/T mixture, when reaching it, will mainly consists of thermal molecules that have to be converted again into suprathermal particles, e.g. into thermal atoms. That may be done either at a hot metal surface (an atomizer) or in a gas discharge. As it has been shown basing on the results of model experiments, a membrane system of

C.H. Wu (ed.), Hydrogen Recycling at Plasma Facing Materials, 25–34.

quite a reasonable size and with a fairly moderate power consumption may isolate more than 99% of D/T, drastically reducing its freezing up at the cryogenic panels [18].

2.2. *Active particle control in Large Helical Device (LHD).*

Several different options of divertor are supposed to be tested at LHD [5]. Every such a divertor might have a membrane pumping system. In contrast to the previous example, in the case of LHD the membrane is supposed to be installed just into the divertor zone. Thus the membrane is a plasma facing component, pumping the energetic hydrogen generated by the working plasma itself. LHD is not a tritium machine and hydrogen is planning to be removed from the divertor for the two reasons:

(1) Helical machine can operate in the *steady state regime,* and pumping is necessary to provide for *the particle balance* at the NBI or pellet injection;

(2) Effective pumping of hydrogen from the divertor region is expected to dramatically increase the divertor plasma temperature, and this, in turn, is supposed to result in a significant *enhancement of the core plasma temperature.*

Pumping with a superpermeable membrane may, in principle, solve the both tasks due to the steady state operation and the highest conceivable pumping speed.

Upgraded helical divertor is aimed for the second purpose. The idea of pumping consists of removing of CX atoms generated by the plasma from the particles reemitted by the divertor plate with using either membrane or absorption panel (the latter is possible since this particular divertor is planned to operate in pulse regime). The following operation parameters are expected: CX atom energy and flux density are ~ 2 keV and ~ $4\times10^{15}cm^{-2}s^{-1}$ correspondingly, heat flux during the shot is ~ 1.3 W $cm^{-2}s^{-1}$, shot duration is 10 s and pause is 300 s. It is believed that 10-20% of all the hydrogen particles reemitted by the divertor may be pumped in such a way.

2.3. *Requirements for the permeation/absorption panels for particle control*

What physical factors possibly may damage superpermeation?

It was theoretically and experimentally shown [1,3,6,9] that, under conditions of interest, *radiation enhanced diffusion* does not in fact affect superpermeation and enhanced absorption and neither does *reemission induced by energetic hydrogen.* At the same time, *sputtering of the plasma facing surface* and the *growth of a polyatomic carbon film* may present a real problem.

If a superpermeable membrane is initially symmetrical and if nothing has been undertaken to dynamically maintain the impurity monolayer, sputtering of only the membrane upstream side will result in the development of an unfavorable asymmetry and in an ensuing degradation of superpermeation (as well as in a dramatic decrease of enhanced absorption). This phenomenon have been observed many times in ion beam (e.g. [3,9]) and plasma experiments [12].

If a graphite film is thicker than the implantation depth, hydrogen does not reach the metal bulk: it is accumulated in the implantation zone until induced reemission secures the balance. *Thus a sufficiently thick graphite film at the membrane surface must cause the destruction of superpermeation* and that was directly observed by Winter at al [14].

How do the above examples of possible fusion applications look from this point of view?

There is neither sputtering nor deposition of carbon in the case of short-way *separation of D/T and He* when the membrane is installed in the pumping duct. Several membrane systems combined with an atomizer with a pumping speed of about 1000 l/s were successfully tested [4,5], and *there are no fundamental limitations on fabrication of a system of any required size.* A membrane system combined with a gas discharge (instead of atomizer) seems to us being even more promising and a work in this direction is in progress now [12].

Superpermeation technology of this type might be included in the ITER Project which is under reviewing now.

The situation seems to be rather favorable also in the case of a *high recycling divertor*: the low temperature divertor plasma must not physically sputter the pumping panel. On the other hand, the dense divertor plasma will capture the sputtered carbon atoms preventing the surface from the deposition of carbon.

Still in the case of *low recycle divertor* (including the upgraded divertor at LHD) the situation looks more problematic. CX atoms of keV scale energy will sputter the pumping panel. The low density divertor plasma is expected to be rather "transparent" for the sputtered C atoms and so C will reach the pumping panel.

Thus two problems have to be solved, if one wants to employ the superpermeation technique under conditions of such a kind:

- *Superpermeable membranes being able to preserve a favorable asymmetry in spite of sputtering should be developed.*
- *The conditions and regimes for the superpermeable membrane operation under the carbon flux should be found.*

3. Experiments on membrane asymmetry under the sputtering.

Ion driven permeation through a 0.1 mm Nb membrane was observed using the scheme presented in Fig.1. There are three factors one could employ to control a membrane asymmetry. (1) 0.6 keV ion beam itself could sputter the membrane upstream side in ultra high vacuum (UHV). (2) Pd could have been deposited *in situ* at the membrane downstream side preliminary cleaned by Ar^+ ion beam. (3) The chemical state of upstream side might have been *in situ* controlled by admitting chemically active gases in the course of ion bombardment. The membrane

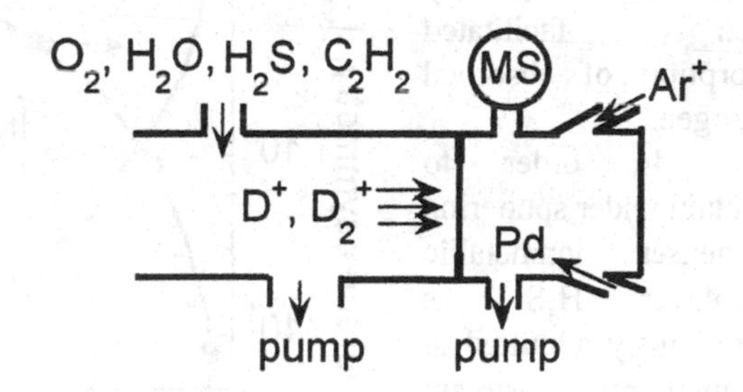

Figure 1. Scheme of ion beam-membrane experiment.

might be heated by an external source of light through the glass window.

Pd deposition results in an about tenfold increase of ion driven permeation (Fig.2) due to the increase of membrane asymmetry; as a result superpermeation is reached in spite of sputtering in UHV. An independent measurement of the membrane asymmetry has shown its hundred fold increase from a very unfavorable magnitude (~0.1) to a rather favorable one (~10) despite the permanent sputtering. When the Pd film thickness reaches a few tens of monolayers the film remains stable and no repeated deposition is required, if temperature does not exceed 500-600°C.

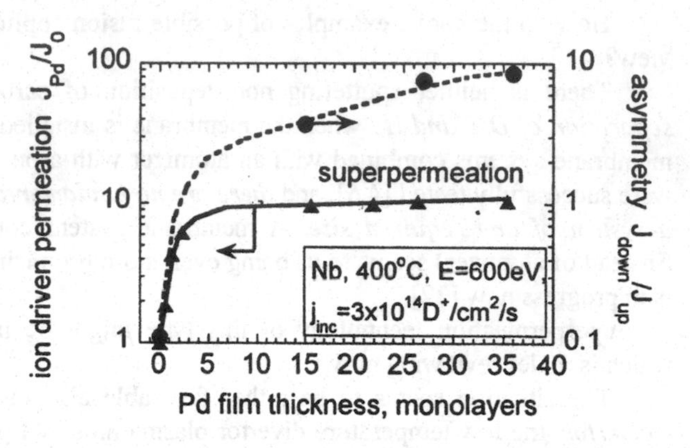

Figure 2. Effect of the deposition of Pd on the Nb membrane downstream side. Permeation flux with a Pd coating (J_{Pd}) is normalized to the permeation flux without the coating (J_0). Asymmetry is determined as a ratio of permeation (J_{down}) and reemission (J_{up}) fluxes.

The main reason of the effect of Pd is that Pd surface can be easily maintained clean (in contrast to Nb surface), and the clean surface ensures a very much facilitated desorption of absorbed hydrogen.

In order to maintain under sputtering a denser nonmetallic monolayer, H_2S was continuously admitted at the membrane upstream side during ion bombardment (Fig.1) while the downstream side was covered with a Pd coating (a few tens of monolayers). That resulted in an additional

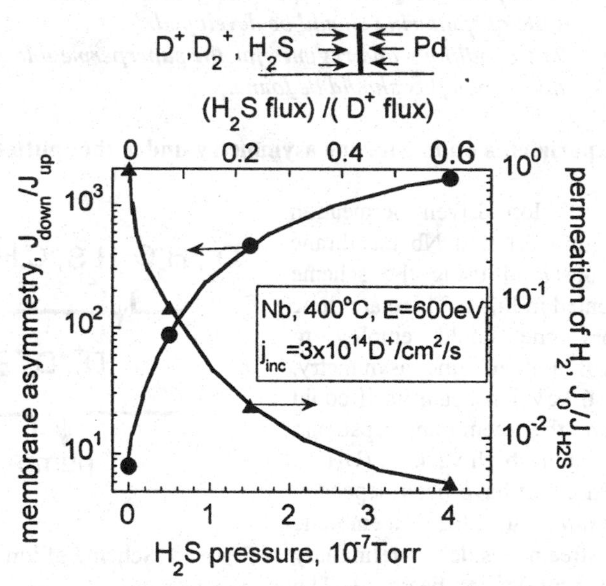

Figure 3. Combined effect of S at the upstream side and Pd at the downstream side on the Nb membrane asymmetry and molecular permeation. Molecular permeation is normalized to the permeation without H_2S.

asymmetry increase and in the suppression of molecule permeation by orders of magnitude (Fig. 3). H_2S is an "exotic" species for fusion devices but H_2O, O_2 and C_2H_2 have a similar effect [9,19].

4. Experiments on the superpermeable membrane operation under the carbon flux.

The goal of this experiment was to investigate the effects produced by carbon being deposited onto a Nb membrane sample at its interaction with energetic hydrogen and, in particular, to model the operation of superpermeable membrane located in the divertor zone of Large Helical Device [15]. The first results obtained by the moment are presented.

4.1. Experimental

Hydrogen plasma surrounded a resistively heated tubular Nb membrane of 1.5 cm diameter, 15 cm length, and 0.2 mm wall thickness (Fig. 4). The plasma ($n_e = 10^9 – 10^{10}$ cm^{-3}, T_e in a few eV, $p \approx 10^{-2}$ Torr) generated by a discharge (3 A × 80 V) with hot W cathode had the shape of a hollow cylinder confined by a magnetic field mainly in the space between a target and a grid anode. A resistively heated partly transparent grid target (56 cm^2) made of graphite contacted the plasma, surrounding it from the outside. A cylindrical Mo shield was placed behind the target. Two main groups of suprathermal hydrogen particles were impinging on the membrane: thermal atoms (typically ~ 5×10^{16} H/(cm^2 s)), and the atoms originating from ion neutralization at the target (an average energy of the latter may have been controlled by target biasing and was 0.15–0.25 of the incident H_2^+ and H_3^+ ion energy, while their flux estimated from the ion current and reflection coefficient typically was 10–20 times smaller than the flux of thermal atoms). An additional flux of energetic ions (~ 3×10^{15} (H_2^+, H_3^+/(cm^2 s)) could have been obtained by membrane biasing. Besides, thermal atoms may have been generated without plasma at hot cathode filaments.

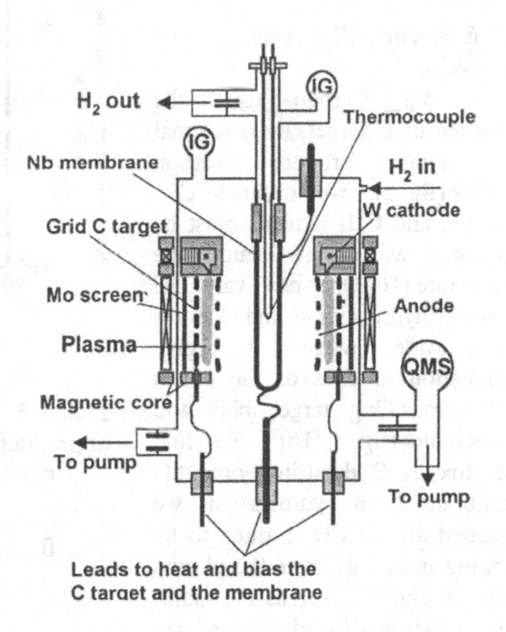

Figure 4. Schematic of the experimental device.

The device is designed as a UHV one. Its upstream and downstream chambers are continuously pumped by two TMP's through calibrated diaphragms (100 and 30 l/s for H_2, respectively), which permits measuring the permeation flux. A differentially pumped quadrupole gas analyzer permits measuring mass spectra during the plasma operation.

Membrane temperature is measured with a thermocouple, and target temperature by its electrical resistance.

Heating of Nb sample in UHV at $T \geq 600$ °C results in dissolution of the original thick oxide film, with only a monolayer film remaining [3,11]. The AES analysis shows that this film usually consists of O and C [11]. If the C flux onto the surface is negligible, the film keeps stable under suprathermal hydrogen particles of an energy of up to a few tens of eV [2,12], with no special efforts needed to maintain the membrane superpermeability to such particles [12].

4.2. Experimental results and discussion

4.2.1. Carbon flux. In the presence of C target, even thermal H atoms produce carbon species (fig. 5): we observe CH_4 directly, and CH_3 radicals must be generated with approximately the same rate [16].One can vary the fluxes of hydrocarbons and C atoms over wide ranges to give domination to ones or the others with controlling target bias and temperature (fig. 4, [16]). To find the flux of C deposited onto the membrane from hydrocarbons we assumed all the CH_3 formed to be sticking uniformly over the plasma facing surfaces (whereas the major part of CH_4 was found to have been pumped).

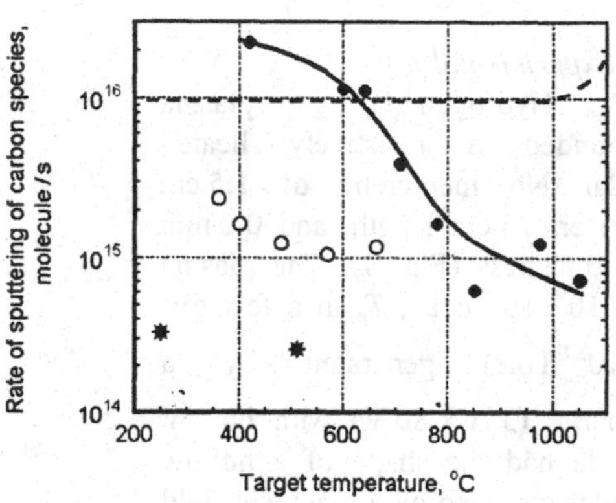

Figure 5. Temperature dependence of the rate of production of carbon-containing species.
- CH_4 (and the same is for CH_3) at a floating potential on the target;
- CH_4 (and the same is for CH_3) at a target bias: –300 V (target current is 80 mA);
- CH_4 (and the same for CH_3) with only the filament on;
- - - C atoms at a target bias –300 V (80 mA of H_2^+, H_3^+), on supposing the coefficient of physical sputtering to be 1×10^{-2} per proton [17].

4.2.2. Effect of membrane temperature on C film growth. With only the cathode on, superpermeability to thermal hydrogen atoms with a typical *temperature independence* of permeation flux was observed (fig. 6) and *was quite stable in spite of the C flux.*

Therefore, chemical sputtering prevents the formation of polyatomic C film, the original monolayer remaining responsible for the interface processes.

A stable temperature independent *plasma driven* permeation (superpermeation) also was observed at not very high C fluxes (fig. 6). At higher C fluxes, the ability of plasma facing surface to withstand the growth of polyatomic C films depends on *T*. That was demonstrated as follows (fig. 7). The membrane underwent isochronal (1.5 hrs) plasma expositions at different *T*. After every exposition, the state of the surface was examined at a standard temperature 520 °C by the atomic driven permeation (ADP) due to hot cathode. Before the exposition at next *T*, the standard state of the membrane (including the ADP magnitude) was restored by heating for a few minutes at 1200 °C.

The observed ADP reduction may result from a decrease in either the asymmetry factor or the implantation coefficient. The asymmetry was shown not to be decreasing, whence it was the growth of a polyatomic film barring the absorption of atoms that was responsible for the effect of plasma. The low-temperature branch of temperature dependence may be associated with the chemical sputtering of C by thermal H atoms [16], while the high temperature branch with the diffusion of C into metal producing a bulk carbide not hindering hydrogen absorption (see below). Thus it is easier to prevent the growth of polyatomic (not carbide) C films at a

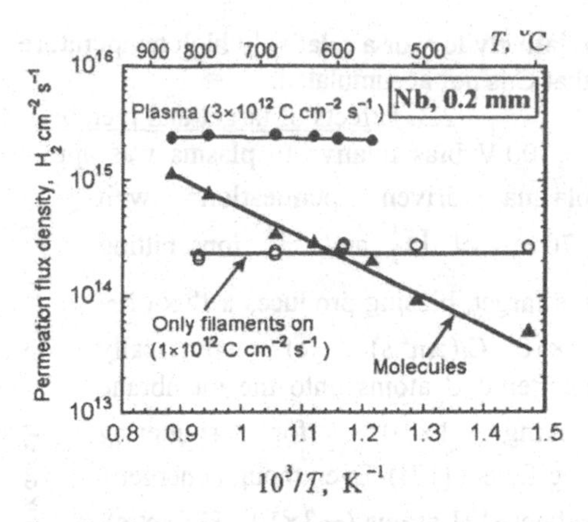

Figure 6. Temperature dependence of molecule-, atom- and plasma-driven permeation in the presence of a flux of carbon (as indicated in the brackets). In the case of plasma, the membrane and target are at a floating potential.

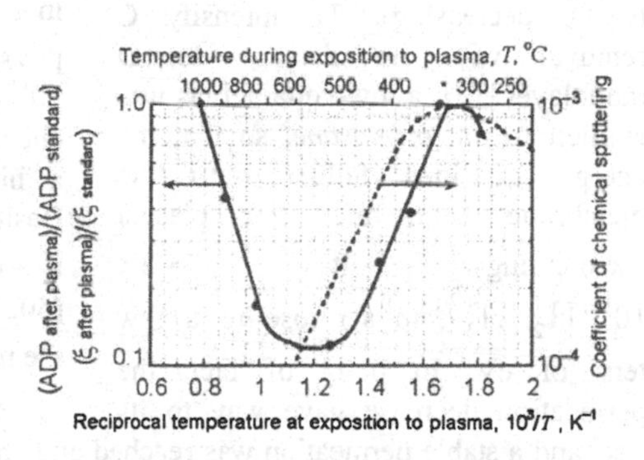

Figure 7. Effects of the isochronal (1.5 h) membrane expositions to a C-contaminated H plasma (C flux $\sim 10^{13}$ cm^{-2} c^{-1}) on the permeation of thermal atoms (ADP) and on the probability of their absorption, ξ, as functions of membrane temperature during the exposition. The coefficient of chemical sputtering [16] by thermal H atoms of the deposited C film is also presented

relatively low or a relatively high temperature. Still a low temperature has the advantage that C is not accumulated.

4.2.3. Effects of target and membrane biasing. In the next experiment (fig. 8), a – 300 V bias relative to plasma was applied to the target after the establishment of plasma driven permeation; with 170 mA of H_2^+ and H_3^+ ions hitting the target, biasing produces a flux of ~ 6×10^{13} C/(cm^2 s) of physically sputtered C atoms onto the membrane (taking 1×10^{-2} for sputtering coefficient [17]). Even though energetic reflected H atoms (~ 3×10^{15} H/(cm^2 s), 50–100 eV) were getting at the membrane alongside of C, the permeation started decreasing due to the ξ decrease that occurs with the growth of polyatomic C film, as the deposition of C surpasses its removal (a special test showed the asymmetry not to be decreasing). To intensify C removal with no damage to O monolayer [12], a low-voltage bias was applied to *the membrane*, so that the energy of ions falling onto the membrane from the rare plasma surrounding it (~ 3×10^{15} H_2^+, H_3^+/(cm^2 s)) rose to a few tens of eV. In point of fact, the permeation decrease gave way to its rise, and a stable permeation was reached and was keeping during the whole observation time[1]. About 10^3 C monolayers might have been deposited during that time; had that actually taken place, such a film would have been virtually impenetrable to all the groups of H particles.

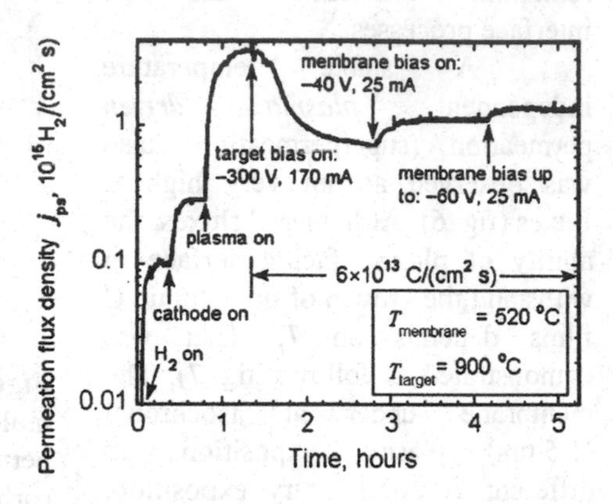

Figure 8. The evolution of plasma driven permeation caused by target and membrane bias. Biasing the target results in a dramatic carbon flux increase, with physically sputtered C atoms (6×10^{13} C/(cm^2 s)) dominating over chemically sputtered hydrocarbons due to a high target temperature (see Fig. 5). Besides, biasing of the target gives rise to a flux (~3×10^{15} H/(cm^2 s) of energetic (50–70e V) reflected H atoms bombarding the membrane.

4.2.4. The role of carbides of the Group Va metals. Although membrane biasing can remove C film, membrane heating at T = 1000-1200 °C was usually employed to restore the due state of superpermeable membrane after C deposition. Carbon will not be removed from the sample at such temperatures, and therefore it must be diffusing into bulk metal: Thousands of C monolayers would be transported in this way inside metal

[1] However permeation was not completely restored with biasing the membrane. The most likely explanation is that the tubular membrane stands out of plasma (fig. 4) and is being only partly sputtered.

during a typical experimental campaign to exist only in the form of carbide there (C solubility in Nb is negligible [7]). In fact, a layer-by-layer AES analysis of extracted samples discovered a 1 μm (!) near-surface NbC/Nb_2C layer. So one has to conclude that, unlike the modifications of pure C (graphite, a:C-films, diamond-like films), *NbC (and perhaps VC, TaC as well) do not stop hydrogen transport in the superpermeation regime*. Hence: (1) the surface barrier inhibiting hydrogen reemission keeps at Nb transformation into NbC (perhaps due to an oxicarbide monolayer), and, (2) a very large magnitude of $D \times S$ inherent to Nb is conserved in NbC (D and S are diffusivity and solubility of H in Nb correspondingly).

5. Conclusions

- Surface composition of PFM is responsible for the large scale effects on hydrogen absorption, permeation and reemission; these effects may be employed for the active particle control in fusion devices.
- There are sufficient physical grounds right now to employ the permeation technique for the short way separation of D/T and He and for the particle control in *high recycle* divertors.
- An employment of permeation/absorption technique in a *low recycle* divertor looks rather attractive but more problematic because of sputtering and possible deposition of carbon.
- It is experimentally shown that superpermeable membrane may keep its favorable asymmetry under the sputtering even under UHV conditions.
- It is experimentally demonstrated that at definite conditions superpermeable membrane may steadily operate under a concurrent flows of suprathermal hydrogen and C.

Acknowledgements

This work was supported by the International Science and Technology Center, Project 1110, and by the Russian Foundation for Basic Researches, Project 98-02-18248.

References

1. A. I. Livshits, *Vacuum* 29 (1979) 103.
2. R. A. Causey, R. A. Kerst and B. E. Mills, *J.Nucl.Mater.* 122&123 (1984) 1547.
3. A. I. Livshits, M. E. Notkin and A. A. Samartsev, *J.Nucl.Mater.* 170 (1990) 74.
4. A. I. Livshits, M. E. Notkin, A. A. Samartsev, A. O. Busnyuk, A. Yu. Doroshin and V. I. Pistunovich, *J.Nucl.Mater.* 196–198 (1992) 159.
5. A. Livshits, N. Ohyabu, M. Notkin *et al.*, *J.Nucl.Mater.* 241–243 (1997) 1203.
6. A. I. Livshits, *Sov.Phys.Tech.Phys.* 20 (1975) 1207; 21 (1976) 187.
7. Gase und Kohlenstoff in Metallen, eds. E. Fromm and E. Gebhardt (*Springer*, Berlin, 1976).
8. E. S. Hotston and G. M. McCracken, *J.Nucl.Mater.* 68 (1977) 277.
9. A. I. Livshits, M. E. Notkin, A. A. Samartsev and I. P. Grigoriadi, *J.Nucl.Mater.* 178 (1991) 1.

10. M. W. Roberts and C. S. McKee, Chemistry of Metal – Gas Interface, (*Claredon Press*, Oxford, 1978).
11. M. Yamawaki, N. Chitose, V. Bandurko and K. Yamaguchi, *Fus.Eng.Design* 28 (1995) 125.
12. A. Livshits, F. Sube, M. Notkin, M. Soloviev and M. Bacal, *J.Appl.Phys.* 84 (1998) 2558.
13. J. Park, T. Bennet, J. Schwarzmann and S. A. Cohen, *J.Nucl.Mater.* 220_222 (1995) 827.
14. J. Pillath, J. Winter and F. Waelbroeck, *J.Nucl.Mater.* 162–164, (1989) 1046.
15. N. Ohyabu, A. Komori, H. Suzuki *et al.*, *J.Nucl.Mater.* 266-269 (1999) 302.
16. E. Vietzke and A. A. Haasz, Chemical Erosion, in: "Physical Processes of the Interaction of Fusion Plasmas with Solids", eds. W. O. Hofer and J. Roth, *Academic Press*.
17. W. Eckstein and V. Phillips, Physical Sputtering and Radiation Enhanced Sublimation, in: "Physical Processes of the Interaction of Fusion Plasmas with Solids", eds. W. O. Hofer and J. Roth, *Academic Press*.
18. _A._I._Livshits, M._E._Notkin, V._I._Pistunovich, M._Bacal and A._O._Busnyuk, *J.Nucl.Mater.* 220-222 (1995) 259.
19. A._I._Livshits, M._E._Notkin, A._A._Samartsev and M._N._Solovyov, *J. Nucl. Mater.* 233-237(1996)1113

DEUTERIUM PUMPING WITH SUPERPERMEABLE MEMBRANE IN THE DIVERTOR OF JFT-2M TOKAMAK

Y. NAKAMURA[a], S. SENGOKU[b], Y. NAKAHARA[a], N. SUZUKI[a], H. SUZUKI[a], N. OHYABU[a], A. BUSNYUK[c], A. LIVSHITS[c]
[a]*National Institute for Fusion Science, Toki 509-5292, Japan*
[b]*Japan Atomic Energy Research Institute, Ibaraki 319-11, Japan*
[c]*Bonch-Bruyevich University, 61 Moika, 191186 St. Petersburg, Russia*

1. Introduction

There has been big progress in the last decade towards the demonstration of high performance plasmas in nuclear fusion research. This progress has often been coupled to improved wall materials and better conditioning techniques for impurity control and lower recycling. Wall coating techniques such as carbonization, boronization and beryllium evaporation not only reduce oxygen contamination and metal influx from the wall but also provide reduced hydrogen/deuterium recycling around the wall [1], which is responsible to prevent a confinement degradation. Due to finite wall capacities, however, these techniques are limited to transient pumping and may not be effective for long pulse operation in superconducting machines such as Tore Supra, LHD and ITER. The future direction towards long-pulse or steady-state experiments requires the development of a new type of divertor pumping system.

The superpermeable membranes may be employed for hydrogen pumping in various places in fusion devices. For instance, metal membranes are proposed for pumping of D/T fuel and its separation from He ashes in ITER [2]. Another possible application is to install the membranes just into the divertor region for strong active pumping of suprathermal hydrogen being present here in order to provide for the high-temperature divertor plasma operation [3]. In applying the membranes to fusion devices, especially, in the divertor where the membranes directly face plasmas, there are several subjects which should be investigated: (a) limitation of permeation flux density, (b) reliability of membrane operation in the presence of chemically active gases, (c) sputtering of nonmetal film at the inlet surface by suprathermal hydrogen particles, (d) deposition of nonmetal (e.g. B, C) and metal (e.g. Fe, Ti) impurities. Some of them have been already investigated experimentally [4-6].

In addition to that, it is very important to apply a membrane pumping system to a real fusion device and to investigate the permeation flux density through the membrane which is located in the divertor region. Therefore we design a prototype of membrane

C.H. Wu (ed.), Hydrogen Recycling at Plasma Facing Materials, 35–40.

pumping system to evacuate deuterium particles in the JFT-2M tokamak divertor and investigate the outcoming flux of deuterium atoms produced by the divertor plasma. In this paper, we present the results on first application of membrane pumping to a fusion device.

2. Experimental Arrangement

A small pumping system with superpermeable membranes was installed into the divertor region of JFT-2M tokamak as shown in Fig. 1(a). The JFT-2M is a medium size tokamak (major radius R = 1.31 m, minor radius a < 0.35 m, elongation $\kappa < 1.7$, magnetic field $B_t < 2.2$ T). Details of this machine were given elsewhere in conjunction with a review of confinement and fueling studies [7]. A large number of hydrogen atoms are produced by the divertor plasma through atomic and molecular processes. The incident flux of energetic hydrogen particles to the membranes, which are located at the position of 15 cm from the center of divertor plasma, is provided by these hydrogen atoms (mainly Frank-Condon atoms).

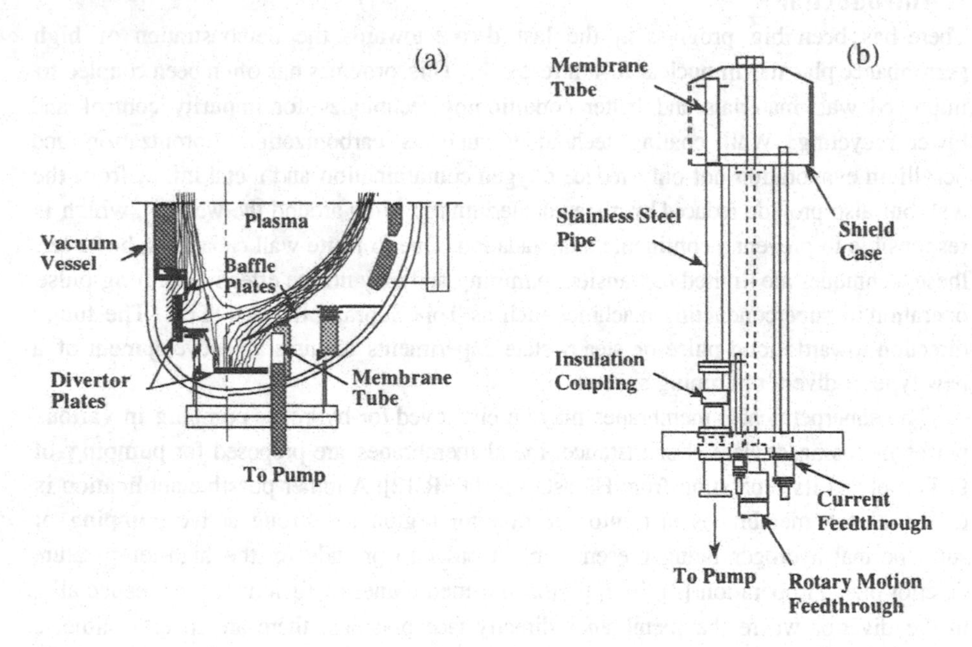

Figure 1. Schematic views of (a) the JFT-2M closed divertor showing the location of tubular membrane and (b) a prototype membrane pumping system. The magnetic field lines are indicated in the scrape-off layer.

The structure of membrane pumping module is shown in Fig. 1(b). This module has two Nb tubular membranes with the diameter of 1.5 cm, the length of 9.4 cm and the thickness of 0.02 cm. The membrane tubes were connected to a pumping system (turbo-molecular pump with the pumping speed of 500 l/s) through stainless steel pipes and the hydrogen particles permeated through the membranes were evacuated. The tubular membranes were electrically isolated with a ceramic from the plasma chamber and resistively heated by an alternating current through the membranes. The membrane temperature could be raised up to 1200 °C with an ohmic current of 250 A. To make clear that the incident hydrogen flux to the membrane originates from the divertor plasma, the membrane tubes are surrounded with a double cylindrical shield box whose outer cylinder is able to rotate by a rotary motion feedthrough.

In membrane experiments, prior to tokamak discharges, the membranes were heated by ohmic current (150 A) up to high temperature (~870 °C) and the membrane temperature was sustained until rising of the toroidal magnetic field. The circuits of membrane as well as the thermocouples was cut off during tokamak operation on account of the Lorenz force due to the interaction between the membrane current and the magnetic field. In this phase the membrane temperature decreased gradually and it was about 650 °C in the course of plasma discharge according to our calculation. Within the interval of tokamak operation (7 minutes), the membranes were heated up again with the same current to release the hydrogen atoms absorbed during the discharge. This time sequence was repeated in the course of membrane experiment.

The tokamak machine was mainly operated in an ohmically heated discharge with a lower-single-null (LSN) divertor configuration. The tokamak plasma was initiated with a fueling gas of D_2 and then gradually proceeded to the divertor configuration. When the divertor configuration is established at around 0.4 s, the plasma flows and reaches the divertor plate, and the pressure in the lower divertor chamber (P_{div}) increases due to the localization of recycling. The divertor configuration was maintained almost unchanged for 0.5 s. For divertor plasma, the density was around 5 x 10^{18} m^{-3} and the temperature 10 - 20 eV. In the course of membrane experiment, there were several discharges additionally heated with neutral beam injection (NBI) with the power of 320 kW. An upper-single-null (USN) divertor configuration with open geometry, where there was no plasma in the closed divertor region, was attempted to compare the membrane permeation flux between two divertor configurations.

3. Experimental Results

Figure 2 shows the time history of the pressure at the downstream side in the course of membrane pumping experiments. Observed evidences for membrane pumping are summarized as follows: The pressure rise at the downstream side is observed only when the shield is open and the membrane faces the divertor plasma (e.g. shots 87011-87013 in Fig. 2(b)). In case of closing the shield to prevent the membrane from facing the plasma, we observed no increase in the downstream pressure, that is, no permeation flux through the membrane as indicated in Fig. 2(a). On the contrary, when the membrane faced the divertor plasma, the downstream pressure increased on a large scale after reheating the membrane as seen in Fig. 2(b). This pressure increase is attributed to the permeation of deuterium atoms originated from the divertor plasma. In the phase of

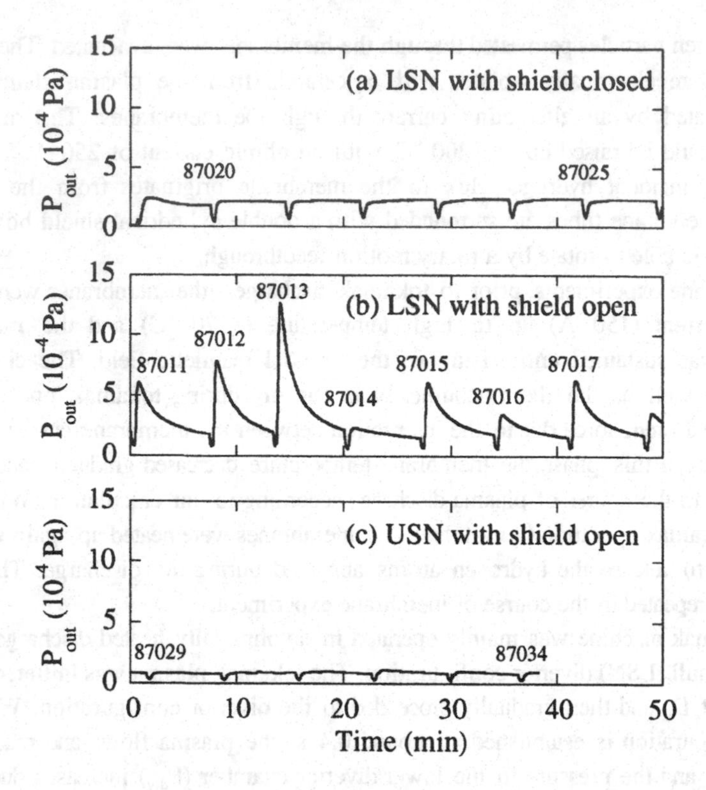

Figure 2. Time history of the pressure at the downstream side of membrane in the course of membrane pumping experiment: (a) with membrane shielding in LSN, (b) without membrane shielding in LSN for the sequential shots of a density scan experiment (87011-87013), disruption occurred before (87014) and after (87015-87017) the establishment of divertor configuration, (c) without membrane shielding in USN. LSN and USN represent the lower- and upper-single-null divertor configurations, respectively. The pressure increase is observed only when the membrane faces plasmas.

decreasing the membrane temperature during tokamak operation, the incident deuterium atoms are absorbed and confined in the metal. With increasing the membrane temperature, a major part of the absorbed particles are released at the downstream side. Another evidence for plasma driven permeation was revealed in the plasma disruption that occurred prior to establishing the divertor configuration (shot 87014 in Fig. 2(b)) and the USN divertor discharges (Fig. 2(c)). As expected, no pressure increase was observed in the downstream chamber because there was no plasma in the divertor region near the membrane module. Thus a deuterium pumping with superpermeable membrane has been demonstrated for the first time in fusion devices.

The particle permeation rate (pumping rate) was estimated by integrating the gas flow through the constant conductance C_0 (= 2.6 l/s) with time. The pumping rate was reduced by shortening of discharge duration due to plasma disruption (e.g. shots

87015-87017 in Fig. 2(b)). For full duration discharges, it was also dependent upon the type of plasma discharges (Ohmically heated, with NBI, with additional gas puffing). The maximum pumping rate of 2.8 x 10^{17} D/s was obtained in the discharge with a strong gas puffing into the divertor chamber. Figure 3 shows the dependence of the pumping rate on the divertor pressure measured by a Penning gauge. One can see that the pumping rate increases in proportion to the divertor pressure, which is closely related to the divertor particle flux. In the USN divertor discharges, the pressure in the lower divertor chamber is much less than that for the LSN divertor discharges because there is no plasma flow from the main plasma to the divertor plate. For discharges with NBI heating, both main and divertor plasma densities increase by 20-30 % and consequently the divertor pressure goes up due to the increase of divertor particle flux. Then one can expect that the increase of plasma density in the divertor region results in the increase of suprathermal atoms, which are responsible for the membrane permeation. Another important discharge with a strong gas puffing was carried out for the purpose of obtaining a dense and cold divertor plasma, which is effective for reducing the heat load of divertor plate during NBI heating. In these discharges, the plasma density in the divertor region increases by two or three times, although the electron temperature decreases with increasing the gas puffing. This divertor plasma serves as an efficient generator of suprathermal deuterium (mainly atoms) and the permeation flux increases proportionally to the divertor pressure. This proportional dependence is greatly advantageous in applying the membrane pumping to the closed divertor in which neutral particles may be compressed by orders of magnitude in future machines.

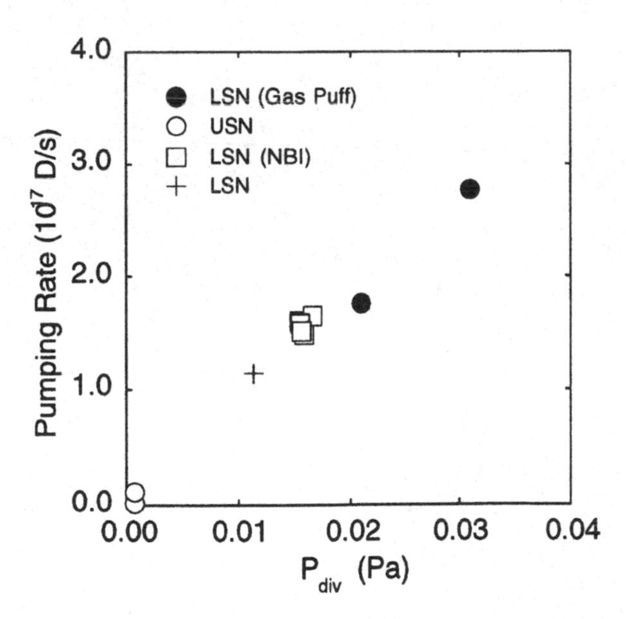

Figure 3. Dependence of pumping rate on divertor pressure. The pumping rate is proportional to the divertor chamber pressure and depends on the magnetic configuration, additional heating (NBI) and gas puffing.

4. Conclusion

A prototype membrane pumping system with superpermeable Nb tubular membranes has been installed into divertor region in the JFT-2M tokamak and the pumping capability has been investigated. A deuterium pumping through the membrane has been demonstrated in tokamak environment for the first time. The pumping flux increases in proportion to the divertor pressure and the maximum permeation flux density amounts to 7.3 x 10^{19} D/m^2s with the divertor pressure of 0.03 Pa. Application of membrane pumping to future large machines, where the divertor pressure may be greatly increased by the compression of neutral particles into the divertor, seems to be rather promising. Further investigation concerning metal membrane heating method would be required for steady state operation of membrane pumping system in long-pulse machines.

References

[1] J. Winter, Plasma Phys. Control. Fusion 36 (1994) B263.

[2] A.I. Livshits et al., J. Nucl. Mater. 196-198 (1992) 159.

[3] N. Ohyabu et al., J. Nucl. Mater. 220-222 (1995) 298.

[4] A. Livshits, N. Ohyabu, M. Notkin et al., J. Nucl. Mater. 241-243 (1997) 1203.

[5] H. Suzuki et al., J. Plasma Fusion Res. SERIES 1 (1998) 402.

[6] A. Livshits, N. Ohyabu, M. Bacal, Y. Nakamura et al., J. Nucl. Mater. 266-269 (1999) 1267.

[7] S. Sengoku and the JFT-2M Team, J. Nucl. Mater. 145-147 (1987) 556.

V.I. PISTUNOVICH

WALL PUMPING IN TOKAMAK WITH LITHIUM

Nuclear Fusio Institute, RRC "Kurchatov Institute", Moscow, Russia

1. INTRODUCTION

The most important issues for steady-state operation of magnetic fusion devices are plasma facing components. Usual materials as graphite, beryllium, tungsten don't decide this problem. Liquid lithium may be considered as possible plasma facing material in magnetic fusion devices. The lithium capillary target experiments in Russia with the electron beam on SPRUT-4 DEVICE, experiments simulating the disruption conditions in a tokamak, and on tokamak T-11M and calculations of the lithium ions behavior in tokamak have been performed at last years. This concept assumes the evaporational-radiational cooling of highest loaded elements.

The beam-plasma discharge in crossed electric and magnetic fields in the SPRUT-4 device was characterized by densities of (10^{12}-10^{14}cm^{-3} and electron temperatures of (2-5) eV. The lithium capillary target has been successfully tested at a specific thermal load of electron beams up to 50MW/m^2. The exposition time was 1000s for a load of (1-5) MW/m^2, 100s for (15-20) MW/m^2 and 5s for (45-50) MW/m^2, fig. 1.

The effects of disruption discharges in tokamak have been simulated by means of magnetized plasma flows in quasi-stationary plasma accelerator QSPA at an energy density up to 5 MJ/m^m and at a pulse duration 0,5 ms, fig. 2. It was found that lithium plasma cloud adsorbs the largest part of the coming energy and re-radiate it, so that only (1-3)% reaches the lithium capillary porous surface..

Experiments have shown the opportunity of lithium capillary targets to withstand high impulse loadings. An experiment on T-11M tokamak has been started with movable lithium capillary porous limiter.

The calculations to base the ITER divertor concept with liquid lithium have been done. 2D modeling of the ITER divertor with lithium target have been performed

C.H. Wu (ed.), Hydrogen Recycling at Plasma Facing Materials, 41–49.

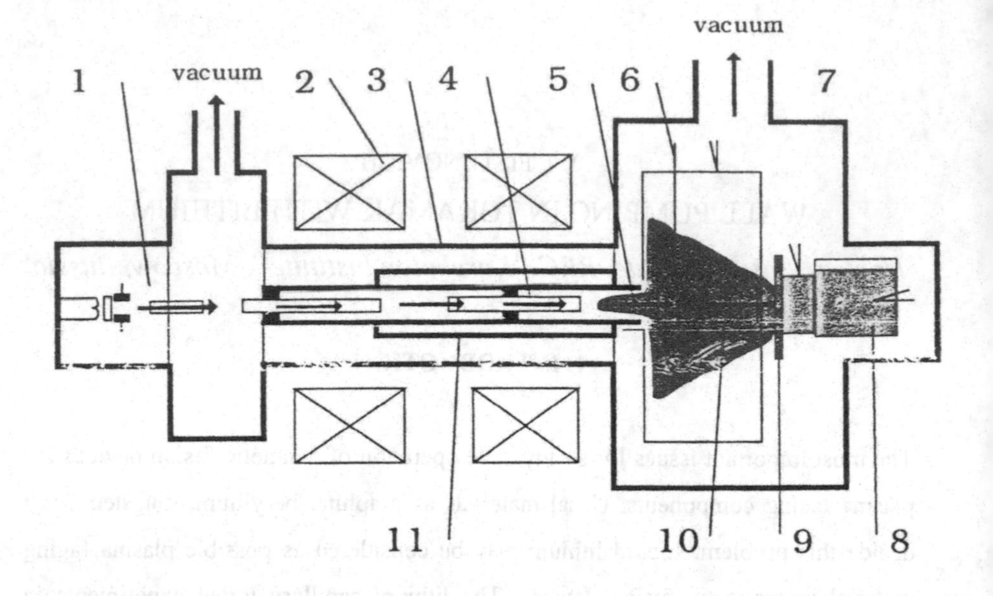

1.Electron beam source
2.Magnetic system
3.Vacuum vessel
4.Defferential vacuum resistance
5.Electron beam
6.Lengmur's probe
7.Lithium vapor condenser
8.Target
9.Capillary surface
10.Viewing window
11.Water cooling condenser

Fig. 1. SPRUT

QSPA JANUARY 1996

Scheme of the target - plasma flow interaction experiment.

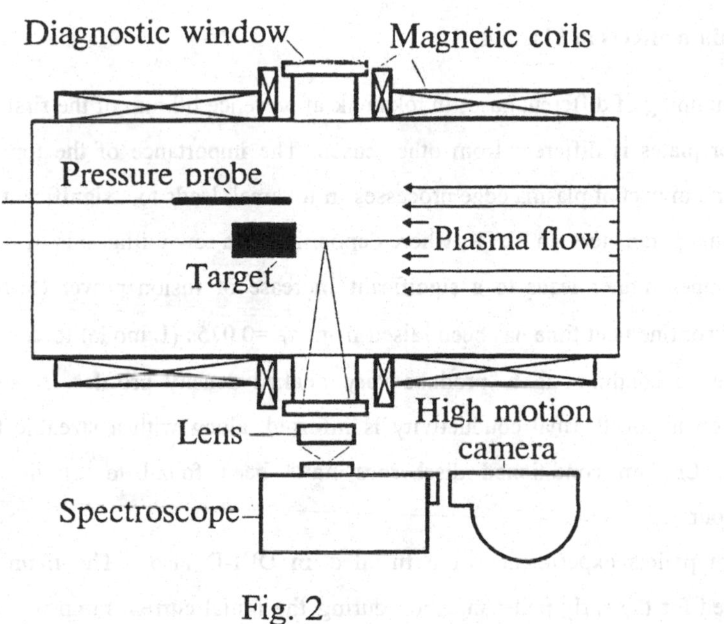

Fig. 2

using the DDIC95 code. The calculations have shown that thermal loads in the divertor plates are reduced down to 1,3MW/m^2 . The main power entering the divertor is radiated on the baffles in the divertor. On the basis of radial Li-ion density distributions in the divertor layer one-fold calculations of ITER plasma parameters inside the separatrix have been done.

The limiting Li-ion fluxes arriving from the peripheral zone to the main reactor plasma have been determined. The calculations take account of helium recycling in the reactor.

Similar calculation has been done with neon, for comparison.

2. Lithium effects in Plasmas

Wall pumping of different gases in tokamak at presence lithium of the first wall and divertor plates is different from other cases. The importance of the topic by the profound impact of plasma edge processes in tokamak leads to a significant increase of plasma parameters. In TFTR, where deposition of a few milligrams of lithium on the bumper limiter leads to a significant increase of fusion power (5MW). The energy confinement time has been raised from τ_E =0.075s (L mode) to τ_E =0.33s by using edge conditioning to produce very peaked density profiles. A significant reduction in ion thermal conductivity is inferred, along with a sizeable transport barrier. Lithium conditioned discharges have been found to limit sawtooth behaviour.

Lithium pellets experiments were fulfilled in DIII-D also. The main changes observed for the early pellet injection during the initial current ramp was a modes increase (23%) in the gas puffing needed to achieve the target density, and a reduction in the oxygen content. In high performance negative central shear (NCS) discharges lithium injection reduced the required duration of intershot helium glow discharge cleaning, showed some tendency to eliminate locked modes and produce less disruptivity at high current. Reductions in the central oxygen content were also observed in NCS discharges with lithium conditioning.

Tokamak Alcator C-Mod operates with metal walls (molybdenum) in contrast to the carbon furfaces of the other machines. Lithium pellet injection leads to higher neutron rates on Alcator C-Mod, but this result is not understood else.

On TdeV tokamak the experiments were to assess the deuterium wall retention with and without lithium conditioning, and to study the effect of lithium conditioning on impurity levels. The result of this experiment was that only a small and transient reduction in hydrogenic recycling (measured by H_{α}light) was found as a result of the lithiumization, along with a small reduction of CO.

There was a negligible effect on the plasma parameters, even through an estimated amount of lithium equivalent from 40 to 100 monolayers was injected in 15 shots. An overall ratio of hydrogen atoms/injected lithium atoms of 0.1 to 0.3 was estimated.

3. Interaction of Li with hydrogen plasma

There has been growing interest in wall conditioning with lithium layer deposition since the first laboratory experiment [1] and lithium pellet injection of TFTR [2]. To date the lithium conditioning has been done in several devices TFTR [3], DIII-D [4], JIPP-TIIV [5], TdeV [6], Alcator C-Mod [7], Heliotron-E [8] and T-11M [9]. In most cases, the main effect has been substantial decrease in oxygen and carbon impurities along with low hydrogen recycling. In addition, small-scale laboratory study [10] revealed various chemical activities of the lithium layer, supporting the observations in fusion devices.

An exposure of the lithium surface to a hydrogen plasma leads to an enormous uptake of hydrogen in the lithium layer. The hydrogen uptake ceases when the

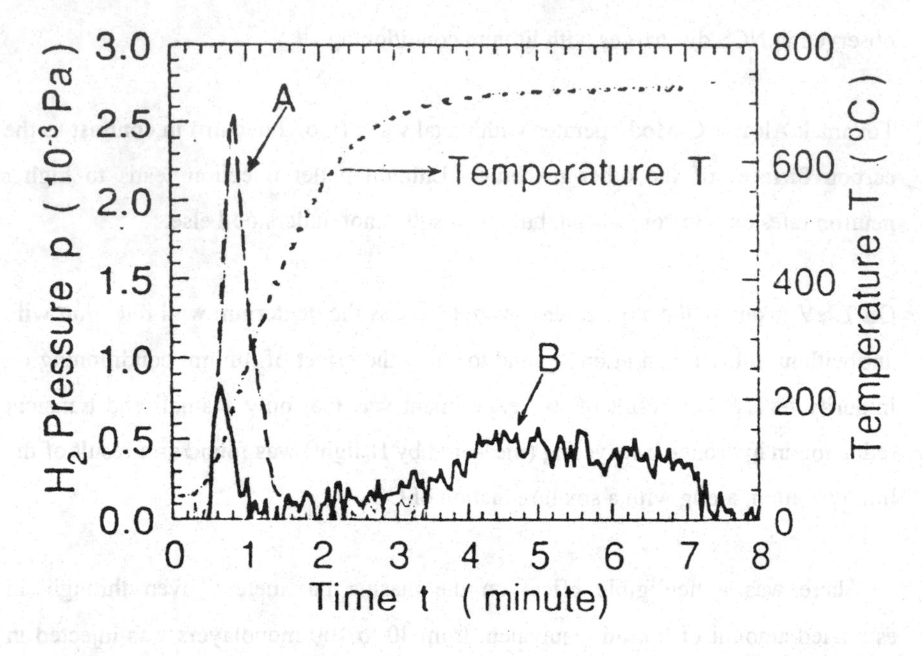

Fig. 3. Thermal desorption of hydrogen molecule from lithium-deposited wall. The solid lines A and B indicate hydrogen pressures and the dashed line the test wall temperature.

lithium layer is saturated with hydrogen at the atomic ratio Li:H=1, suggesting lithium hydride (LiH) formation.

Thermal desorbtion of hydrogen molecules from the lithium layer saturated with hydrogen is observed at 200 °C, which is much lower then the melting point (680 °C) of LiH. Thus were remains a question about the chemical state of hydrogen atoms in the lithium layer. Fig. 3 (Sugai) shows examples of thermal desorption spectra (TDS) [11]. The curve labeled "A" shows a clear peak at T~200 °C just around the melting point (179 °C) of lithium. The time integration of hydrogen release rate gives the total hydrogen inventory, which was the same order of magnitude as the hydrogen uptake estimated from the pressure decrease during the hydrogen glow discharge. Thus most of hydrogen atoms contained in the lithium layer are easily desorbed at relatively low temperatures.

However, the TDS seems to depend of the experimental history and circumstances. The curve "B" in Fig.3 was obtained as follows. The test wall with lithium layer containing H_2 is heated at a slow ramp rate (0.25 °C/s) up to 300 °C, kept there for 60min, again exposed to the hydrogen glow at room temperature for 10min, and finally heated at a fast ramp rate to 700 °C as shown in Fig.3. In this case, hydrogen molecules are released not only at low temperature (<200 °C) but also at high temperatures around the melting point (680 °C) of lithium hydride. A thin lithium layer on graphite was found to reduce both physical sputtering (in a helium plasma) and chemical sputtering. LiOH is more easily desorbed than H_2O. From the results, a provisional model speculates that the lack of recycling effects on AlcatorC-Mod may be due to the oxygen, which is normally present as $M0_2O_3$, bound up by lithium.

4. Modelling and lithium conditioning on large tokamaks.

At present the working model for lithium effects in plasmas is at the conceptual stage: there are some general ideas about the processes at work. Detailed modelling that would permit a quantitative comparison of these general ideas with confinement experiments is just beginning. Thus, we are some what removed from having a predictive model that could be extrapolated to future devices, such as ITER. The model development problem is separable into core and wall/scrape-off layer sub problems [12]. The core transport problems requires, at present experimental specifications of the boundary conditions and of the density profile, and these can be provided by existing diagnostics. The SOL problems require more basic laboratory and other fundamental data (e.g., more reflection data and sputtering yields for realistic surfaces). It is necessary to simulate the observed lithium dept distribution in samples taken from confinement experiments. Further, detailed comparison of existing models with spectroscopic observations is required. The ITER issues are discussed:

a) Steady state erosion of lithium coatings and development of ideas to extend the lifetime of lithium coatings;
b) Exploration of the use of lithium during ITER startup scenarios;
c) Comparison of the relative merits of lithium and beryllium;
d) Extension of lithium confinement enhancement to regimes other than those found in TFTR.

5. Summary

The results illustrate the potential for lithium conditioning to improve the performance of future fusion reactors to a significant degree. The realization of this benefit will require continued study leading to a better understanding of confinement and of the associated interdisciplinary issues in solid state, atomic, molecular and surface physics.

References

1. T.Isozumi, S. Yoshida and H. Sugai, Kaku Yugo Kenkun 60 (1988)304.
2. J.A. Snipes et.al., J. Nucl. Mater. 196-198 (1992) 686.
3. D.K. Mansfield et.al. Phys. Plasmas 3 (1996) 1892; J.T. Hogan, C.E. Bush, C.H. Skinner, Nucl. Fusion v. 37 (1997) 705.
4. J.L. Jackson et.al. J. Nucl. Mater. 241-243 (1997) 655.
5. H. Sugai et.al., J. Nucl. Mater. 220-222 (1995) 254.
6. B. Terreault, J. Nucl. Mater. 220-222 (1995) 790.
7. J.T. Hogan et.al. Nucl. Fusion v.37 (1997) 705.
8. K.Kondo et.al. Annu. Meet. Of the Jap. Soc. Of Plasma Sei. and Fusion Research (Kyoto, March 1996).
9. V.A. Evtikhin et.al. IAEA-CN-69/FTP/27, Proc. 17th Int. Conf. of Fusion Energy, 1998.
10. H. Sugai et.al., J. Nucl. Mater. 220-222 (1995) 254.
11. H. Toyoda et.al., J. of Nucl. Mater. 248 (1997) 85.

REFLECTION AND ADSORPTION OF HYDROGEN ATOMS AND MOLECULES ON GRAPHITE AND TUNGSTEN

E. Vietzke, M. Wada*, M. Hennes

*Institut für Plasmaphysik, Forschungszentrum Juelich GmbH, Euratom Association, Trilateral Euregio Cluster, 52425 Juelich, Germany; *on sabbatical leave from Department of Electronics, Doshisha University, Kyoto, Japan.*

Abstract

The reflection coefficients of hydrogen/deuterium atoms and molecules on graphite and tungsten are experimentally determined by using a mixed thermal beam of atoms and molecules with temperatures above 2000 K. The velocity distributions of released atoms and molecules measured by the time-of-flight are used to distinguish the elastically reflected particles from desorbed particles. The released deuterium atoms and the elastically reflected molecules show the energy distribution of the incident beams. Most of the adsorbed deuterium is released as molecules with a Maxwell-Boltzmann velocity distribution of the target temperature. The reflection coefficients for deuterium atoms and molecules (above 2000 K) on graphite and tungsten ranges from 0.7 to 0.9. The reflection coefficient for cold deuterium molecules on hot tungsten is zero.

1. Introduction

The hydrogen recycling on plasma-facing materials largely affects the fuelling efficiency of fusion plasmas. Recycling includes both the reemission of implanted or adsorbed hydrogen and the reflection processes. Particle reflection is an important parameter in modeling the recycling, since reflected particles have higher energies than the reemitted ones. For modeling the divertor plasma it is essential to know the reflection coefficient of hydrogen isotopes at low energies.

For energetic ions, the reflection process is an ion-atom scattering between the incident ion and the target atoms which leads to charge transfer, energy loss and momentum transfer. Many features of the reflection in the keV range can be reasonably understood in the single-collision model [1,2] in which the incident particle penetrates the solid in a straight trajectory, experiences a collision at some depth, and leaves the solid again in a straight trajectory.

C.H. Wu (ed.), Hydrogen Recycling at Plasma Facing Materials, 51–58.

Computer simulation has been very successfully applied to the reflection. All simulations are based on the binary collision approximation.[3,4]. Many results are in good agreement with the observed reflection, i.e. the validities of interatomic potentials and electronic energy losses are satisfactory. A problem arises by applying the binary collision approximation at low impact energies as it will be discussed below. Experimental reflection data of hydrogen and deuterium on graphite exist for energies down to 30 eV [5, 6] and they agree with TRIM calculations[7]. Experimental reflection data of hydrogen and deuterium on W exist only for energies above 5 keV [8]. Their values are smaller by a factor between 2 to 3 compared to the calculated ones.

Very little is known about the reflection and adsorption of hydrogen atoms and molecules at lower energies. In model calculations [7], the reflection coefficients below one eV decreases to zero due to the influence of the surface binding energy. However, if the incident atom finds no possible way to dissipate the kinetic energy the reflection coefficient could be close to one. In this low energy range, the most effective mechanism in energy transfer is the excitation of phonons in the adsorbing surface [9]. In the study of model calculation of low temperature plasmas the reflection coefficient of hydrogen atoms is used as a fitting parameter, and mostly a value close to one is employed [10,11].

2. Experimental and data evaluation

To determine the reflection and adsorption coefficients of atomic and molecular hydrogen (H)/ deuterium (D) on graphite and tungsten, a mixed beam of atomic and molecular hydrogen (deuterium) with temperatures of 1900 and 2530 K is directed with an incident angle of 45° to the surface and the amount of released H^o (D^o) and H_2 (D_2) are measured by line-of-sight mass spectroscopy. The velocity distributions measured by the time-of-flight (TOF) signals are used to distinguish the elastically reflected particles having the distribution of the incident beam from the desorbed particles that show the distribution corresponding to the equilibrium with the target temperature.

Details of the experimental set-up and the data evaluation are described in a previous paper [12]. Briefly, 4 mm wide samples were bombarded by a mixed beam of $5x10^{19}$ H^o/m^2s and $4.7x10^{19}$ H_2/m^2*s (or $4.3x10^{19}$ D^o/m^2s and $5x10^{19}$ D_2/m^2s) in the case of 2530 K source (tungsten tube) temperature and an incident angle of 45°. A beam temperature of 1900 K is used to determine the reflection of the molecules alone with $7x10^{19}$ H_2 or D_2/m^2*s and 2% atoms. The emitted particles were periodically chopped by a motor driven disk with a trapezoidal transmission pulse (HWFM = 13 μs, repetition time 700 μs) and were directly detected after a flight path of s = 15 cm by a differentially pumped line-of-sight quadrupole mass spectrometer (QMS) aligned at 90° with respect to the incident beam direction.

The reflection and adsorption coefficients are determined by the assumption that the retention of atomic and molecular hydrogen in the material is negligibly small (in [13] the retention in graphite is estimated to 4% at the beginning of hydrogen atom exposure and decreasing with further fluence, in tungsten it should be even smaller by this long exposure). The quantitative estimation of the reflection coefficient is complicated by two facts. The incoming beam intensity has been calculated and an angular distribution has been predicted from measurements at limited number of data points [12]. The angular distribution in the backward direction can not be measured. In this paper, from the distributions for an incident angle of 45° and 60° the specular reflection component is estimated.

The graphite samples are made from pyrolytic graphite (Union Carbide, HPG) by machine milling and baked before the experiment at 2200 K. The tungsten (W) samples are cut from 0.05 mm thick foils and ultrasonic cleaned in methanol before installing. The W sample is heated up to 1300 K before each exposure to remove all adsorbed gases. Experiments with sputter cleaned W surfaces are in progress.

3. Results

3.1 H/D REFLECTION ON GRAPHITE

The results of reflection and absorption of hydrogen/deuterium atoms and molecules on graphite have been recently published [12]. Here, only a summary is given and for comparison some results for C are presented.

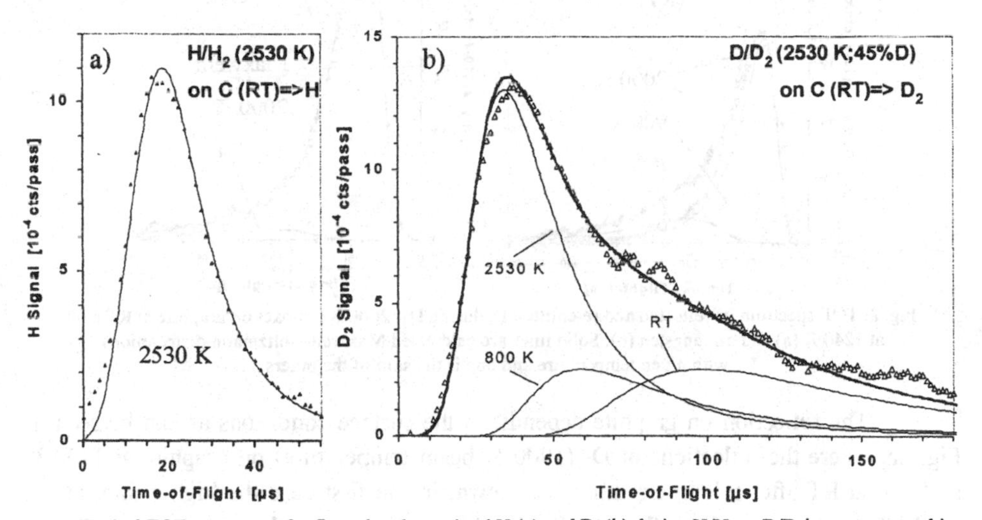

Fig. 1. Typical TOF spectrum of reflected and reemitted H (a) and D_2 (b) during H/H_2 or D/D_2 impact on graphite at RT under 45°. Thin solid lines are calculated Maxwell-Boltzmann distributions with given temperatures and the thick one the sum of others.

Figure 1 shows a typical TOF spectrum of reemitted D (left) and D_2 for a mixed incident beam of D and D_2 on graphite at room temperature after a long-term exposure. The temperature of the beam is 2530 K. For all observed results obtained with the beam temperature higher than the target temperature, the D (H) spectra consist of only one component with the incident beam temperature. The D_2 (H_2) spectra consists of two or three components. The largest one has the incident beam temperature and the slowest shows the target temperature. In some distribution a third component has to be taken into account to fit the experimental result. For simplicity a Maxwellian has been assumed, in the case of Fig. 1b with 800 K temperature. It is obvious that the beam temperature component is formed from elastic reflections without energy loss, and the target temperature component corresponds to the group formed by desorption of adsorbed D or D_2. In [12] is shown that the incident atoms contribute to the target temperature components of the emitted D_2 by using a beam with much higher atomic component. The medium energy component could be formed from D_2 reflection with loss of energy or an exoergic desorption due to recombination which is similar to the effect observed in hydrogen desorption after permeation from tungsten [13]. The measured angular dependence of the re-emitted particle flux, as described in section 2, suggests that the atoms and molecules are reflected with a nearly cosine angular distribution [12].

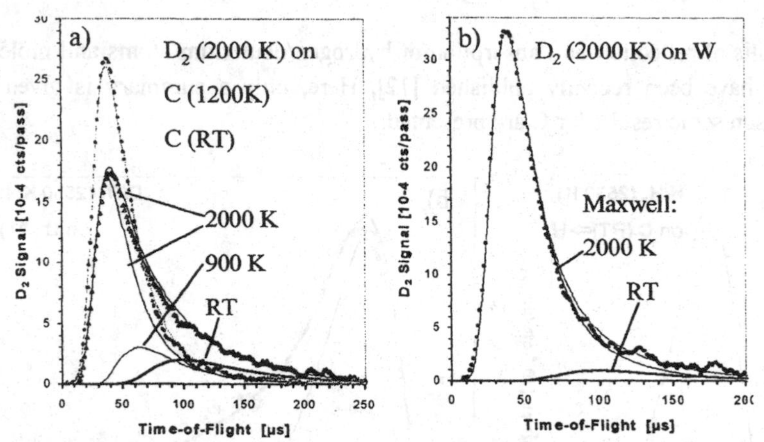

Fig. 2. TOF spectrum of reflected and re-emitted D_2 during D_2 (2000 K) impact on graphite at RT and at 1240 K (a) and on tungsten (b). Solid lines are calculated Maxwell-Boltzmann distributions with given temperature and one is the sum of the others.

The reflection on graphite depends on the surface conditions as can be seen in Fig. 2a, where the reflections of D_2 (1900 K beam temperature) on graphite at 1200 K and that at RT after a long exposure are shown. In the first case the higher intensity is partly due to the specular reflection. Similar high reflection peaks are obtained directly after baking the target to 1400 K [12]. After a longer irradiation these peaks are reduced (without a detectable increase of the RT component of D_2) due to the modification of the surface by the hydrogen atom bombardment. Since this temperature is above the thermal desorption temperature of implanted hydrogen or that of an a-C:H film the hydrogen in the surface layer can cause this change in the reflection behavior. The reflection coefficients obtained under different conditions [12] are collected in Table 1.

3.2 H/D REFLECTION ON TUNGSTEN

The elastic reflection of the deuterium molecules from W at RT is rather high as seen in figure 2b using a 1900 K D_2 beam. By comparing this results with those on C at RT in figure 2a it is clear that the reflection contains a large component of the specular reflection.

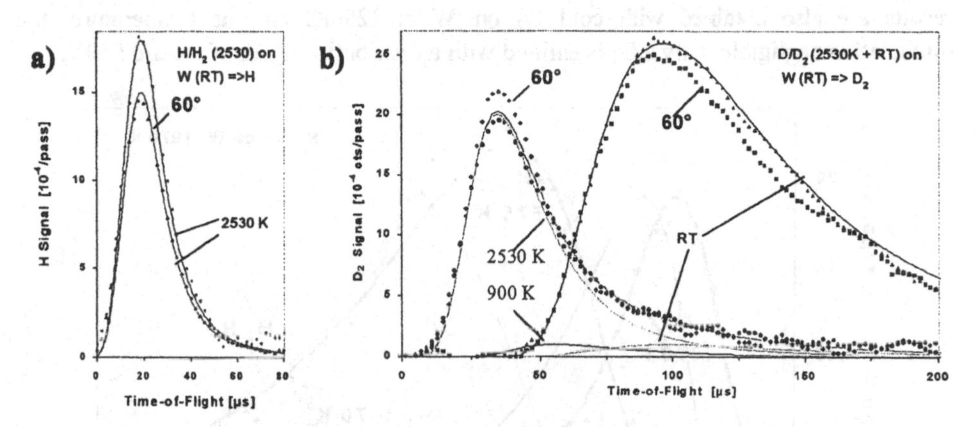

Fig. 3. TOF spectrum of re-emitted H and D_2 during H/H_2 or D/D_2(2530K and RT) impact on tungsten at RT for incident angles of 45° and 60°. Thin solid lines are calculated Maxwell-Boltzmann distributions of the given temperatures, the thick one the sum of others

Figure 3 shows the TOF spectrum of the re-emitted H, D (left figure) and D_2 from tungsten at room temperature exposed to a mixed incident beam of D and D_2 (or H and H_2) at two different incident angles with respect to the target normal, 45° and 60°. The temperature of the beams are 2530 K and RT. We clearly see the re-emitted H signals from an irradiation with a 2500 K beam are roughly 25% larger for the 60°case (re-emission angle = 30°) indicating that the atomic re-emission occurs by a nearly cosine distribution. In contrast to this result, the D_2 signals for the RT beam as well as that for the 2500 K one are nearly equal for both incident angles, which means the signals for both beams at the incident angle of 45° contain a large components of specular reflection. From this comparison around 30 % of the reemitted D_2 of the RT beam and 20 % in the 2500 K peak are assumed to be components due to the specular reflection.

Remarkable in the remission of D_2 at 2500 K irradiation are the small contributions from components of the RT and 900 K to the distribution meaning the adsorption of D and D_2 are small. There is no change in this behavior at higher target temperature and also a longer exposure to very high fluences up to $6*10^{23}$ D/m^2 and $7*10^{23}$ D_2/m^2 results only in a negligible reduction of the atomic elastic reflection. The obtained reflection coefficients are collected in Table 1.

3.3 REFLECTION OF COLD H_2/D_2 ON W AT ELEVATED TEMPERATURE

Completely different is the reflection behavior of cold hydrogen molecules on W at elevated temperatures. The reemission of cold (RT) H_2 from W at 1970 K is shown in Fig. 4. There is no or negligible elastic reflection of H_2. Atomic H is emitted with a target temperature distribution and H_2 with an Maxwell distribution of about 575 K. For orientation the H_2 distribution for 1970 K and RT are also marked by thin lines. Similar results are also obtained with cold D_2 on W at 1250K. At this temperature, the dissociation negligible, only a D_2 is emitted with a distribution again of around 550K.

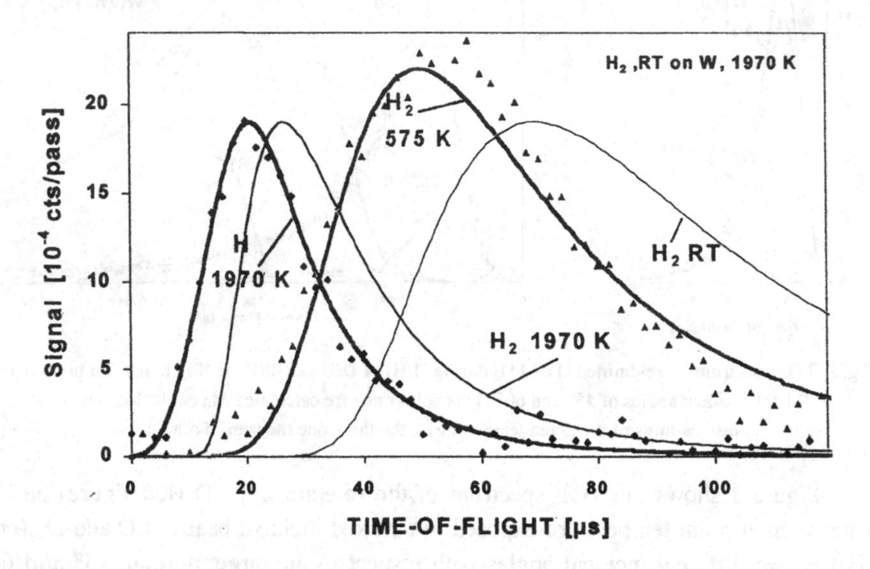

Fig. 4: TOF spectrum of re-emitted H and H_2 during cold H_2 (room temperature) impact on tungsten at 1970 K. Solid lines are calculated Maxwell-Boltzmann distributions of the given temperatures

4. Discussion

An important observation is the reflection of overthermal hydrogen/deuterium atoms and molecules without any energy loss, i.e. they are elastically reflected with the velocity distribution of the incident beam. This result can be associated with the mechanism that an incident particle will be reflected if it has no possibility to dissipate its energy [9]. We have to state that only the momentum is exchanged with the surface, but no energy transfer is possible to the target.

The other point of view is also relevant: the reflection of low energy hydrogen on surfaces is not only governed by a kinetic mechanism as in the high energy range, but surface conditions play a important role as it appears in the dependence on the modification of the graphite surface by atomic hydrogen. The elastic reflection on H-rich graphite surfaces after long hydrogen atom exposure is reduced for both, atoms and molecules, which has been clearly observed in a long term reflection experiment [12]. This observation has a direct relevance to the modeling of the recycling for low energetic hydrogen on graphite since the surface of plasma-facing graphite components is mostly a hydrogen saturated layer or even an a-C:H film.

Change in surface condition is clearly the reason for the completely different behavior of the overthermal hydrogen/deuterium atoms and molecules on cold W surfaces and cold hydrogen/deuterium molecules on hot W. In the first case there is only less adsorption due to the dissipating of the kinetic energy for the incoming hydrogen and due to possible complete hydrogen coverage of the surface. In the second case, the hot surfaces with a short residence time of adsorbed particles are hydrogen free and an complete adsorption occurs for incident cold H_2 or D_2. After the dissociative adsorption the atoms are in thermal equilibrium with the surface whereas the molecules are released with a distribution far less than the surface temperature. A reason could be the molecules undergo only one encounter with the surface without coming to an equilibrium.

As already mentioned the quantitative estimation of the reflection coefficient is a rather difficult task. With the assumptions discussed almost all measured distributions could be fitted within an error of ± 0. 15 for R_D and R_{D2}.

5. Summary and Conclusion

The main results of the reflection and adsorption of overthermal hydrogen/deuterium atoms and molecules (≈2500 K) on graphite and tungsten are:

- The high energy component of released atoms and molecules show the energy distribution of the incident beams with a cosine angular distribution, i.e. they are elastically reflected. The reflection coefficient for overthermal atoms and molecules on room temperature graphite and tungsten ranges from 0.7 to 0.9. At high reflection probability, there is always found an enhanced specular reflection.
- The adsorbed hydrogen/deuterium is released as molecules with Maxwell-Boltzmann velocity distribution of the target temperature and of one at around 900 K.
- Within the experimental errors the reflection coefficients for hydrogen atoms and molecules are the same as those for deuterium.

Because the reflected atoms and molecules show a cosine distribution they are supposed to by elastically scatter at the surface, and only momentum but no energy of these particles is transferred to the target.

The obtained reflection coefficients follow the trend which has been calculated for hydrogen/deuterium ions above 30 eV, namely an increasing value with decreasing energy [5,6]. From the obtained result it seems also possible that the reflection coefficient for the intermediate energy range of some eV can be interpolated between 30 eV D^+

results and the overthermal D°, whereby the acceleration of the low energetic ions in the ion-surface potential up to their neutralization has to be taken into account.

Completely different is the reflection behavior of cold hydrogen molecules an hot surfaces. The reflection coefficient is zero and surface conditions determine the properties of the hydrogen desorption.

In actual fusion devices, the situation is much more complicated. On most of the plasma-facing surfaces hydrogen-rich carbon films are deposited often within islands. In such cases the lifetime of such films have to be estimated to decide which of the reflection coefficient should be applied.

Table 1: Reflection coefficients for D and D_2 (above 2000 K) on graphite at room temperature and at 1400 K, on tungsten at room temperature and for cold D_2 on hot tungsten surface at an incident angle of 45°. The error is estimated to be smaller than ± 0.2.

	D/D_2 (>2000 K) on			cold D_2 on
	D-rich C, RT	C above 1400 K	W, RT	on W, > 1200 K
R_D	0.65	0.9	0.9	--
R_{D2}	0.7	0.9	0.9	0

References:

[1] G.M. McCracken, N.J.Freeman, J.Phys. B2(1969)661.
[2] W.Eckstein, J.P.Biersack, Z.Phys. A 310 (1983) 1
[3] J.P.Biersack and W.Eckstein, Appl. Phys. 34 (1984) 73
[4] W.Eckstein, Computer Simulation of Ion-Solid Interaction, Springer Berlin 1991.
[5] R. Aratari and W. Eckstein, J. Nucl. Mater. 162-164 (1989) 910.
[6] M. Mayer, W. Eckstein, B.M.U. Scherzer, J. Appl. Phys.77 (1995) 6609.
[7] W. Eckstein and J.P. Biersack. Appl. Phys. A38 (1985) 123.
[8] W. Eckstein and H. Verbeek, 'Data on light ion reflection', IPP report 9/32, Garching 1979.
[9] K.D. Rendulic and A. Winkler, Surf. Science 299/300 (1994) 261.
[10] W.P. Biel, Ph.D. thesis, University Düsseldorf 1996.
[11] C. Gorse, Chem. Phys. 161 (1992) 211.
[12] E. Vietzke, M. Wada, M.Hennes, J. Nucl. Mater. 266-269 (1999) 324.
[13] P. Hucks, K. Flaskamp and E. Vietzke, J. Nucl. Mater. 93+94 (1980) 558.
[14] G. Comsa and R. David, Surf. Science Rep. 5 (1985) 145.

DETACHED RECOMBINING PLASMAS IN RELATION TO VOLUMETRIC HYDROGEN RECYCLE

S. TAKAMURA, N. OHNO, N. EZUMI, D. NISHIJIMA,
*S. I. KRASHENINNIKOV and *A. YU. PIGAROV
Department of Energy Engineering and Science, Graduate School of Engineering, Nagoya University, Nagoya 464-8603, Japan
**Plasma Science and Fusion Center, Massachusetts Institute of Technology, Cambridge, Massachusetts 02139, U. S. A.*

1. Introduction

Hydrogen recycling in fusion devices takes place not only on the material surface, but also volumetrically in a diffusive plasma-gas interface. In ITER, a detached recombining plasma is considered to be a promising scheme for divertor plasma operation in terms of reduction of the plasma heat load onto the divertor target plate, in which the plasma particle flux from the core through SOL diminishes due to plasma recombination near the target in the divertor chamber. Figure 1(a) shows the schematic concept in the heat and particles removal in such a dissipative divertor. The real configurations incorporated by a vertical target to have a good confinement of neutrals is shown for ITER-EDA original version in (b).

The hydrogen gas generated volumetrically by plasma recombination makes a flow with excited atoms and molecules which gives important effects on the hydrogen recycling processes and also on the plasma detachment self-consistently. The electron-ion recombination (EIR) including three-body and radiative processes makes highly-excited atoms associated with a strong emission represented by Lyman or Balmer series radiation. Some of these emission lines are optically thick in a high density environment of hydrogen atoms. It may decrease the plasma recombination. In some cases the hydrogen molecules contribute significantly to the plasma detachment in which the hydrogen molecules should penetrate into the plasma against the emanation of fast Frank-Condon hydrogen atoms made of breakup of hydrogen molecules in the

C.H. Wu (ed.), Hydrogen Recycling at Plasma Facing Materials, 59–65.

plasma. Vibrationally excited hydrogen molecules may be generated on material surfaces. They are thought to enhance the above-mentioned molecular activated recombination (MAR).

In the above context, first of all we summarize the experimental observation of plasma detachment due to EIR, secondly, the so-called MAR in high density helium plasma with hydrogen gas puffing, and thirdly, the detachment of pure hydrogen plasmas associated with MAR. Finally, we will discuss the relation of such plasma detachment processes to the hydrogen recycling under the condition of divertors in fusion devices.

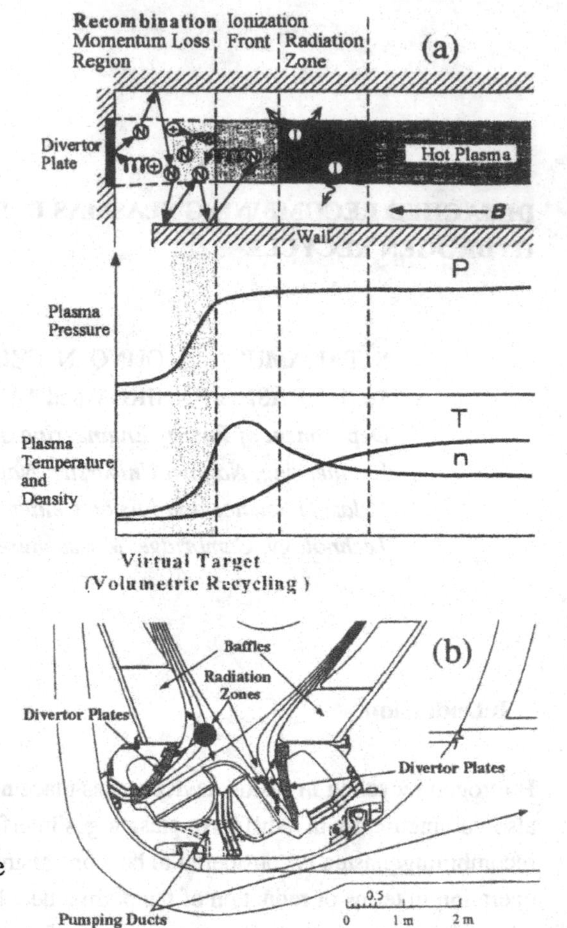

Figure 1. (a) Schematic diagram of dissipative detached recombining plasma concept. (b) Real divertor configuration of ITER-EDA original version where the vertical target is employed to have a good confinement of recycled hydrogen neutrals

2. Experimental Setup

The experiment was performed in the linear divertor plasma simulator NAGDIS-II [1-5] as shown in Fig. 2. Helium plasmas are produced by the modified TP-D type discharge. Sometimes we can choose hydrogen as a working gas. The neutral pressure P in the divertor test region can be controlled 1.0 to 20 mtorr by feeding a secondary gas and/or changing the pumping speed. The change of P in the divertor test region has no effect on the plasma production in the discharge region due to 3 orders of magnitude pressure difference between the discharge and the divertor test regions. Spectra of visible and infra-red light emissions are detected. Several sets of fast scanning probes are installed over the long plasma column to measure the local plasma parameters.

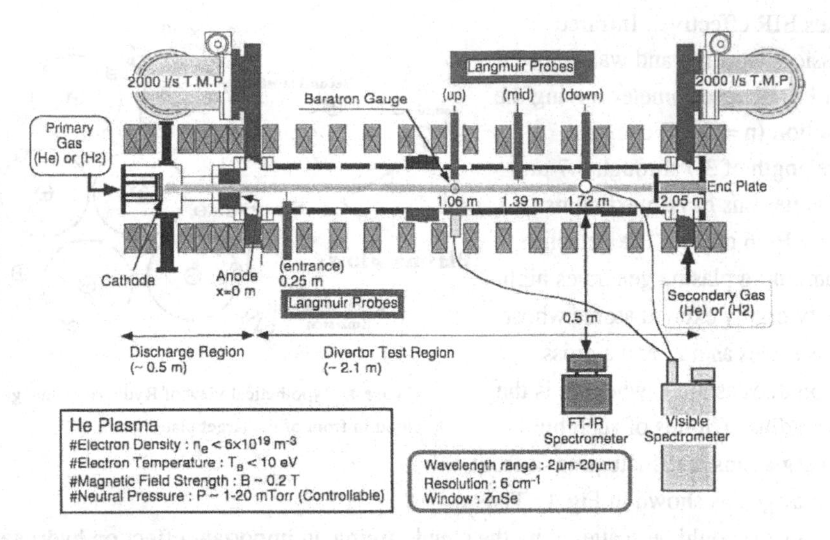

Figure 2. **Linear divertor plasma simulator in Nagoya University; NAGDIS-II.**

3. Electron-ion Recombination

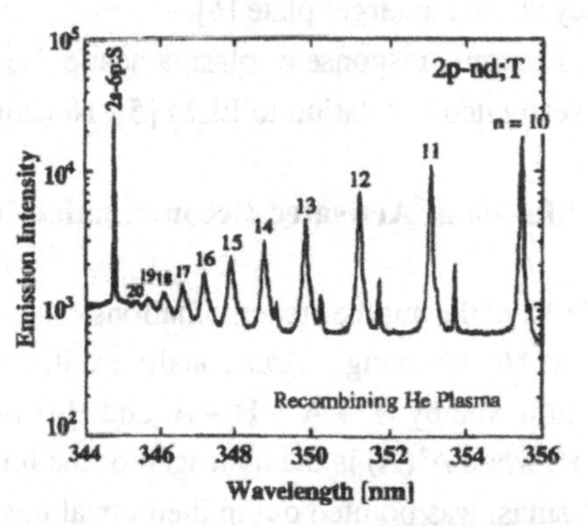

Figure 3. **Ultraviolet light emission spectra from He plasma with a He gas puff.**

Plasma detachment is achieved by feeding the secondary gas near the target. Here, we investigate the He plasmas with He gas puff. Figure 3 shows the spectrum of ultraviolet light emission from 344 nm through 356 nm observed in the recombining region with sufficient secondary He gas puff. Without the secondary He gas puff at P ~ 1.0 mtorr, the plasma column had T_e of 5eV and n_e of 0.8 x 10^{19} m^{-3}. The emission spectrum from this recombining regions of the plasma shows a very prominent continuum radiations below about 344 nm which is connected to the HeI(2p-nd; T)triplet series limit. We can distinguish the line emission up to n ~ 21. This experimental result indicates that EIR occurs in the detached region and contributes to the reduction of ion flux and heat load to the target. The charge-exchange process is essential to remove the plasma energy to have a cold plasma leading to three-body recombination [1]. The population distribution among the highly excited states (n > 5) follows the Saha- Boltzmann relation and gives the T_e of less than 0.5 eV. The value is consistent with that obtained from the wavelength dependence of continuum radiation intensity [5]. Such a low T_e associated by high-density around 1 x 10^{19} m^{-3}

makes EIR effective. Infrared emission appeared and was detected with FT-IR spectrometer, giving the transition (n = 5→ 4) over the wavelength of 3.7 through 4.7 μm. The emissions have maximums around P~ 6 mtorr. The detached recombining plasma generates high density highly excited atoms whose radius scales as n^2a_0 and a cross section does as $n^4a_0^2$, where a_0 is the Bohr radius. Clouds of such big Rydberg atoms are fluctuating in front of the target, as shown in Fig.4. The plasma flow could be scattered by the cloud, giving an important effect on hydrogen recycle on the target plate [6].

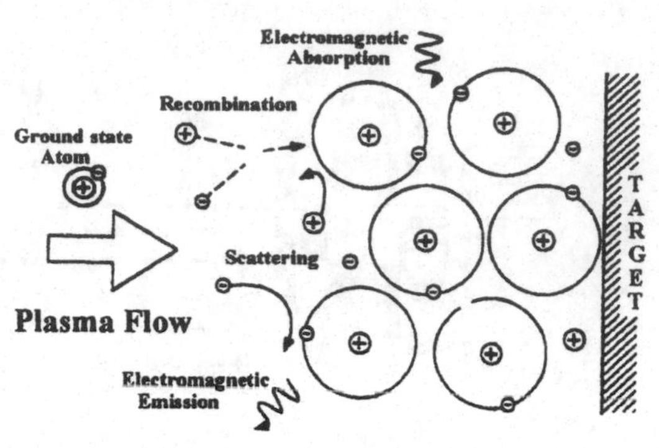

Figure 4. Hypothetical view of Rydberg atomic gas cloud in front of the target plate.

Dynamic response of plasma heat pulse on detached recombining plasma has been investigated in relation to ELM [5]. Nonequilibrium processes might be expected .

4. Molecular Activated Recombinations in Helium and Hydrogen Plasma

We have the another recombinations channels associated with molecular reactions, that is, MAR involving a vibrationally excited hydrogen molecule such as $H_2(v) + e \rightarrow H^- + H$ followed by $H^- + A^+ \rightarrow H + A$, and $H_2(v) + A^+ \rightarrow (AH)^+ + H$ followed by $(AH)^+ + e \rightarrow A + H$, when A^+(A) is the hydrogen or the impurity ion (atom) existing in divertor plasmas, was pointed out in theoretical investigations and modelling [7, 9]. The rate coefficient of MAR is much greater than that of EIR at relatively high T_e above 0.5 eV so that the volumetric recycling is very much influenced by the effect of MAR. In NAGDIS-II device MAR has been clearly observed for the first time in a fusion divertor relevant conditions [2]. A small amount of H_2 gas puffing into a He plasma strongly reduced the ion particle flux along the B field, although the conventional EIR processes were quenched. Careful comparison of the observed He Balmer spectra with CRAMD code [9] indicated that the populations distributions over the atomic levels with relatively low principal quantum numbers can be well explained by taking the MAR effects into account.

Although the rate coefficient of MAR, K_{MAR}, is quite high compared with that of EIR, K_{EIR} ($\sim n_eT_e^{-4.5}$), the reduction rates of plasma density are $-K_{MAR}n_en_{H2}$ and $-K_{EIR}n_en_i$, respectively so that the former becomes dominant only when the H_2 density is high in the plasma. When the hydrogen molecules penetrate into the plasma, the molecules dissociate and decay with the penetrating path. In addition, the fast Frank-Condon atoms generated by the H_2 molecule dissociation emanate from the interior and collide with the incident H_2 molecule, suppressing the penetration as shown schematically in Fig. 5 [10]. The spatial decay of H_2 density is described simply by

$$n_{H2} = f\exp\left(-\frac{1}{v_{H2}}\int n_e \langle\sigma v\rangle_{dis} dY\right), \quad \text{and} \quad f = \frac{m_{H2} v_{H2}}{\alpha \cdot g(n_e, T_e) m_H v_H} \qquad (1)$$

where f is the spatial reduction of H_2 influx due to collision with fast H atoms in which α is the probability of H atom collisions with H_2 in the plasma column with the radius of R, $\alpha = 1 - \exp(-R/\lambda)$, and g is the number of H atoms per an incident H_2 molecule:

$$g(n_e, T_e) = 2\cdot\left(1 - \exp\left(-\frac{1}{v_{H2}}\int n_e \langle\sigma v\rangle_{dis} dY\right)\right) \qquad (2)$$

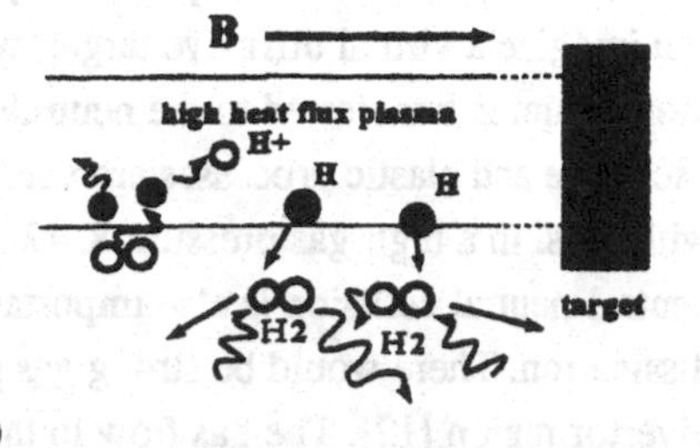

Figure 5. Schematic picture of H_2 penetration into high-flux plasma.

Under the NAGDIS-II condition, the distribution of hydrogen molecules obtained by the above estimation is shown in [3]. The effect of penetration process on the longitudinal structure of MAR detached plasma is studied in the experiment. The ratio of ion flux $\Gamma(down)/\Gamma(up)$ was found to increase with the plasma density at the entrance [3], which was discussed by the above argument.

An interesting byproduct of MAR in He plasma with H_2 gas puffing is singly charged helium-hydride molecular ion HeH^+ which is the simplest closed-shell heteronulcear molecule. It is the molecule formed from the most abundant atoms H and He in universe so that its presence in stellar atmospheres, interstellar clouds et al. is expected, and is considered to be responsible for apparent absorption bands in the spectra of a number of early-type stars [11], in which vibrational transitions of HeH^+ (3.346 and 3.607 μm) are examined. The FT-IR installed in NAGDIS-II indicated that the signals from 3.58 to 3.62 μm becomes large with H_2 gas injections into He plasma from 4mtorr (without H_2 gas) to 10 mtorr.

The experiment on the pure hydrogen plasmas has shown the clear reductions of the ion particle flux and heat load to the target in hydrogen plasmas with the hydrogen gas puff due to MAR[4]. Detailed analysis of Balmer series spectra with the CRAMD-code shows that MAR mainly appears as a weak dependence of Balmer series spectral emission intensities on the plasma condition, being consistent with experimental results. In addition, the hydrogen plasma conditions necessary to obtain plasma detachment through either MAR or EIR in tokamak divertors are discussed in terms of the plasma density, the electron temperature and the hydrogen gas pressure.

5. Effects of Detached Recombining Plasmas on Hydrogen Recycle and Retention

Hydrogen recycle does not always occur on the material surface, the first wall and the divertor target plate by the surface recombination of plasmas. In detached plasma condition which is considered as the main operating mode in ITER, the volumetric plasma recombination takes place away from the divertor plate, as shown in Fig.1. We can imagine a virtual diffusive target, where the plasma is extinguished, and its momentum is transferred to the neutrals. The ion-neutral collisions through charge-exchange and elastic processes are very strong so that the neutrals are strongly coupled with ions. In a high-gas pressure $K = \lambda/L \leq 1$ where K is the Knudsen number, the neutral-neutral collision is also important in terms of plasma/neutral momentum dissipation. There would be strong gas recirculation, possibly gas turbulence, in the divertor region [12]. The gas flow in the divertor could give the effects on the refueling into the core region, the impurity transport and the dust formation. The energy transport through the recombining plasma and gas phase are affected by the opacity of hydrogen resonance line (Ly-α, Ly-β) which has been suggested in the tokamak divertor experiments [13].

We have pointed out that the highly-excited Rydberg atoms generated by EIR could give a deceleration of ion flow to the target. There could be interesting dynamics of Rydberg gas which could be studied by a pseudo-acoustic wave propagation. Many-body effects would affect the energy levels close to the ionization limit which could be studied with IR spectra. It has been noted that vibrationally excited molecules would enhance MAR process. Hydrogen molecules recycled on the material surface may be vibrationally excited so that the wall recycling may give an important influence on MAR enhancement [14]. We have already discussed about the H_2 penetration into the plasmas in relation to MAR.

Finally we note that the detached recombining plasma with a low-Te, high-n_e and n_n has a very similar condition to the processing plasma in gas discharge for semiconductor manufacturing. When the graphite is employed in fusion divertors, the chemical sputtering produces many kinds of hydro-carbon radicals C_xH_y, while the processing plasma with the gas of silane yields Si_xH_y which becomes a precursor of unfavorable dust particulates with the size of micrometers [15]. In fusion divertor relevant conditions the dust-like spheres made of hydrocarbon are redeposited on the plasma-facing component [16]. Recently, hydrogen (tritium) retention in hydrocarbon dusts have been identified in the divertors of large tokamaks and is worried due to the large amount of quantity. There would be similar and also different processes in nucleation and coagulation between the fusion divertors and the processing plasmas.

6. References

1. Ezumi, N., Mori, S., Ohno, N., Takagi, M., Takamura, S., Suzuki, H., and Park, J. (1997) Density threshold for plasma detachment in gas target, *J. Nucl. Mater.* **241-243**, 349-252.
2. Ohno, N., Ezumi, N., Takamura, S. et al. (1998) Experimental evidence of molecular activated recombination in detached recombining plasmas, *Phys. Rev. Lett.* **81**, 818-821.
3. Nishijima, D., Ezumi, N., Kojima, H., Ohno, N., Takamura, S., Krasheninnikov, S.I., and Pigarov, A.Yu. (1999) Two-dimensional structure of the detached recombining helium plasma associated with molecular activated recombination, *J. Nucl. Mater.* **266-269**, 1161-1166.
4. Ezumi, N., Nishijima, D., Kojima, H., Ohno, N., Takamura, S., Krasheninnikov, S.I., and Pigarov, A.Yu. (1999) Contribution of molecular activated recombination to hydrogen plasma detachment in the divertor plasma simulator NAGDIS-II, *J. Nucl. Mater.* **266-269**, 337-342.
5. Ohno, N, Ezumi, N., Takamura, S., Krasheninnikov, S.I., and Pigarov, A.YU. (1999) Dynamic response of detached recombining plasmas to plasma heat pulse in a divertor simulator, *Phys. Plasmas* **6**, 2486-2492.
6. Nextarov, E.S. et al. (1996) Role of highly excited atoms in plasma flow deceleration in gas target, *Plasma Phys. Reports* **22**, 390-394.
7. Krasheninnikov, S.I., Pigarov, A.Yu., and Sigmar, D.J. (1996) Plasma recombination and divertor detachment, *Phys Lett.* **214**, 385-291.
8. Post, D.E. (1995) A review of recent developments in atomic processes for divertors and edge plasma, *J. Nucl. Mater.* **220-222**, 143-157.
9. Pigarov, A.Yu. And Krasheninnikov, S.I. (1996) Application of the collisional-radiative, atomic-molecular model to the recombining divertor plasma, *Phys. Lett.* **222**, 251-257.
10. Fielding, S.J., Johnson, P.C., and Guilhem, D. (1984) Recent results from the DITE bundle divertor - gas refueling efficiency in the divertor chamber, *J. Nucl. Mater.* **128&129**, 390-394.
11. Miller, S., Tennyson, J., Lepp, S., and Dalgarno, A. (1992) Identification of features due to H_3^+ in the infrared spectrum of supernova 1987A, *Nature* **355**, 420-422.
12. Knoll, D.A., et al. (1996) Simulation of an ITER-like dissipative divertor plasma with a combined edge plasma Naviet-Stokes neutral model, *Contrib. Plasma Phys.* **36**, 328-332.
13. Wenzel, U., Behringer, B., Buechl, K., Herrmann, A., and Schmidtmann, K. (1999) Characterization of the hydrogen emission in divertor II of ASDEX -Upgrade, *J. Nucl. Mater.* **266-269**, 1252-1256.
14. Reiter, D., May, Chr., Baelmans, M., and Borner, P. (1997) Non-linear effects on neutral gas transport in divertors, *J. Nucl. Mater.* **241-243**, 342-348.
15. Shiratani, M., et al. (1996) Simultaneous *in situ* measurements of properties of particulates in rf silane plasmas using a polarization-sensitive laser-light-scattering method, *J. Appl. Phys.* **79**, 104-109.
16. Hirooka, Y. and Conn, R.W. (1993) A review of materials erosion and redeposition research at UCLA for the development of plasma-facing component in ITER, in "*Atomic and Plasma-Material Interaction Processes in Controlled Thermonuclear Fusion*", Elsevier Science Pub. 429-453.

Hα spectroscopic study of hydrogen behavior in a low temperature plasma

Bingjia Xiao, Kobayashi Kazuki and Satoru Tanaka
Department of Quantum engineering and Systems Sciences
Faculty of Engineering, The University of Tokyo

Abstract: In high recycling or detached divertor plasma, the plasma temperature can be lowered to several eV and even below 1 eV. MAP [1], a linear steady plasma facility with plasma temperature of ~ 10 eV and density of ~ $10^{18}/m^3$, can in some extent, be a simulator to low temperature divertor plasma. This research concentrates on the numerical simulation on the hydrogen transport in MAP. Hα spectra are specially modeled and compared with experiment. It is observed if the dissociated atoms' energy could be reduced to about 0.45 eV, the modeled spectrum have a much better agreement with the experiment. This could be explained caused by high vibration excited molecular population in the case studied. It is also shown for hydrogen reflection from carbon target, the yield might be much smaller than that predicted by the existing database [2]. Calculation also shows large energy sinks which would decrease the electron energy considerably. The considerable decrease of electron temperature along the flow direction of plasmas would cause a strong decrease of Hα emission but it was only observed in the near wall region. This would probably be due to the overestimation of energy loss of electron from molecular dissociation.

1. Introduction:

The transport of neutrals, the atomic process and plasma wall interactions play key roles in edge plasma physics. The density, energy (velocity) of the atomic and molecular fractions of hydrogen released from the plasma facing components determine the impact of the neutrals to the properties of edge plasma and thus the heat and particle fluxes to the target.

Hydrogen plasma or neutral atom incidences on the material can induce both atomic and molecular forms of hydrogen release from the material surface. The atomic form release is via reflection or ion-induced desorption. The processes may be well modeled by some codes, e.g. TRIM. But in the low energy case, e.g., below several ten eV, there is limited verification of the obtained data. When the hydrogen is released as molecular form, it would be broken-up via the collisions with electrons and ions. The break-up atoms of molecular hydrogen occupy large fractions of the hydrogen atoms in the edge plasma. Molecules may be released from the surface in vibrationally excited states. The released molecules can also be vibrationally excited via the collisions with the plasma. These vibrationally excitations would impact the dissociation channels and give rise to dissociated products with different energy. Former experimental study of MAP plasma showed that Hα spectrum observed in MAP plasma can be decomposed into a low energy component with the energy about 1 eV and a high energy component [1]. High energy component may be considered the contribution of reflected hydrogen atoms but the low energy component can only be contributed by the dissociated atoms. In order to understand the processes more clearly, we perform now the numerical simulation via DEGAS 2 [3,4]

2. Experimental

A linear steady plasma facility, MAP, is introduced in [1]. Recently MAP was extended with a new (second) chamber. The plasma column in the second chamber is with the size of

C.H. Wu (ed.), Hydrogen Recycling at Plasma Facing Materials, 67–73.

40 cm(length) and ~5 cm (diameter). The electron temperature is ~ 13 eV and density about 3. $\times 10^{17}/m^3$ (see also Fig. 1 where the density and temperature were measured near the carbon target by Langmuir probe), the neutral pressure in the chamber is kept constant as 0.25 mTorr via a diffusion pump. Hα spectra at different positions to the target were observed via a telescope and a monochrometer with resolution of 0.012 nm (see [1] for the detail). All the spectra recorded show a narrow broadening and the decomposed spectra shows a component with the neutral energy less than 1 eV. Fig. 2 shows the intensity distribution of Hα along the axis. It is shown that Hα density decreases very much in the near target region but the variation is much smaller in the region far from the target. In Fig. 4, Hα spectrum observed at position at 5 mm from carbon target is shown (note that the experiment spectrum in the wavelength region less than 6561.7 A° does not represent Hα)

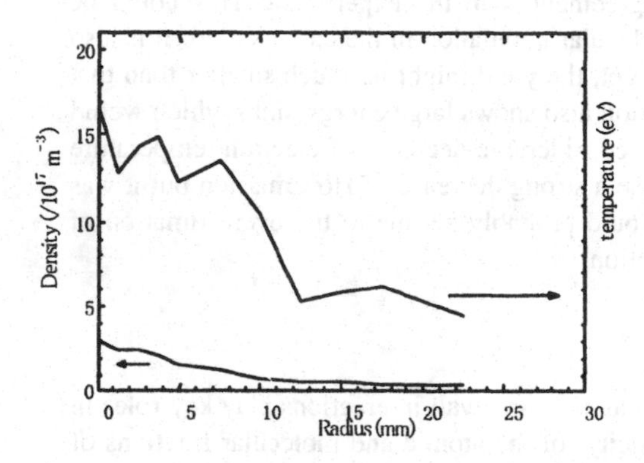

Fig. 1 electron temperature and density

Fig.2 Hα intensity along the axis

3. Numerical simulation via DEGAS 2

The simulation was performed by DEGAS 2 code [3,4]

DEGAS 2 is a successor of DEGAS [5], which is a widely used neutral transport code. It was developed in PPPL to improve the speed and flexibility of DEGAS calculation. At given geometry and plasma parameters, neutrals are modeled in a Monte-Carlo fashion as they encounter charge exchange, ionization, dissociation (for molecules) and recombination collisions with the plasma and the interactions with material surfaces as well. Then the final results are indeed the neutral distribution solution of Boltzmann equation.

When a hydrogen atom decays from n=3 excited state to n=2 state, Hα is emitted. The spectrum can be modeled in DEGAS 2 by recording the wavelength of Hα emitted within the observation volume (detector). In the modeling, we set several detectors to record the spectra in order to have an information of emission in different regions. The wavelength observed is Doppler shifted according to the velocity of the atom which emits the photon, $\lambda = \lambda_0(1 - \vec{v}\cdot\vec{x}/C)$, where λ_0 is the wavelength center of Hα, $\vec{v}$ the velocity of the atom and $\vec{x}$ the position unit vector relative to the detector. In the calculation, the intensity is scored according to the atomic process such as excitation and de-excitation, etc., and also the molecular processes which give rise to the generation of n=3 excitation state of hydrogen atoms. The local Hα emission rate from ground state atoms is according to the density of the n=3 population data from the calculation by a radiation collision model and the Einstein emission rate. Hydrogen atoms with excitation state of n=3 can also be directly from the molecular dissociated products and thus the photon emission is determined directly from these pathways.

In most of the cases in divertor plasmas, the neutrals' density is not high enough to thermalize the velocity distribution. Therefore, hydrogen atoms from different pathways have different energy or velocity and contribute to different wavelength regions of the observed spectrum. For instance, the reflected atoms from the ion incidence on material walls will have energies, $E_{refle} = R_E(E_i + \phi)$, where E_i is the ion energy before the sheath, ϕ the sheath potential or biased potential of the target and R_E is the fraction of incident energy retained by the reflected atoms, which is dependent on the incident angle, energy and the target material. For the atoms resulting from the dissociation of H2 molecules, the energies may be varied in the range of 0.2 – 7.8 eV according to different pathways.

3.1 Geometry and background.

As schematically shown in Fig. 3, in MAP, the diameter of the chamber is 50 cm which is much larger than that of the plasma column (5 cm). The background pressure is almost constant in experiment. Therefore we only perform the neutral transport calculation in the plasma column and we consider that the neutrals escape from the plasma when they reach the plasma circumference. The background neutrals are taken as a hydrogen molecular puff (molecular temperature was assumed in the order of room temperature, here it is assumed as 0.04 eV) from the plasma circumference with a uniform rate which is adjusted according the neutral pressure from the experiment and the simulation. Here the puffing rate is about $4.\times10^{21}/m^2s$. The simplification may cause some problems because there exists the neutral atoms outside the plasma column which would pass through the calculation region. Because of the large area of the chamber wall and the large chamber volume, it is reasonable to take them as the strong neutral sink and to neglect the influence of neutral atoms outside the plasma column. So this simplification would not deviate from the real case much.

The plasma parameters, mainly electron temperature and density are taken according to probe measurements. Radial distributions are taken into account (see Fig.1) while the axial distribution is assumed uniform.

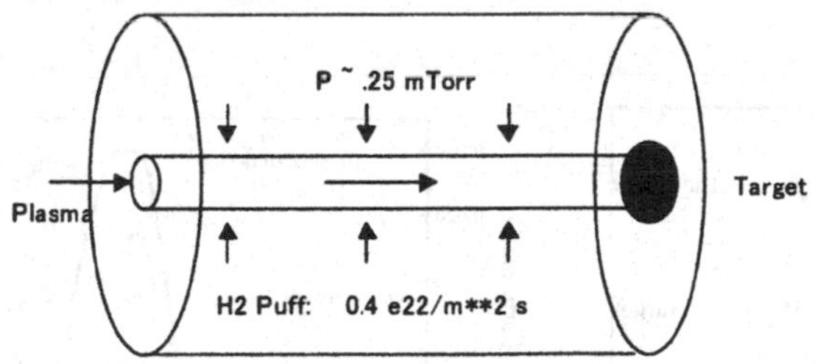

Fig. 3 Schematic show of the model of MAP plasma.

3.2 Atomic and molecular physics considered

For the atomic and molecular physic processes, we take into account the following reactions:

A. $e + H^* \rightarrow e + H^+ + e$ ionization

B. $e + H2 \rightarrow H + H + e$ molecular dissociation

C. $e + H2 \rightarrow H2^+ + e$ molecular ionization

D. $e + H2^+ \rightarrow e + H^+ + H$ molecular ion dissociation

E. $H^+ + H \rightarrow H + H^+$ charge exchange

F. $e + H^+ \rightarrow H^*$ recombination

It is believed that the neutral collisions with ions, except charge exchange, can be neglected because of their much smaller cross sections. Other atomic and molecular reactions are not considered here both because of their much smaller reaction rates and because our calculation with them included verified their negligible contributions to the spectra and also the transport. Elastic collision is negligible due to the thin structure of the plasma and the low neutral density in the chamber. In the case of MAP with much narrow radial dimension, low density and low temperature, reactions E and F can also be neglected for the contribution to photon emission. The reason we take them into account here is that they are much active in most of other cases of divertor plasma.

3.3 Surface physics

Hydrogen ions impinge on the material surface with energies, $E_{reflc} = R_E(E_i + \phi)$. In the experiment, the target was isolated from the chamber wall so ϕ is the floating potential of the target against the plasma. According to the sheath theory [6], the floating potential is ~ 3 T_e and the plasma flux is 0.5 n_eC_s where C_s is the sound speed of ions which is taken to be $\sqrt{k(T_i + T_e)/m_i}$.

Hydrogen ions and atoms are reflected from the wall surface as neutral atoms and desorbed as molecules when they impinge on the solid surface. It can also induce physical and chemical sputtering. Because the sputtering yield is relevant small in MAP case(less than 5 % of the incident flux), we neglect the effects of sputtered species to the plasma and neutral transport. So in the steady state, we assume that part of the incident particles would be reflected as neutral atoms, with the rest are desorbed as neutral molecules. The dependence of reflection yield and energy reflection coefficient on the incidence energy used in this paper is from the 'refl.dat' in old DEGAS code which is calculated by VFTRIM code [2] and is implemented in the DEGAS 2 package. The angular distribution of reflected particles is assumed cosine distribution. For the desorbed molecules, the energy was assumed to be equal to the temperature of the target which in our case, is about 0.04 eV.

4. Results and Discussion

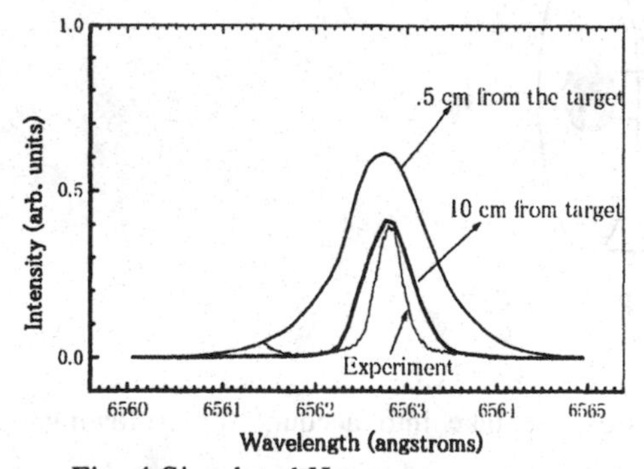

Fig. 4 Simulated Hα spectra

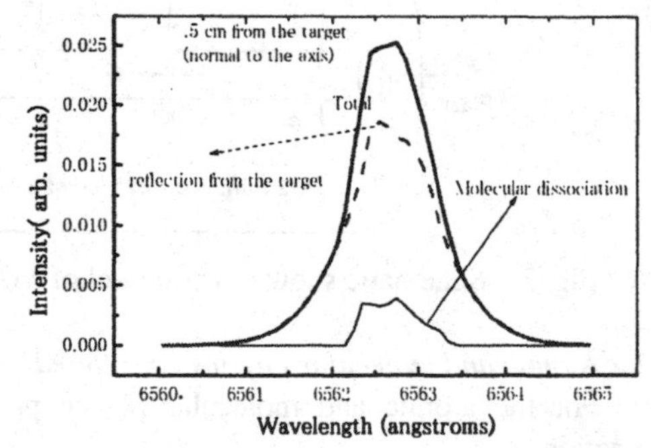

Fig.5 Main component of Hα close to the target (not convoluted by the instrumental function)

Fig.4 gives the simulated Hα spectra at the positions of 5 mm and 10 cm distant from the target (in all the cases in this paper, the watching directions is normal to the axis, in other words, parallel to the target surface.). It is also shown in Fig.4 Hα spectrum observed by the experiment at the position of 5 mm distant from the target. The calculation was performed underlying the databases included in the current DEGAS 2 package. The dominant impacts to the current results are of the dissociation products' energy and the reflection yield. The energy of the dissociated atoms is assumed 3 eV for the reaction A shown in 3.2, which contributes most to the spectra especially at the place far from the target. For the reflection yield, in our case with the ion energy ~ 40 eV toward the floating target, it is set ~ 0.4. From Fig. 4, we can see the modeled spectra obviously deviates from the experimental. The modeled spectra are much broader. This suggests that the neutrals, which emit Hα in our case, could have much lower energy than 3 eV. Meanwhile, in comparison to the spectrum by experiment, the simulated spectrum has much higher contribution from the high-energy component (the wing region).

There might be the possibilities of the geometry and plasma parameters to affect the spectrum shape. For the clarification, we set some extreme conditions, for instance, with the plasma radial extension much smaller than the real value or set the plasma parameters much lower than those from the probe experiment. All of these efforts can not help us to mitigate the discrepancy of the broadening of modeled spectra. To some extent, it can reduce the high energy component because its intensity is directly determined by the plasma flux and thus parameters.

In order to match the results with those from experiments, we assume the energy of dissociation products for reaction A as 0.45 eV. Reaction A in section 3.1 is ground state molecular dissociation considered in the database. According Frank-Condon theory, this channel can only give rise to hydrogen atoms with energy of 2 - 3 eV to be generated. . Because dissociation of vibrational excited molecules currently can not be modeled and also reaction A is the dominant source for neutral hydrogen atoms, we make our assumption here is only trying to match the experiment. Reflection yield is assumed 40% of that predicted by the database ('refl.dat' for hydrogen incidence on carbon [2]). The simulated spectra are shown in Fig.6 and Fig. 7. Fig. 6 gives the comparison with the experiment. The simulated spectrum agrees much better with that by experiment. Fig. 7 illustrates the Hα intensity distribution along the axis. The photon emission intensity in the near wall region decreases very much. The tendency generally agrees with the experiment observation (Fig. 2).

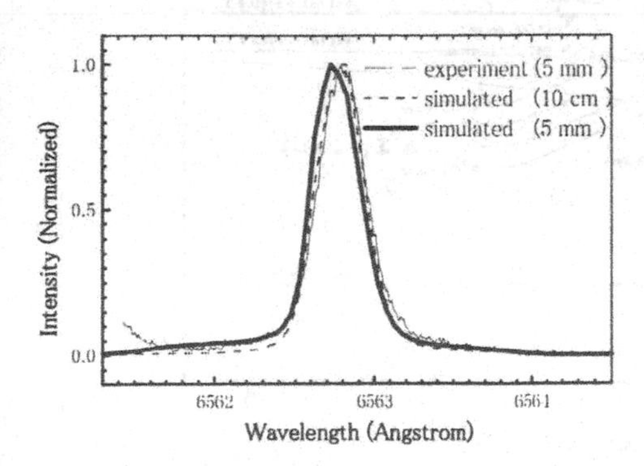

Fig. 6* Hα spectrum

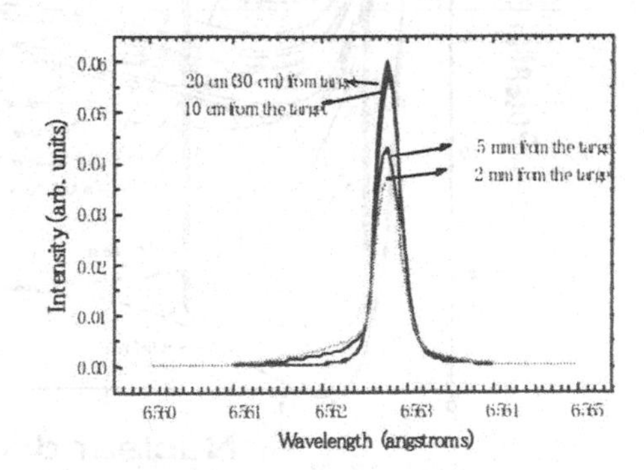

Fig. 7* Illustration of Hα intensity distribution along the axis

Simulation assumption: dissociated atoms' energy as 0.45 eV, reflection yield: about 0.08

From the experiment and the simulation, we can see that Hα emission is mainly from the low energy (less than 1 eV) hydrogen atoms. However, neutral atoms are mainly from the dissociated atoms which in the most cases have energies about 3 eV. The possible explanation of our observation is that there may exist high population of vibrationally excited hydrogen molecules. These vibrationally excited states of molecules can give rise to low energy atom dissociation. Fig.8 schematically shows the potential curves of hydrogen molecules and ions [7]. Both via the repulsive state $b^3\Sigma$ of H_2 and the states $X^2\Sigma$ and $2p\sigma$ of H_2^+, it is possible to obtain (excited) atoms with low energies(~0.5 eV) and high energies (~10 eV) as indicated in Fig. 8 by arrows. In MAP plasma, the temperature is about 10 eV, which is in the turning point to provide higher probability for the electron induced molecular excitation. Moreover, the molecules can be desorbed from the walls with the states vibrationally excited. Also, because of the thin structure of MAP plasma, there is higher probabilities for H_2 (v=0) to be vibrationally excited and escape from the plasma column before being dissociated. This would bring about higher population of vibrationally excited H_2 around the plasma and they can re-enter the plasma to be dissociated. The vibrationally excited H_2, especially v>=4, can also give rise to the dissociation, H_2 (v>0) + e → H + H^- and then H^- + H^+ → H which have larger cross sections (~ 2 orders) than the volumetric recombination of H^+ with electrons. Because of ion temperature of MAP plasma is relatively low (less than 1 eV), this could partly contributes to the population of low energy atoms. Low energy H atoms have shorter collision mean free path than those with high energy have. So Hα is mainly contributed by the low energy atoms in our observations and we conclude that there is high population of vibrationally excited H_2 in our case.

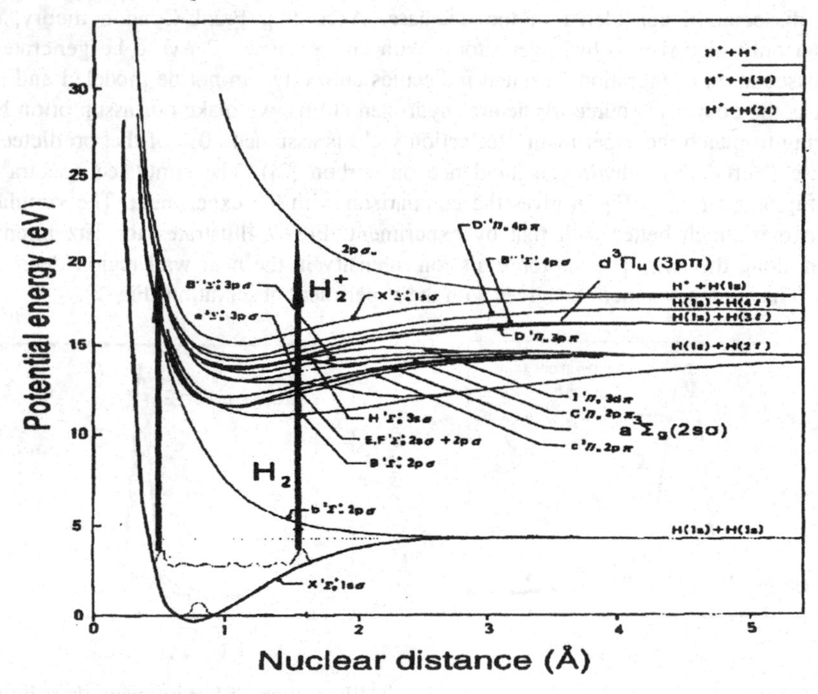

Fig.8 Potential curves of hydrogen molecule
(copied from A. Pospieszczyk et al., Journal of Nuclear Materials 266-269 (1999) 138-145)

Fig. 9 shows the electron energy loss rate due to the collision with neutral. The energy loss is about 0.1 MW/m^3. Such kind of level energy loss would cause an electron temperature decrease of 20~ 30 eV from the plasma entrance to the target according to the simple balance of electron energy balance. The previous probe measurements showed the electron temperature in the first chamber is very close to that in the second chamber made. Furthermore, we tried to take into this level of electron temperature change into account in Hα modeling, results showed Hα intensity decreased very much along the plasma axis and this can not be verified by experiment. Because the dominant collision in our case is molecular dissociation, so most probably the data [8,9] overestimate the electron energy loss in the dissociation processes.

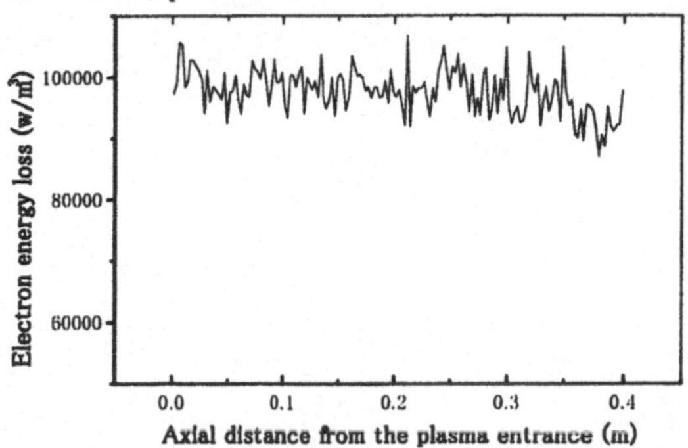

Fig. 9 Electron energy loss

5. Conclusions

From this study, in MAP plasma, there is high population of hydrogen neutral atoms with low energy (less than 1 eV) and they are most possibly the dissociated products of vibrationally excited hydrogen molecules. This phenomenon was also observed by TEXTOR experiment [5]. Our observation also find the reflection yield in the case of low energy ion incidence may be much less than that predicted by the current database at least for carbon targets. So we should pay enough attention when we perform the modeling of high recycling low temperature edge plasma and neutral transport based on the current databases. Detailed consideration of the molecular physics may be necessary.

6. Acknowledgement

We are very grateful for Dr. Daren Stotler to permit us to use DEGAS 2 code and also for his many instructions and help. This work is partly supported by JSPS grant-in-aid.

References:

[1] K. Kobayashi, S. Ohtsu and S. Tanaka, J. Nucl. Mat. 266-269 (1999) 850-855

[2] David N. Ruzic. Nuclear Instruments and Methods in Physics Research. Vol. B47 (1990) pp.118-125.

[3] Daren Stotler et al., 12th topical meeting on high-temperature plasma diagnostics, Princeton, NJ, 1998

[4] Daren Stotler and Charles Karney, Contrib. Plasma Physics, 34(1994) 2/3, 392-397

[5] A. Pospieszczyk et al., J. Nucl. Mat. 266-269 (1999) 138-145

[6] P. C. Stangeby, in *Physics of Plasma-Wall Interactions in Controlled Fusion*, edited by D. E. Post and R. Behrisch (Plenum Press, New York, 1986), NATO ASI Series, p41

[7] D. Heifetz D. Post, M. Petravic, J. Weisheit and G. Bateman, J. Comp. Phys. 46(1982) 309

[8] "Elementary Processes in Hydrogen-Helium Plasmas", by R.K. Janev, W.D. Langer, K. Evans & Jr, D.E. Post, Jr, Springer Series on Atoms and Plasmas, Springer-Verlag (Berlin, Heidelberg, New York), 1987.

[9] R. K. Janev and J. J. Smith, Atomic and Plasma-Material Interaction Data for Fusion (Supplement to the journal Nuclear Fusion) 4 (1993)

METAL SURFACE MICRORELIEF FORMED DUE TO SPUTTERING BY MONO- AND VARIABLE ENERGY IONS OF HYDROGEN PLASMA

A.F.BARDAMID[1], V.T.GRITSYNA[2], V.G.KONOVALOV, N.B.ODINTSOV[3], D.V.ORLINSKIJ[4], A.N.SHAPOVAL, A.F.SHTAN[1], S.I.SOLODOVCHENKO, V.S.VOITSENYA, K.I.YAKIMOV[1].

IPP NSC KIPT, Kharkov 310108, [1]Kiev University, [2]Kharkov University, Ukraine;
[3]SRC Prometei, St-Petersburg, RF; [4]RRC Kurchtov Institute, Moscow, RF

1. Introduction

It is known from experiments on several tokamaks (PLT, ASDEX, JFT-IIM) that energy distribution of charge exchange atoms (CXA) is very wide, from tens eV up to tens keV. Calculation for ITER gave qualitatively similar long-tail distribution of CXA. Evidently, that consequences of such energy-distributed atoms bombardment will differ from those for the case when surface is bombarded with ions having some fixed energy, as it is usually realized in simulation experiments.

In this paper results are presented on bombardment of polycrystalline stainless steel mirrors with ions having different fixed energy (0.3, 0.65, 1.43 keV) and with energy-distributed ions (0.1-1.43 keV) of deuterium plasma.

2. Experiment

As an ion source the steady ECR discharge in deuterium was used. The water-cooled sample holder was installed in a free streaming plasma that flows out of the mirror-type magnetic configuration. Plasma density was maintained at the level $\sim 10^{10} cm^{-3}$ and electron temperature $\leq 5 eV$. Samples were supplied with negative voltage fixed or variable in time. Correspondingly, the bombarding sample surface ions had a fixed energy or their energy distribution was wide, as a rule, between 0.1 and 1.5 keV. The effective energy distribution, obtained using current-voltage characteristics of the holder with copper sample installed and with taking into account the secondary electron yield, is shown in Fig.1b. The chemical composition of plasma ion component was not measured but the comparison of sputtering rate measurement at given accelerating voltage gave the concentration of gas impurities at the level of several per cents.

C.H. Wu (ed.), Hydrogen Recycling at Plasma Facing Materials, 75–81.

The square shape (22x22x3mm) stainless steel (of 316 type) mirror samples were exposed to plasma ions in many identical steps. After every exposure the mass loss of sample was measured with accuracy 10^{-5}g together with measurement of spectral reflectance, R(λ), at normal incidence in the range 250-650 nm. Thus, the dependence of reflectance on some average value of sputtered layer thickness was obtained. The evolution of mirror surface morphology due to sputtering was investigated by means of scanning electron microscope (SEM).

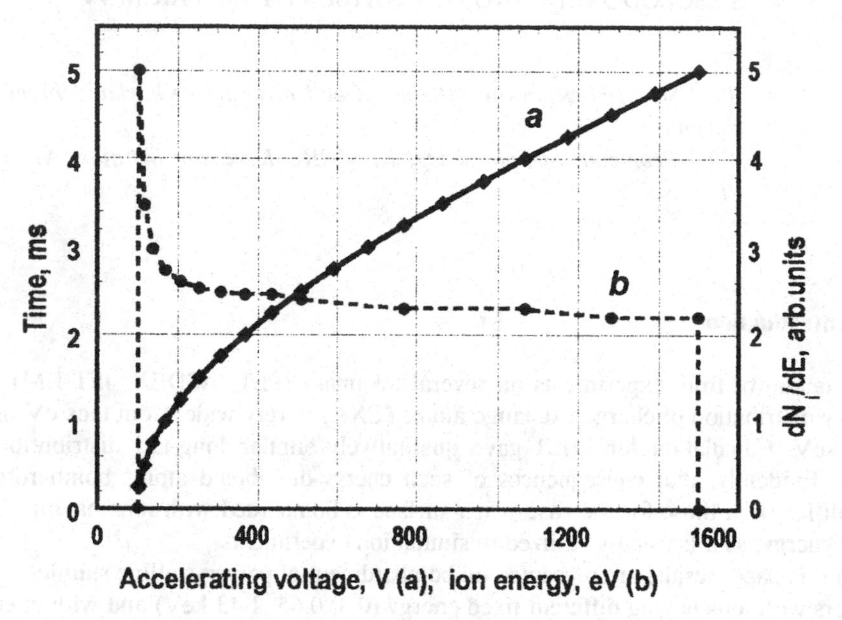

Fig.1. Time dependence of accelerating voltage, *a*, and ion energy distribution function found with taking into account the current-voltage characteristics and secondary electron emission, *b*.

3. Results

Fig. 2 shows the dependence of R versus mean value of thickness of sputtered layer (h) for three fixed energy of bombarding ions and for the energy-spreaded ion distribution. It is seen that in all cases the mirror reflectance decreases with increasing thickness of sputtered layer (i.e., total ion fluence). However, the rate of reflectance degradation strongly depends on the ion energy: the higher energy the faster reflectance degrades. The case with wide ion energy distribution is between curves for 0.3 and 0.65 keV cases.

Qualitatively similar results were obtained earlier for copper mirrors as was shown by Bardamid *et al.* [1] with only one difference: in the case of copper the curve for energy-spreaded ions was positioned between 0.65 and 1.5 keV curves. Much more important difference between these metals was the strong difference in the reflectance

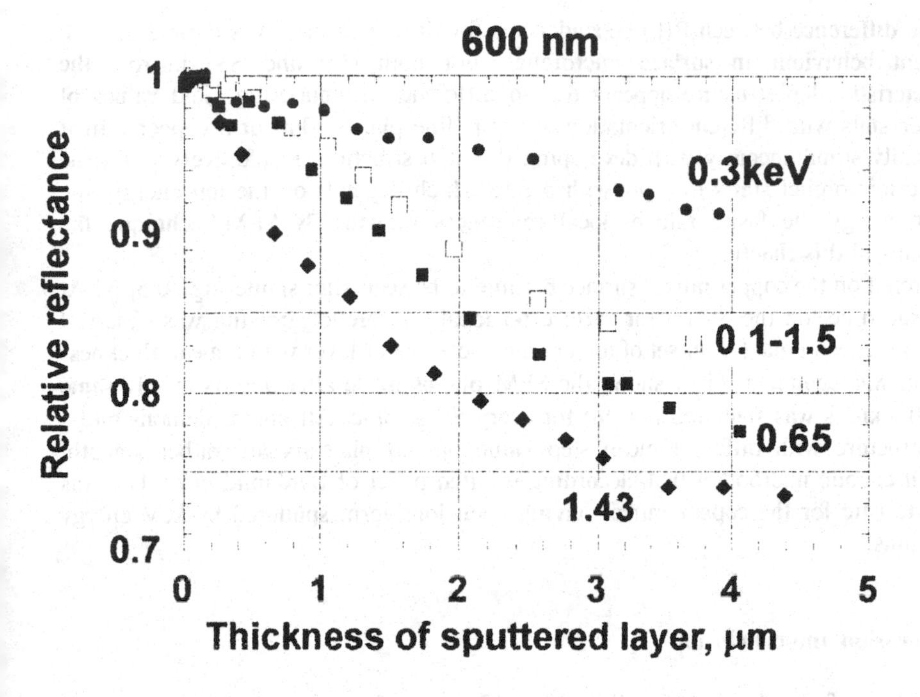

Fig. 2. Relative reflectance versus thickness of sputtered layer for SS mirrors bombarded with different energy ions of deuterium plasma.

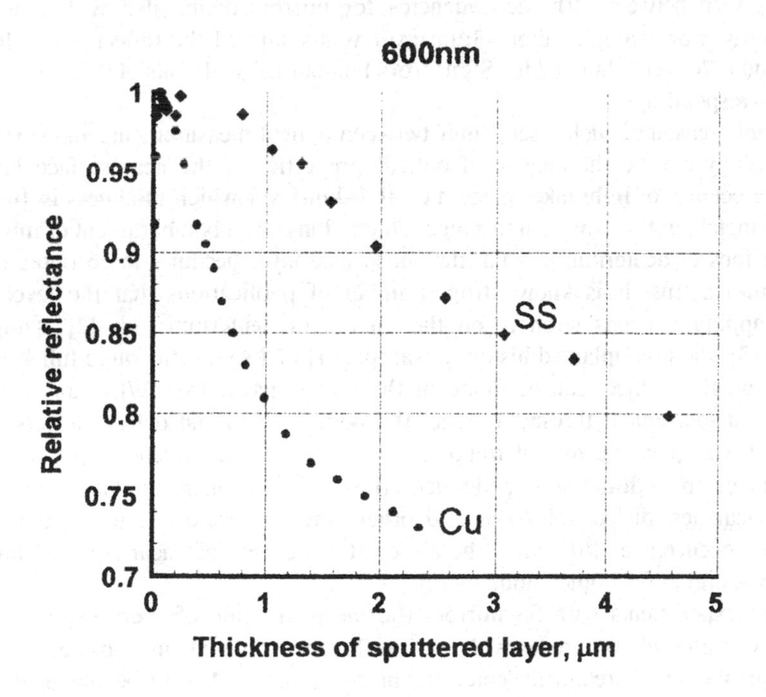

Fig.3. Long-term sputtering effect on degradation of relative reflectance (at normal incidence for λ=600nm) of copper and stainless steel mirrors fabricated of polycrystalline materials.

one can conclude that compare to copper, SS is much more resistant in preserving the reflectance being subjected to the long-term bombardment with keV energy range ions

The difference between R(h) dependencies for these two metals is correlated with different behaviour in surface microrelief. For both (Cu and SS) mirrors the characteristic step structure appears due to difference in sputtering yield values of microcrystals with different orientation of crystalline planes. But for a copper mirror practically simultaneously with developing the step structure, inside every grain the chaotic microrelief starts to grow with a rate which depends on the ion energy: the higher energy the faster rate of local roughness growth. By SEM technique the appearing of this chaotic
microrelief on the copper mirror surface became to be seen after sputtering ~0.5μm. At the same time, on the SS mirror surface no local microrelief growing was observed using same technique for all set of ion energies, even after layer with a mean thickness ~4.5μm was sputtered. Fig.4 shows the SEM photos made after approximately same layer thickness was sputtered (~4μm) for every SS sample. All photos demonstrate a step structure with different mean step value but all plateaux are rather smooth, without chaotic microrelief that, according to cited paper of Bardamid *at al.* [1], was characteristic for the copper mirror having been long-term sputtered by keV energy range ions.

4. Discussion and conclusion

It follows from photos of Fig.4 that the reflectance for normal light incidence (as measured in experiment) has not to have big difference. At the same time, a significant difference is seen between R(h) dependencies for mirrors bombarded with ions of different energy. For example, after ~3μm layer was sputtered the reflectance values 0.93, 0.81 and 0.76 were obtained for SS mirrors bombarded with ions of 0.3, 0.65 and 1.43 keV, correspondingly.

The possible reason of such discrepancy between optical measurements and surface micromorphology can be the change of optical properties of the near surface layer where the reflectance of light takes place, i.e., the skin-layer which thickness is 10-20 nm for most metals in this wavelength range. Due to long-term bombardment of mirror surface with ions of deuterium plasma, the subsurface layer becomes to be implanted with deuterium atoms. It is known from number of publications that the level of hydrogen trapping depends strongly on the energy of deuterium ions [2], sample temperature [3], the preimplanted history of sample [4] and so on (the quite full list of publications on this subject can be found in the review paper [5]). With ion energy prevaling the displacement threshold value, the point defects and defect clusters are created which can play the role of trapping sites for hydrogen atoms. Thus, in the subsurface layer in addition to mobile deuterium, some deuterium atoms will be trapped in vacancies, dislocation loops and other defects created due to metal atom displacement occurrence [6], and therefore the degree of near-surface layer modification is higher for higher atom anergy.

In present experiments with SS mirrors the mean duration of every exposure to plasma ions was around 60 minutes with the ion fluence ~$2{\cdot}10^{19}$D/cm^2. For keV range ions the total amount of retained deuterium at such fluence has to be much above $2{\cdot}10^{16}$at/cm^2 [7], taking into account that optical measurements were carried out during 1-2 hours after finishing exposures. According to [8] the dense dislocation network was observed already at fluence $1\cdot10^{18}$D/cm^2 after bombardment of 316 SS speciment at room temperature. And what is important, the gradual decrease of density of retained deuterium atoms caused by diffusion without heating does not lead to disappearing the defect structure which has formed under ion bombardment. Even the high temperature (up to 873K) annealing of molybdenum samples resulted only in the modification of the defect structure but not in its full recovering [9].

Thus, it follows from comparing published data and experimental conditions that properties of the subsurface layer have to be changed significantly due to long-term exposure to ions of deuterium plasma in the present case. Accumulation of different kinds of defects, i.e., platelets, vacancies, dislocation loops, interstitial loops, voids, highly strained regions, etc. with volumetric density of the order $1\cdot10^{18}$/cm^3 can influence the electrical conductivity of the near-surface layer and therefore can change the reflectance coefficient of light. This effect has to be stronger with increasing the energy of deuterium ions. In the case of long-term bombardment of copper and some other metals in similar experiments (Al, Mo, W,Ta, [10]) the change of optical constants of mirror material could be masked by developing the microrelief on the mirror surface. On our knoledge, nowadays there were no publications concerning effects of hydrogen implantation on the optical properties of metals.

Coming from what was said above, the following experiment could give an answer the question relatively influence of hydrogen retaining in the subsurface layer on the optical properties of stainless steel mirror. With energy of hydrogen ions near 1.0keV, the mean ion range in stainless steel is ~12nm [8] and sputtering yield ~$1.5\cdot10^{-2}$at/at. Thus for reasonable fluence $(1\text{-}4)\cdot10^{19}$D/cm^2 there will be sputtered ~(1.5-6.0)$\cdot10^{17}$at/cm^2, which corresponds to ~(0.02-0.08) μm layer thickness. With such thickness of sputtered layer the surface will continue to be smooth enough, and therefore the change of reflectance (if would be found) can be assigned to changing the optical constants of a subsurface layer.

References

[1] Bardamid, A.F., et al. (1998) Ion energy distribution effects on degradation of optical properties of ion-bombarded copper mirrors, *Surface Coatings & Technology* **103-104**, 365-369.

[2] Blewer, R.S., Behrisch, R., Scherzer, B.M.I., and Schulz, R. (1978) Trapping and replacement of 1-14 keV hydrogen and deuterium in 316 stainless steel, *J. Nuclear Materials* **76&77**, 305-312.

[3] Wilson, K.L., Pontau, E. (1979) The temperature dependence of deuterium trapping in fusion reactor materials, *J. Nuclear Materials* **85&86**, 989-993.

[4] Besenbacher, F., Bottiger, J., Laursen, T., and Moller, W. (1980) Hydrogen trapping in ion-implanted nickel, *J. Nuclear Materials* **93/94**, 617-621.

[5] Pisarev, A.A. and Chernikov, V.N. (1987) Interaction of hydrogen with radiation defects in metals, in A.P.Zakharov (ed.), *Interaction of hydrogen with metals*, Nauka, Moscow, pp. 233-264 (in Russian).

[6] Langley, R., (1984) Hydrogen trapping, diffusion and recombination in austenitic stainless steel, *J. Nuclear Materials* **128/129**, 622-628.

[7] Wilson, K.L. and Baskes, M.I. (1978) Deuterium trapping in irradiated 316 stainless steel, *J. Nuclear Materials* **76&77**, 291-297.

[8] Thomas, G.J. and Wilson, K.L. (1978) Microstructure of low energy deuterium implanted stainless steel, *J. Nuclear Materials* **76&77**, 332-336.

[9] Sakamoto, R., Muroga, T., and Yoshida, N. (1994) Microstructural evolution in molybdenum during hydrogen ion implantation with energies comparable to the boundary plasma, *J. Nuclear Materials* **212-215**, 1426-1430.
[10] V.S.Voitsenta, A.F.Bardamid,

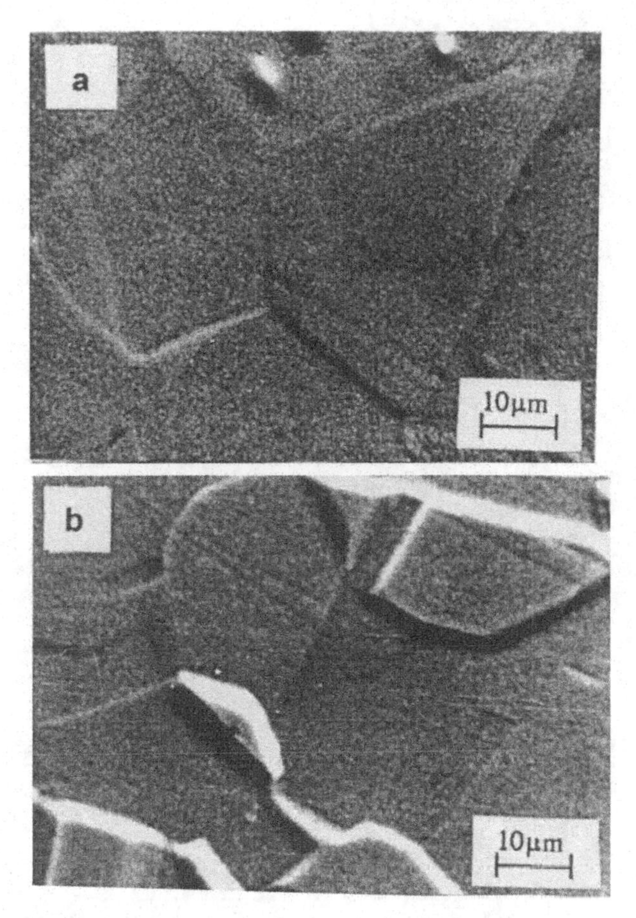

Fig. 4 SEM images of SS surfaces after sputtered of the ~4μm thickness layer by D^+ ions with energy 0.3 keV (A0 and 1.43 keV (b).

RECYCLED HYDROGEN IN EXCITED STATE

Tetsuo TANABE
Center for Integrated Research in Science and Engineering,
Nagoya University, Furo-cho, Chikusa-ku, Nagoya 464-8603, Japan

Abstract

Released hydrogens from plasma facing materials are constructed of reflected, desorbed (ion induced, electron and photon stimulated), and reemitted particles. Most of the previous hydrogen recycling studies have focused to particle balance between the boundary plasma and the plasma facing materials, assuming the released particles are all at ground states. However, the released particles, whatever their origins are, are not necessarily at the ground states in their electronic, rotational and vibrational states. Ionization process in the boundary plasma should be influenced by at what state the reflected particles are. For example, the penetration length or ionization length of the reemitted and reflected particles in the excited states are much longer than those particles at the ground state.

Thus the velocity distribution and energy state (electronic state including ion fraction, rotational and vibrational states) of both reemitted and reflected hydrogen are very important not only to investigate the surface kinetics occurring under energetic hydrogen injection but also to understand the role of the reemitted hydrogen and impurities on the plasma materials interactions.

The present paper demonstrates the significance of the velocity distribution and/or energy state of the released hydrogen from the wall on the plasma-surface interactions and gives a few example recently observed in the TEXTOR tokamak.

C.H. Wu (ed.), Hydrogen Recycling at Plasma Facing Materials, 83–94.

1. Introduction

Hydrogen recycling is one of the key issues to ensure steady or long pulsed operations in DT fusion reactors, and various laboratory studies have been extensively done[1]. Up to now, however, most of them have concentrated to particle balance between plasma and materials, whereas energy and momentum balance between them has scarcely been considered.

Energetic hydrogen particles escaping from boundary plasma directly impinge into the solid, deposit their energy by electron excitation, atomic displacement and phonon excitation, and become thermal remaining thermal motion (diffusion, trapping, reemission and permeation). All these deposited energies cooperatively influence the migration of the thermalized hydrogen in the target through, for examples, producing excited or ionized atoms, defect trapping, diffusion enhancement and so on. Thus hydrogen molecules reemitted and/or permeated under the energetic hydrogen injection are not necessarily be equilibrated with the material's temperature, taking higher energy states than expected from the materials temperature, i.e. having higher translational energy, and/or being in excited states electronically, rotationally and vibrationally [2]. In such cases the hydrogen recombination process may not be the rate limiting process.

In addition to the reemitted hydrogen molecules, various particles are emitted from the surface, which include reflected and sputtered particles, secondary electrons and photons as well. If displaced interstitial atoms migrate to the top surface before being thermalized, they are released from the materials surface with excess energy. Depending on their velocities the reflected particles show various electronic states, distributing among ions, excited neutrals and ground stat neutrals. With increasing the incident energy, the fraction of the ion in the reflected particles increases [3].

Thus the energy (velocity) distribution and the energy state (electronic state including ion fraction, rotational and vibrational states) of both reemitted and reflected hydrogen are very important not only to investigate the surface kinetics occurring under energetic hydrogen injection but also to understand the role of the reemitted hydrogen and impurities on the plasma-materials-interactions because they directly correlate to the penetration length or ionization length of the reemitted and reflected particles in the boundary plasma.

The present paper demonstrates the significance of the velocity distribution and/or energy state of the released hydrogen from the wall and gives a few examples recently observed in the TEXTOR tokamak.

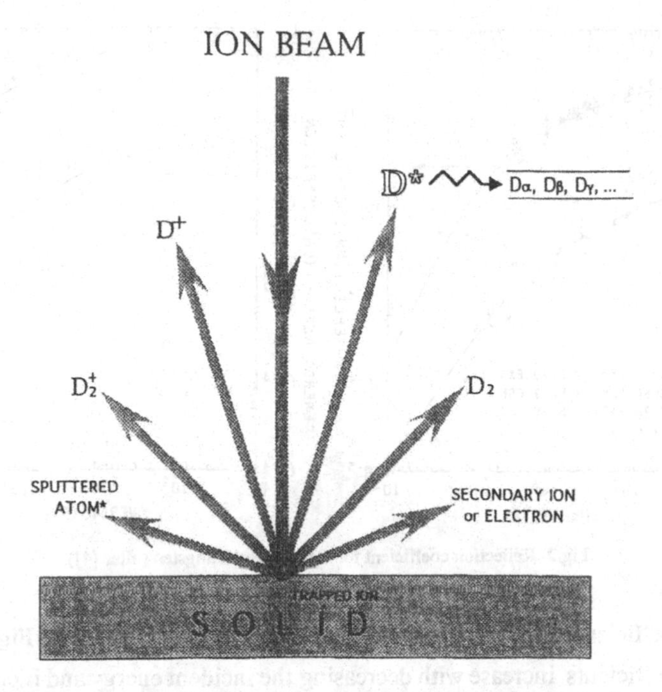

Fig.1 A schematic drawing for released hydrogens from plasma facing materials under hydrogen ion injection, reflected, desorbed (ion induced. electron and photon stimulated), and reemitted particles. Fig.2 Reflection coefficient for Carbon and Tungsten (after [4])

2. Particle balance (Released particle)

Released hydrogens from plasma facing materials are constructed of reflected, desorbed (ion induced, electron and photon stimulated), and reemitted particles as schematically shown in Fig.1. For discussions of particle balance (hydrogen recycling), hydrogen retention (trapping and solution) in the materials should also be taken into account, of which contribution is very large during the transient state like startup phase of discharges and temperature change. In the present work, however, we concentrate on steady state and will not discuss the retention in detail.

2.1. REFLECTION

Reflection can be analyzed based on the assumption of binary collision dominated process. For the incidence energies above several tens eV, the theoretical estimation of

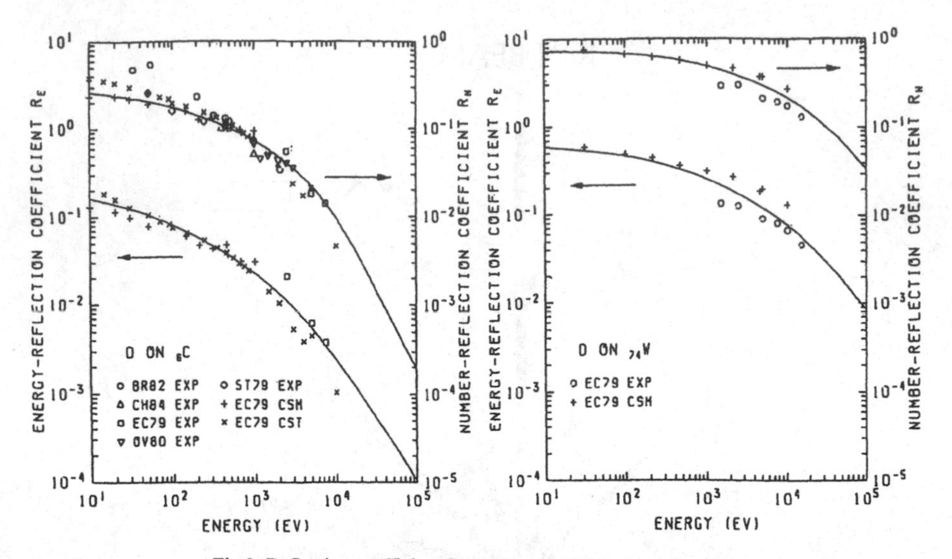

Fig.2 Reflection coefficient for Carbon and Tungsten (after [4])

reflection coefficients agrees well with experimental data as shown in Fig.2 [4]. The reflection coefficients increase with decreasing the incident energy and increasing target mass (Z number) For the incident energies below 100 eV, more than half of the incident hydrogen is reflected at tungsten (W), whereas the reflection at graphite is about 1/5. (See Fig. 2) As discussed in the next section, reflected particles bring some of their incident energy, and hence deposited energy to the target or energy reflection coefficients are quite different for high-Z and low-Z materials. Actually some difference was observed in the deposited energy on the W and C limiter in TEXTOR (See sec. 3)[5] At very low incident energy, however, the assumption of the binary collision is no more reliable and also no experimental data are available due to the difficulty to get a low energy ion beam, too.

The reflected hydrogen is either an ion or atom, of which penetration length into the plasma critically depends on their energy (or velocity) and angular distribution. Generally incident angle at plasma facing surface is very small. The angular distribution of the reflected hydrogen for small angle incidence is important topics both theoretically and experimentally. In addition, it should be noted that the reflected neutral atoms are not necessarily in the ground states. Depending on the characteristics of the reflected particles, i.e. velocity and angular distribution, ions or neutral atoms and electronic state, ionization process in the boundary plasma must be varied and hence the edge plasma should be influenced, which is discussed in Sec.3.

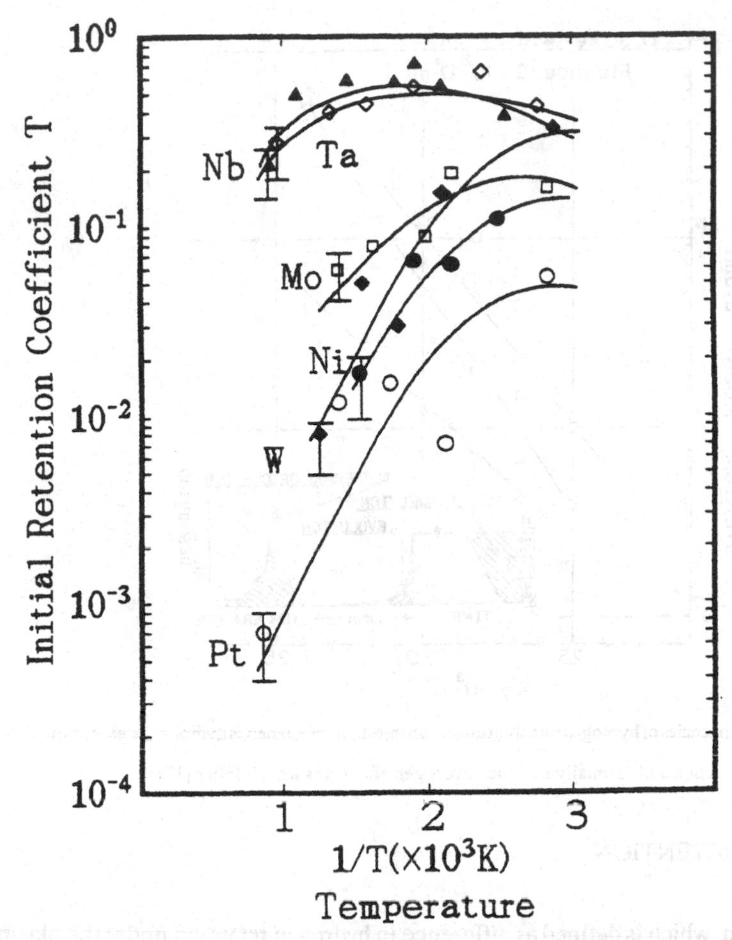

Fig.3 Retention coefficient of atomic hydrogen at higher temperatures (after [9])
In the experiments surface chemisorption sites would be fully occupied by residual hydrogen.

To analyze the reemission process, recombination coefficients for metals have been estimated semi-empirically by several authors [14-16] . All of them have utilized solubility which is basically applicable for thermodynamical equilibrium system. When the reemission is dominated by excited molecules or atoms, as described above, these estimations lose their theoretical base. Again, the release of the excited particles would change the character of boundary plasma because the penetration or ionization length of those particles becomes longer.

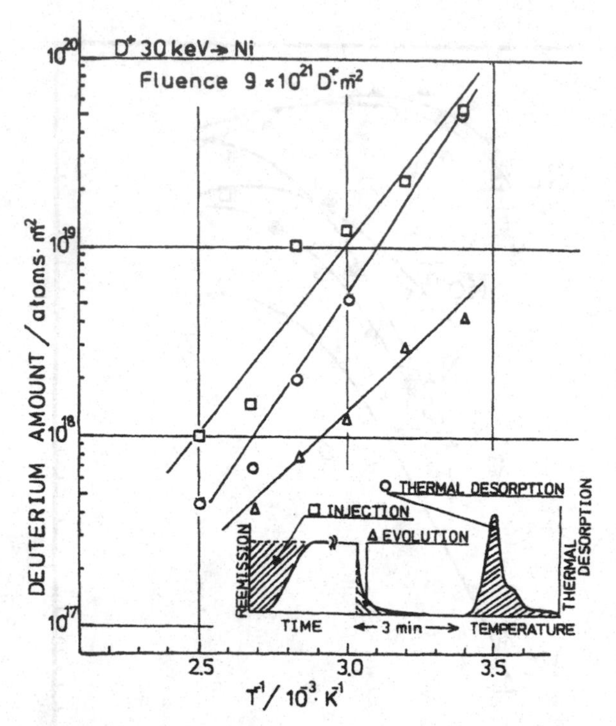

Fig.4 Temperature dependencies of hydrogen retention under ion injection, spontaneous hydrogen release immediately after stopping the ion injection and thermally desorbed hydrogen afterwards for Ni.(after [17])

2.2. DYNAMIC RETENTION

Dynamic retention, which is defined as difference in hydrogen retention under the plasma exposure and that after the exposure, is related to the excited molecular and/or atomic reemission. Fig. 4 shows temperature dependencies of hydrogen retention under ion injection, spontaneous hydrogen release immediately after stopping the ion injection, and thermally desorbed hydrogen afterward for Ni [17]. As seen in the figure, the total retention decreases with increasing the temperature, while the fraction of the spontaneous release to the total retention increases with increasing the temperature and most of the injected hydrogen is spontaneously released at elevated temperatures. This means the decreases of the residence time of the hydrogen. It is very much reasonable that hydrogen atomic reemission starts when the residence time becomes very small, because hydrogen atoms are losing a chance to recombine and to stay at the surface to be thermalized. Such argument may lead to a conclusion that the influence of retained clearly demonstrated that permeated hydrogen through Pd with contaminated surface by sulfur gives much higher translation energy (velocity) than that expected from the materials temperature. Even for non-contaminated surface, hydrogen recombination, one atom on the surface and the other in the subsurface or both in the subsurface, produces the excited molecules rotationally or vibrationaly[13], when the recombined molecules do not stay longer at the surface absorption site to be thermalized. At very high temperatures, as already pointed out, atomic reemission is favorable, because the residence time becomes so short that hydrogen atoms lose a chance to recombine.

Again it should be pointed out that no current recombination theories take such epi-thermal process into account, and a more realistic theory which takes the excited state into account is awaited.

3. Some examples in the TEXTOR tokamak

In recent experiment in TEXTOR[5], new results which clearly indicate the influence of the wall character on hydrogen recycling and heat deposition have been observed, as summarized in the following.

3.1. EFFECT OF REFLECTION ON DEPOSITED ENERGY

Energy deposition and hydrogen recycling with different Z number materials are compared with using a C/W twin limiter in TEXTOR-94. The experimental details have been given in elsewhere[5]. Fig.5 shows the potograph of the C/W twin limiter and the depostion profiles of C, D, Si and W after the plasma exposure [25]. C contamination on the W side was not appreciable owing to large sputtering and most of the W part remained as shiny, except the very edge of the limiter where was seen C deposition with a very sharp boundary. Owing to its large reflection detureium retention in the W part is very small, whereas the significant amount of deuterium is retained in the C side, and in the deposited region of the W side with D/C ratio corresponding the surace tempearture. (The surface temperature is the highest at the center and the lowest at the edge)

Effect of the reflection is also seen in the difference of the absorebed heat. In fig.6 is compared the deposited heat between the C and W sides of the C/W twin limiter inserted in TEXTOR-94 plasma. One can see the difference in the absorbed (deposited) heat flux between the C side and the W side. Although the absorbed heat flux in the C

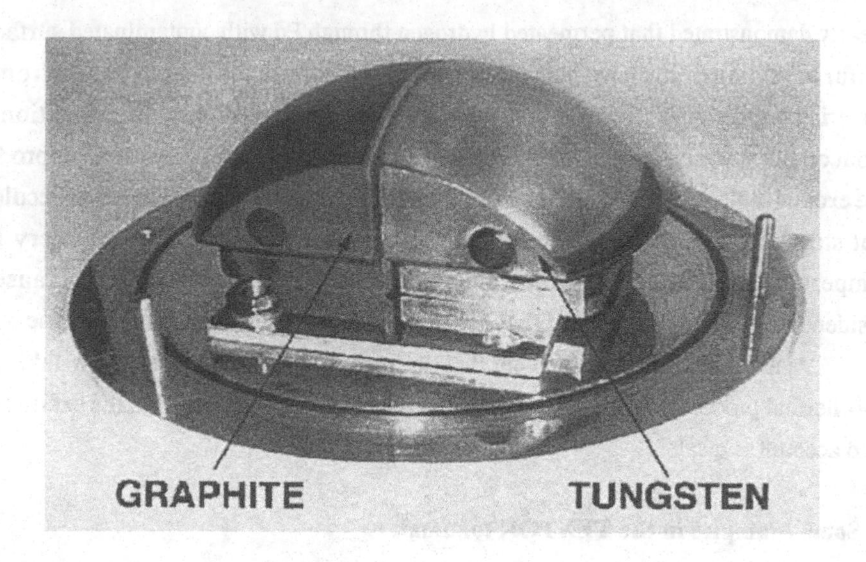

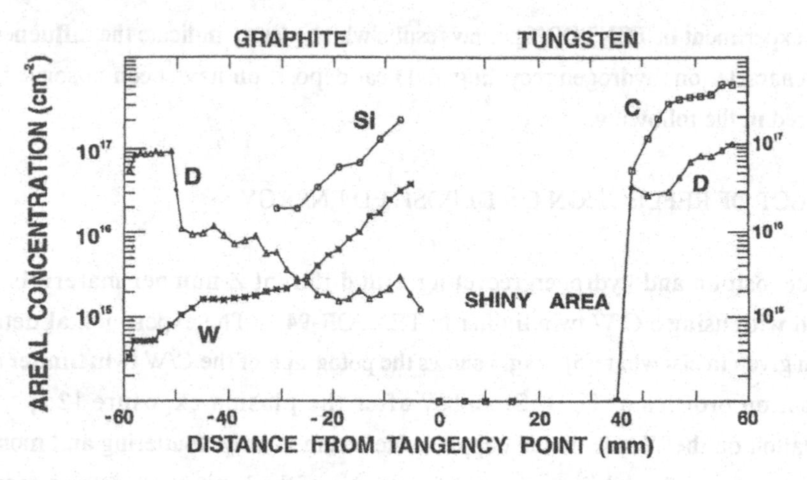

Fig.5 Photograph of C/W twin limiter and C, D and W deposited profies(after [25])

side stayed relatively constant with increasing line averaged density, that in W showed steeper decrease. Such density dependence of the absorbed heat on the density is very likely governed by the reflection phenomena. The reflection coefficient of W increases with decreasing impact energy so is the decreasing electron temperature. Accordingly, the deposited heat on the W side decreases with increasing density. On the other hand the heat load to the C side was not severely influenced by the reflection owing to its small reflection, and showed little dependence upon density. Neverthless, the difference in

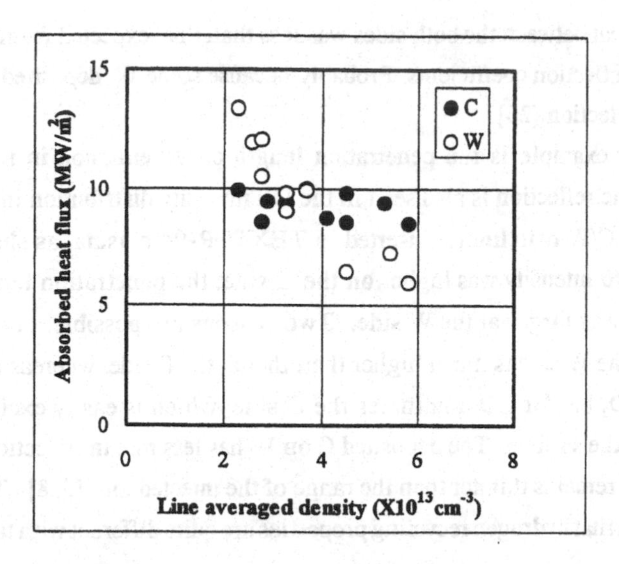

Fig.6 Comparison of deposited heat between C and W sides of C/W twin limiter inserted in TEXTOR-94 plasma (after [5])

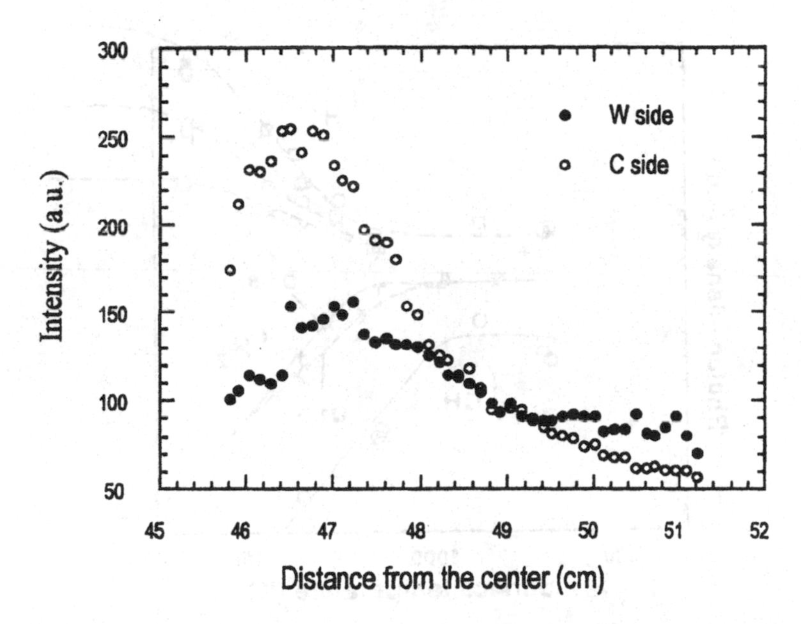

Fig.7 Comparison of D α intensity decay in front of the C and W sides of C/W twin limiter inserted in TEXTOR-94 plasma. The limiter was positioned at 46cm from the plasma center (after [5]).

the deposited heat between the both sides was less than that expected from the difference between their reflection coefficients. Probably because some W deposited on the C side enhcaces its reflection [26].

Another example is the penetration length of Dα emission in to the plasma. Difference in the reflection is also seen in the Dα intensity distribution in front of the C and W sides of C/W twin limiter inserted in TEXTOR-94 plasma. as shown in Fig.7. Although the Dα intensity was higher on the C side, the penetration length (e-folding length) of Dα was larger at the W side. Two reasons are possible, (1) the energy of reflected D at the W side is much higher than that at the C side, whereas (2) reemission as low energy D_2 and/or CD dominates the C side, which is easily excited to give Dα emission near the surface. The deposited C on W has less role in reflection as far as the deposited layer remains thinner than the range of the injected ions [5, 8]. This is the first clear indication that hydrogen recycling properties are quite different with target materials in tokamaks.

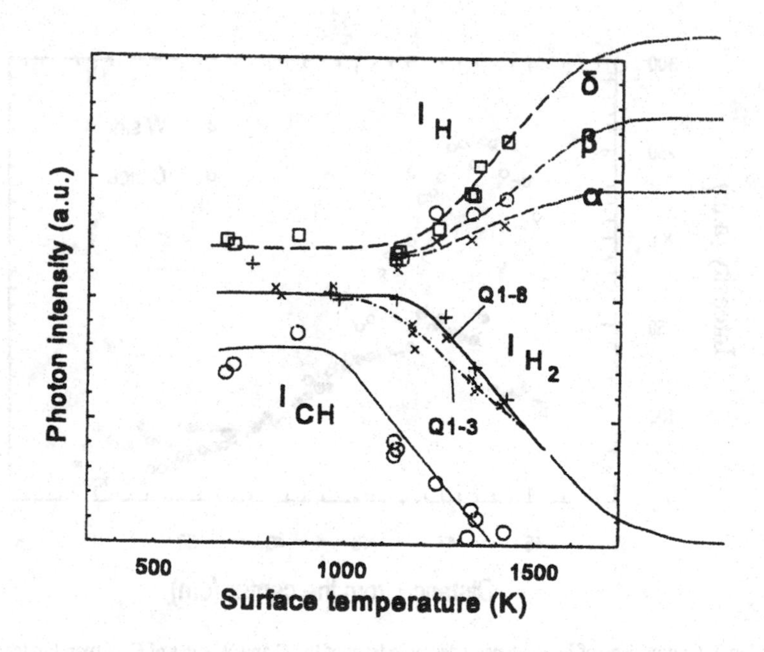

Fig.8 Variations of the H_2-(Q_{1-3}, Q_{1-8} band), CH-, and H δ β α -emission(□,○,×, resp.) as a function of the graphite limiter temperature in the Ohmic part of the TEXTOR-94 discharge. A smaller sum of individual Qi-component is represented by (x) (after [27])

3.2. OBSERVATION OF EXCITED ATOMS AND MOLECULES

Fig. 8 shows variations of the H_2-(Q_{1-3}, Q_{1-8} band), CH-, and H δ β α −emission(□ ,O, × resp.) as a function of the graphite limiter temperature in the Ohmic part of the TEXTOR-94 discharge[27]. Increase of H δ β α-emissions above 1200K corresponds well to the atomic hydrogen emission [21]. At the same time H_2 band emission decreases. From $Q_5/Q_1$of the (0-0) band in creases with the limiter temperature. The cause of these band emission in the boundary plasma, whether plasma origin or materials origin, is not clear. In any case the large fraction of molecular hydrogen may lead to different energy exhaust mechanisms in the boundary plasma or divertor of a fusion plasma.

4. Conclusions

Released hydrogens from plasma facing materials are constructed of reflected, desorbed (ion induced, electron and photon stimulated), and reemitted particles. Most of the previous hydrogen recycling studies have focused to particle balance between the edge plasma and the plasma facing materials, assuming the released particles are all in ground states. However, the released particles, whatever their origins are, are not necessarily at the ground state in their electronic, rotational and vibrational states.

Reflected particles generally carries some of their incident energies as their velocities, which usually give energy reflection. Depending on their velocities, they are distributed among ions, excited neutrals and ground state neutrals. The higher (lower) the velocity of the reflected particles, the more the fraction of the ions (the ground state neutrals) is. Except for very low energies the fraction of excited neutrals are not small, which might contribute to Hα emission in boundary plasmas . Ionization process in the boundary plasma should be influenced by at what state the reflected particles are.

The reemitted molecules are often carrying their recombination energy with a form of rotational and vibrational energies. Even atomic reemission dominates at elevated temperatures but far below a thermal dissociation temperature. This also influences hydrogen recycling process because the less ionization energy is needed for excited molecules.

Not a small amount of hydrogen and/or other volatile impurities are released from the plasma facing wall by particle and photo induced desorption in future tokamaks. Although photo desorption and electron stimulated desorption are well know in surface

5. References

[1] See, for example, Wilson, K. L. *et al.*: Trapping, detrapping and release of implanted hydrogen isotopes, *Suppl. J. Nuclear Fusion,* **1**(1991)pp.31-50, and W. O. Hoffer and J. Roth, *Physical processes of the interaction of fusion plasmas with solids*, (Academic press, 1996)

[2] See for example, Tanabe, T. : *J. Nucl. Mater.* **248**(1997)418

[3] Verbeek, H. : *Backscattering of light ions from metal surfaces, in Radiation effects on Solids Surfaces*, Ed.by M. Kaminsky, (American Chem. Soc. 1976)pp.245-261

[4] Ito, R., Tabata, T., Itoh, N. *et al.*: Data on the backscattering coefficients of light ions from solids, IPPJ-AM-41(1985)

[5] Tanabe, T., Ogo, T., Wada, M., *et al.*, Proc. 5th Inter. Conf. Fusion Nuclear Technology, Sep.19-24, 1999, Rome, Italy

[6] Naujoks, D., Roth, J. and Krieger, K. : *J. Nucl. Mater.* **210**(1994)43

[7] Ueda, Y., Tanabe, T., Philipps, V., *et al.*: *J. Nucl. Mater.* **220-222**(1995)224

[8] Ohya, K., Tanabe, T., Wada, M., *et al.* : *Nucl. Instr. Methods in Phys. research*, **B153** (1999)354

[9] Kiriyama, T., Tanabe, T. : *J. Nucl. Mater.* **220-222**(1995)873

[10] Franzen, P., Vietzke, E., : *J. Vac. Sci. Technol.* **A12**(1994)820

[11] Davis, J. and Haasz, A. A.: *J. Nucl. Mater.* **223**(1995)312

[12] Comsia, C., David, R. : *Surface Sci. Repts.* **5**(1985)145

[13] Schtroeter, L., David, R., Zacharias,H. : *Surf. Sci.* **258**(1991)259

[14] Baskes,M. I.: *J. Nucl. Mater.* **92**(1980)318

[15] Pick, M. A., Sonneberg, K. : *J. Nucl. Mater.***131**(1985)208

[16] Richards,P. M. : *J. Nucl. Mater.* **152**(1988)246

[17] Tanabe, T, Hirano, H., Imoto, S. : *J. Nucl. Mater.* **151**(1987)38

[18] for example, see Tolk, N. H., Traum, M. M., Tully, J. H. and Madey, T. E. : *Desorption Induced by Electronic transitions DIET*, (Springer-Verlag, Berlin, 1983)

[19] Vietzke, E. and Hassz, A. A. : *Chemical Erosion, in Physical Processes of the Interaction of Fusion plasmas with Solids*, Ed. By Hofer, W. O. and Roth, J. (Academic Press, 1996) pp.135-176

[20] Haasz, A. A., Stangeby, P. C. and Auciello, O, : *J. Nucl. Mater.* **111-112**(1982)757

[21] Vietzke, E., Flaskamp, K. and Philipps, V. : *J. Nucl. Mater.* **111-112**(1982)763

[22] Benninghoven,A., Ruedenauer, F. G., Werner, H. W. : *Secondary Ion Mass Spectrometry, Basic concepts, Instrmental Aspects, Application and Trends*, (John Wiley & Sons, 1987)

[23] Tanabe, T. and Omori, A. : *J. Nucl. Mater.* **266-269**(1999)703

[24] Ohmori, A. and Tanabe, T. : *J. Nucl. Mater.* **258-263**(1999)666

[25] Rubel, M., Philipps, V., Huber, A. et al. : Formation of carbon containing layers on tungsten limiters, Physica Scripta, T81 (1999) 61

[26] Ohya, K., Kawata, J. Tanabe, T. : *J. Nucl. Mater.* **266-269**(1999)629

[27] Pospieszczyck, A., Mertens, Ph., Sergienko,G., *et al.*: *J. Nucl. Mater,* **166-269**(1999)138

MAIN RESEARCH RESULTS IN HYDROGEN THERMO-SORPTIVE ACTIVATION BY METAL HYDRIDES

YU.F.SHMAL'KO

A.N.Podgorny Institute of Mechanical Engineering Problems of National Academy of Sciences of Ukraine
2/10 Pozharsky St., Kharkov, 310046, Ukraine

1. Introduction

The processes of hydrides formation and decomposition at reversible interaction of a number of metals and alloys with hydrogen have a number of specific features. First of all it concerns to increased catalytic activity of hydrides of metals and intermetallic compounds in reactions connected to hydrogen transfer, i.e. hetero-phase isotope exchange [1], catalytic hydriding in gas and liquid phases [2,3], hydrogen saturation of alloys which are difficult for hydriding [4,5], low-temperature hydrogen oxidation with water formation [6], etc. The mentioned effects first of all were explained by hydrogen appearance in monatomic or any other excited state on metal surface or in a gas phase. In a number of experiments there was possible to identify and even "to preserve" for rather long time the H atoms formed during interaction of gaseous hydrogen with hydride forming materials. It has become possible due to spill-over effect of monatomic hydrogen along surfaces of Al_2O_3, SiO_2 and some other oxygen-containing phases. It has been shown, that hydrogen atoms migrating on a surface of these carriers have an effective positive charge [7,8]. The maximum emission of monatomic hydrogen was observed in a during its sorption-desorption using fresh palladium catalyst [7].

In a 1958, by direct mass-spectrometric measurements in conditions eliminating processes of secondary ionisation (for example, ignition of gas discharge), it was shown [9], that hydrogen diffusion through palladium membrane is accompanied by emission of protons and, to a less degree, H_2^+ ions.

Then, the author of work [10] revealed, that hydrogen diffusion through heated up to 773–1273 K steel or palladium anodes into evacuated (down to 10^{-1} Pa) volume with cold (293 K) cathode, is accompanied by passing an electric current in inter-electrode gap. Replacement of steel with hydride-forming palladium results in 2-3 times increase of the current. The current, under the author's [10] opinion, is stipulated by proton emission, i.e. hydrogen, diffusing through the membrane, is desorbed as protons.

In the work [11] it is informed that hydrogen, thermo-desorbed from metals (Fe, W, Ta) being previously saturated with it, is largely dissociated, and the dissociation is not thermal, as with temperature growth in a range of 473–1173 K, the relation of concentrations of monatomic and molecular hydrogen does not increase.

C.H. Wu (ed.), Hydrogen Recycling at Plasma Facing Materials, 95–107.

Approximately at the same time in [12,13] it was shown that photolysis of alkaline and alkaline-earth metal hydrides is accompanied by emission of H^- ions into a gas phase.

The experimental researches carried out by group of the contributors from Institute of Mechanical Engineering Problems of National Academy of Sciences of Ukraine, Sukhumi Physical-Technical Institute and D.V.Efremov Research Institute of Electro-Physical Equipment [14-16] have shown, that in "metal hydride – hydrogen" systems it is observed an essential change of electro-transfer characteristics in a gas phase, in comparison with equilibrium hydrogen. In particular, the ignition of the glow discharge in hydrogen, desorbed from $LaNi_5H_x$, takes place at relative potential gradients E/P 1.5–2 times lower, than in usual molecular one [14,15]. Among particles forming discharge plasma, the fraction of atomic and complex ions (H^+, H_3^+) related to molecular ions H_2^+ [15] is considerably increased. In the certain conditions (hydrogen pressure of 10^{-5}–10^{-3} Pa) in hydrogen desorbed from the indicated intermetallic hydrides the noticeable fraction of H^+ and H_2^+ ions is observed [14]. The data [15,16] show, that hydrogen desorbed from the indicated hydrides is characterised by increased, in comparison with equilibrium one, ionisation efficiency.

The stated facts allow to assert, that sorption-desorption processes in "hydrogen – metal hydride" systems are accompanied by hydrogen activation. The indicated effect is, that the processes of formation and decomposition of binary and intermetallic hydrides are accompanied by appearance near their surface of non-equilibrium state particles (atoms, atomic and molecular ions, excited atoms and molecules).

This report generalises research results obtained in the author's group under the line of the above-mentioned effect of sorption activation, by detail study of energy and charge states of particles in a gas phase of "hydride forming material – hydrogen isotopes" systems, as well as electro-transfer performances in a gas phase of these systems.

2. Mass-spectrometry investigations of sorption activation

As early as 1987 we published the first experimental results [14] on investigation of energy state of hydrogen molecules, desorbed into vacuum from a surface of metal hydrides. Mass-spectrometry measurements [15,16], carried out hereafter for process of molecular ions H_2^+ formation have shown, that hydrogen, desorbed from metal hydrides ($LaNi_5H_x$ at pressure close to atmospheric and room temperature; hydrogen-saturated Zr-V-Fe getter at T=800–900 K and $P=(5–7)\cdot10^{-1}$ Pa) has 30–50 % higher cross-sections of ionisation, and the potential of appearance of H_2^+ ions is accordingly 0.3-0.5 eV lower, than for hydrogen from a cylinder (Fig. 1). The last value is close to energy of the first vibrational level of hydrogen molecule in ground electron state $X^1\Sigma_g^+$. On a basis of observed effects of a decrease of ionisation potential and significant life time of excited states for desorbed hydrogen molecules the supposition was expressed, that thermo-sorption hydrogen activation is accompanied by excitation of vibrational levels of H_2 molecules, formed during recombination of hydrogen atoms going out onto a surface of a hydride.

For refinement of vibrational levels and determination of force field of molecules the research of vibrational spectra of their isotope varieties is widely used. Within the framework of Born-Oppenheimer's approximation it is supposed, that isotope substitution does not change distribution of electron density, equilibrium inter-nuclear distances, function of potential energy, and the force constants. Differences in nuclei masses results only in a change of kinetic energy of their oscillations, that stipulates difference of vibration frequencies of isotope varieties of molecules. Therefore substitution of protium atoms with deuterium or tritium will give maximum isotope effect in differences of vibrational levels of appropriate molecules.

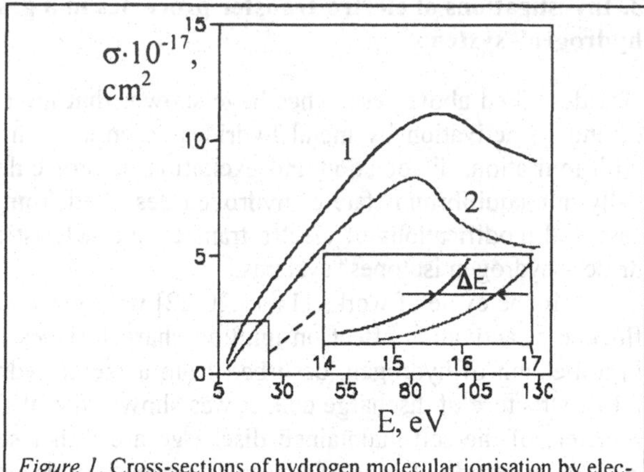

Figure 1. Cross-sections of hydrogen molecular ionisation by electron impact:
1 - H_2 desorbed from (Zr-V-Fe)H_x
2 - Equilibrium H_2 from a cylinder

Oscillation frequency for protium molecule H_2 (at calculations in terms of harmonic oscillator approximation) $\sqrt{2}$ times exceed oscillation frequency for deuterium molecule D_2.

The obtained ionisation efficiency curves for hydrogen, desorbed from a metal-hydride sample (curve 1), and also for thermodynamically equilibrium hydrogen (curve 2), are represented in insertion of Fig. 1. The ionisation efficiency curves for deuterium are similar. From each pair of these curves there was found a shift of appearance potential using Warren's method [16,18]. The obtained results are represented in Table 1.

TABLE 1. Ionisation potentials for hydrogen isotopes (protium and deuterium)

Hydrogen isotope	Adiabatic ionisation potential, eV	Measured ionisation potential for molecules of desorbed gas, eV	Shift of ionisation potential for molecules of desorbed gas, eV	Excitation energy for the first vibrational level, eV
H_2	15.426	14.91	0.520	0.546
D_2	15.467	15.08	0.387	0.386

The ratio of the experimentally measured shifts of ionisation potentials for desorbed protium and deuterium (see Table 1) is equal to 1.35. This value, within the allowable experimental error for the given technique, is close to the value of isotopic effect observed in infra-red spectroscopy for isotope varieties of diatomic molecules, containing protium, or replacing it deuterium, which have the first vibrational level of a ground electron term $X^1\Sigma_g^+$ [19]. The obtained result is a direct experimental confirmation of above-stated hypothesis about vibrational excitation of hydrogen molecules during their desorption from metal hydride surface.

3. Investigations of electro-transfer processes in a gas phase of "metal hydride – hydrogen" systems

The described above researches have shown, that the change of hydrogen energy state during its activation by metal hydrides essentially influences to processes connected with ionisation, dissociation and excitation of molecules. In particular, thermodynamically non-equilibrium state of hydrogen desorbed from metal hydrides should result in essential modifications of electro-transfer characteristics of a gas phase of "metal hydride – hydrogen isotopes" systems.

In the cycle of works [14,15,20-22] we carried out comparative researches of influence of activation effect on ignition characteristics of the self-maintained discharge in a medium of hydrogen, desorbed from a metal-hydride electrode which is included into a structure of discharge cell. It was shown, that the threshold values of ignition parameters of the self-maintained discharge are rather sensitive to a potential of metal-hydride electrode, degree of its saturation with hydrogen, geometry of a discharge cell and strength of the magnetic field.

It has been shown [14], that the ignition of the glow discharge in hydrogen, desorbed from $LaNi_5H_x$ hydride, takes place at relative potential gradients E/P 1.5–2 times lower, than in usual molecular one. To be sure, that the observable effect is stipulated by a modification of properties of a gas phase, instead of any near-electrode process (for example, change of coefficients of secondary electron emission for electrodes surface) we have changed an experimental technique [15]. A metal-hydride sample (the composite material on a basis of $LaNi_5H_x$ hydride with Teflon binder) had a dielectric insulation with electrodes of discharge cell, and its surface area from a side of discharge zone was negligibly small in comparison with electrodes surface area. The discharge cell was made as an element of the gas ion source, installed in a mass-spectrograph.

The results of the experiments are presented in Table 2. The mass-spectrographic investigations of a gas phase have shown, that the full spectrum of masses for gas medium in the discharge cell (in conditions of experiments [15]) contains only positive ions H^+, H_2^+, H_3^+.

Table 2. Burning characteristics and plasma composition for glow discharge in hydrogen

Mode of hydrogen supply into discharge zone	Discharge burning characteristics		Relative intensities of ions' lines		
	Hydrogen pressure, Pa	Discharge burning voltage, V	H^+	H_2^+	H_3^+
Leak-in from a cylinder (molecular hydrogen)	1 – 40	630	2.33	1	0.56
Desorption from $LaNi_5H_x$ hydride	1 – 40	420	7.69	1	2.23

Under hydrogen desorption from a metal-hydride sample the discharge was ignited already at voltage of 300-350 V (at hydrogen pressure range in the discharge chamber of 5–50 Pa), while the reference discharge in a medium of cylinder hydrogen (in the same pressure range) was ignited only at 600–700 V. In accordance with sample leaning with hydrogen during desorption the essential increase of absolute values of intensities of lines of these ions took place. The intensities were maximally high even at long-time evacuation of a sample at 300 K, though the pressure in discharge gap thus decreased down to 1 Pa. It has confirmed our suppositions [14] that the effect of hydrogen sorption activation by metal hydrides is exhibited in a maximum degree during a finishing stage of a desorption, i.e. at transition from (α+β)- to α- region of pressure – composition isotherm.

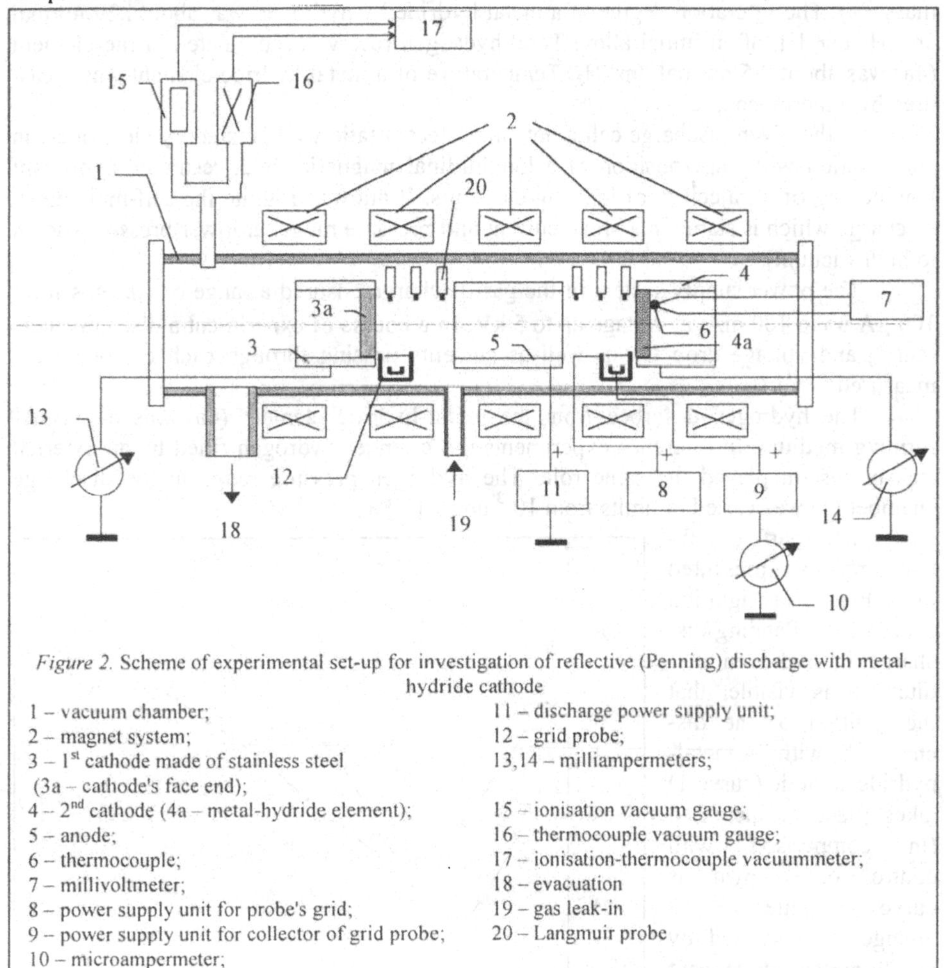

Figure 2. Scheme of experimental set-up for investigation of reflective (Penning) discharge with metal-hydride cathode

1 – vacuum chamber;
2 – magnet system;
3 – 1st cathode made of stainless steel (3a – cathode's face end);
4 – 2nd cathode (4a – metal-hydride element);
5 – anode;
6 – thermocouple;
7 – millivoltmeter;
8 – power supply unit for probe's grid;
9 – power supply unit for collector of grid probe;
10 – microampermeter;
11 – discharge power supply unit;
12 – grid probe;
13,14 – milliampermeters;
15 – ionisation vacuum gauge;
16 – thermocouple vacuum gauge;
17 – ionisation-thermocouple vacuummeter;
18 – evacuation
19 – gas leak-in
20 – Langmuir probe

As it was already noted, performances of ignition and burning of the self-maintained discharge in a medium of hydrogen, desorbed from metal hydrides, are very sensitive to a type of discharge cell, potential of a metal-hydride element, the strength of external magnetic field, etc. So, the ignition characteristics of the discharge in Penning cell with the metal-hydride cathode qualitatively differ from described above ones for the glow discharge with a metal-hydride electrode.

In a Fig. 2 the scheme of experimental installation for investigation of the reflective discharge with the metal-hydride cathode is shown. The system of electrodes

(3,4,5) was mounted inside quartz or Teflon tube and represented Penning discharge cell with tubular electrodes. The anode (5) and cathodes (3,4) were made of a stainless steel. In an end face of the cathode (4), faced to the anode (5), the metal-hydride element (4a) was placed. It was made as a tablet by a diameter of 20 mm and thickness of 4 mm pressed from a powder of saturated with hydrogen $Zr_{55}V_{40}Fe_5$ + 3% B_2O_3 alloy [23], mixed with heat and electro-conducting binder (copper powder in an amount of 40 mass. %). The saturation degree of a metal hydride by hydrogen was about 229 normal cm^3 H_2 per 1 g of an initial alloy. Total hydrogen reserve accumulated in the element (4a) was about 2.5 normal dm^3 H_2. Temperature of a metal-hydride element was measured by a thermocouple.

In the given discharge cell a potential electrostatic wall created by electrodes, in combination with superposition of a longitudinal magnetic field, result in significant lengthening of a trajectory of ionising electrons. It allows to ignite the self-maintained discharge which is stable in a broad current and pressure range, at lower pressure (down to high vacuum),

The power supply system of the gas discharge ensured a range of currents from 100 μA up to 100 mA, at voltage up to 5 kV. In a course of experiment a discharge current I_d and voltage drop U_d, as well as currents passing through each cathode were measured.

The hydrogen desorbed from the metal-hydride element (4a) was a plasma-forming medium. In reference experiments the cylinder hydrogen filled by an external leak-in system played the same role. The hydrogen pressure range in the discharge chamber (1) was varied in limits from 10^{-3} up to 10 Pa.

The part of obtained results is presented in a Fig. 3 as ignition curve of the Penning discharge in hydrogen medium. It is visible, that the ignition of the discharge with metal-hydride cathode (curve 1) takes place at increased (in comparison with control one shown as curve 2) values of the voltage. It is caused by an increased attachment rate of low-energy electrons to the vibrationally excited molecules of hydrogen, desorbed from a surface of the metal-hydride cathode. As a result, near-cathode region became lean by electrons, and the development of electronic avalanches is inhibited. Therefore, for fulfilment of existence condition for the self-

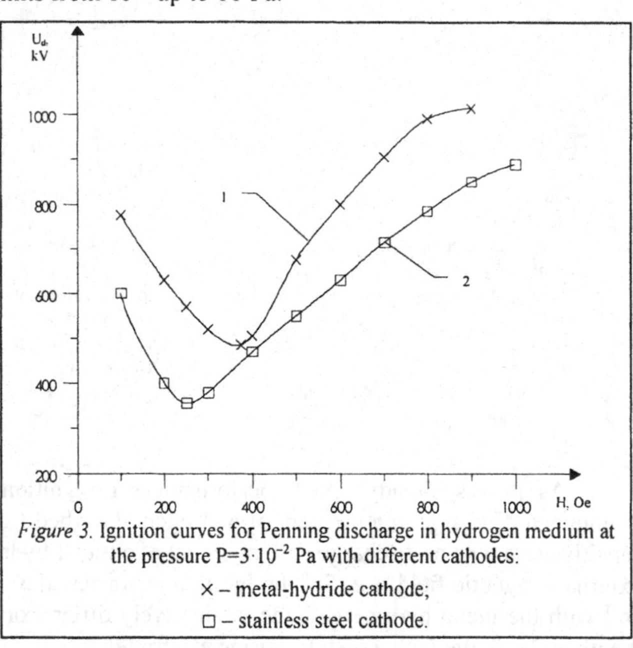

Figure 3. Ignition curves for Penning discharge in hydrogen medium at the pressure $P=3\cdot10^{-2}$ Pa with different cathodes:
× – metal-hydride cathode;
□ – stainless steel cathode.

maintained discharge with the metal-hydride cathode, it is necessary to increase a rate of development of electronic avalanches. It is realised by increase of the voltage.

As these results have given a ground to assume, that in the Penning discharge with metal-hydride cathode the increased yield of negative hydrogen ions is possible, we carried out special researches [24] of formation processes of negative ions in this discharge. They have clearly shown, that the desorption of vibrationally-excite hydrogen molecules from a surface of the metal-hydride cathode unambiguously results in essential increase of H^- concentration in near-cathode region. The dissociative attachment cross-section for low-energy electrons to desorbed from metal hydride surface vibrationally-excited hydrogen molecules is fastly increase with increase of a vibrational quantum number v and reaches the greatest value (about $3\cdot10^{-17}$ cm^2) at $v\geq5$ and energies of electrons about several electron - volts [25]. Thus the concentration of negative hydrogen ions and low-energy electrons at near-cathode region is determined by energy state of hydrogen molecules, desorbed from metal hydride surface.

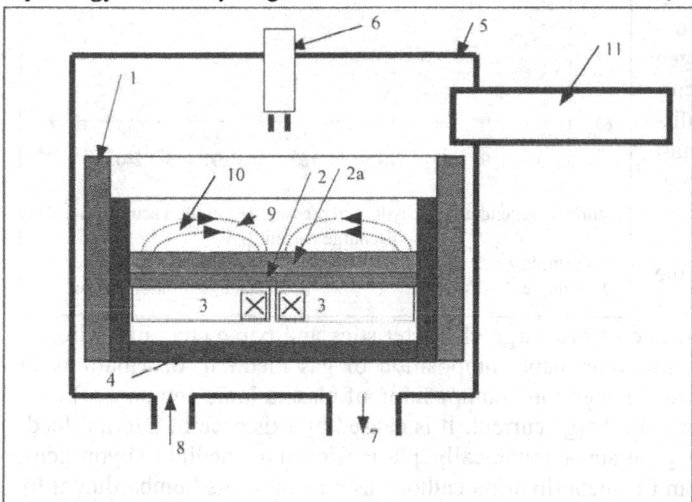

Figure 4. Scheme of experimental set-up for investigation of magnetron discharge with metal-hydride cathode:

1 – anode;	6 – double electric probe;
2 – cathode (2a – metal hydride layer);	7 – evacuation;
3 – magnet system;	8 – leak-in of working gas;
4 – isolator;	9 – force lines of magnetic field;
5 – vacuum chamber;	10 – localisation zone of dense plasma;
	11 – mass-spectrometer.

During further experimental researches the new effect – pressure-consistent mode of gas discharge [22,26] was detected. The experiments were carried out with two types of the discharges magnetron and penning ones – burning in hydrogen medium desorbed from the metal-hydride cathode in closed volume (i.e. without external leak-in of plasma-forming gas and external evacuation). It was shown, that these discharges have a number of specific features exhibiting itself at presence of metal-hydride electrodes. Briefly about research results of Penning discharge have been said above. The magnetron discharge was investigated using the cell, which scheme is presented in a Fig. 4. The metal-hydride element covered a working surface of the cathode as the layer consisting of a mixture of a powder of hydride forming alloy, saturated with hydrogen, and a copper powder in an amount of 40 mass. %. The hydrogen pressure was changed in limits from 55 Pa up to 133 Pa. The discharge current was varied from 50 mA up to 300 mA. Values of a voltage drop, depending on experimental conditions, was from 240 V up to 600 V. A mass spectrum of gas in the chamber, and in some experiments, distributions of plasma density and electron temperature (using double electric probe), were measured in parallel.

In a fig. 5 the experimentally observed dependence of hydrogen pressure on discharge current for magnetron discharge with the metal-hydride cathode is presented. Characteristics of an obtained burning regime of the discharge with the metal-hydride cathode is the unambiguous dependence of discharge characteristics and parameters of hydrogen plasma (hydrogen pressure, component composition of gas medium, distributions of plasma density and electron temperature, composition of plasma ionic component) on a single external parameter – discharge current. It is caused by existence of internal feedback, which enables to generate automatically plasma-forming medium (hydrogen), using a heat released from the metal-hydride cathode as a result of its bombardment by plasma ions. We name such a mode of the electric discharge with the metal-hydride cathode as pressure auto stabilised one [26].

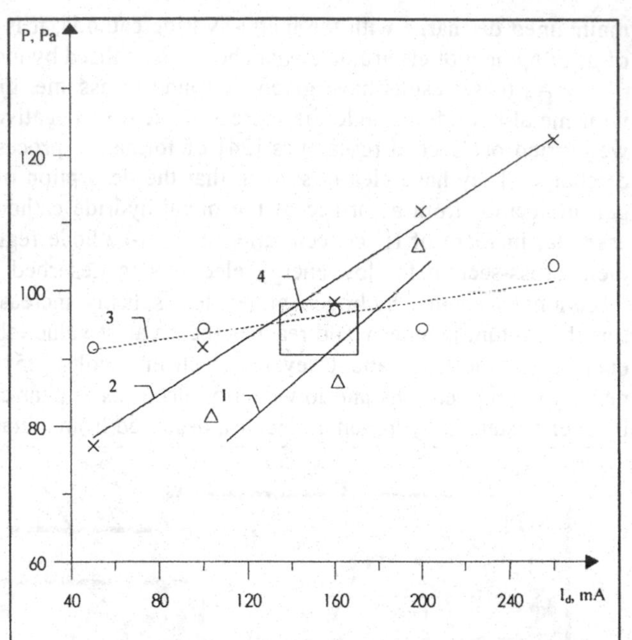

Figure 5. Dependence of hydrogen pressure (P) in the vacuum chamber on discharge current (I_d).
1 – sample $Zr_{55}V_{40}Fe_5$+3%B_2O_3; 3 – hydrogen from a cylinder;
2 – sample $Zr_{33}V_{57}Fe_{10}$; 4 – discharge ignition region.

4. The influence of the effect of hydrogen sorption activation on a composition of gas phase in the systems "metal hydride – hydrogen – gas impurities"

The modification of energy states of hydrogen molecules during of their desorption from the metal-hydride cathode into the region of discharge plasma exerts essential influence upon a composition of gas medium. For a research of this problem we have carried out mass-spectrometry measurements of composition of plasma neutral component in conditions of pressure auto-stabilised magnetron discharge with the metal-hydride cathode [22].

According to mass-spectrometry data, there were constructed the dependencies of the ratios of partial pressures for gas phase components at the presence of a metal-hydride sample (P_i) to similar values for reference discharge (P_i^o) from discharge (I_d) which are presented in a Fig. 6. The analysis of the spectra shows, that values of relative

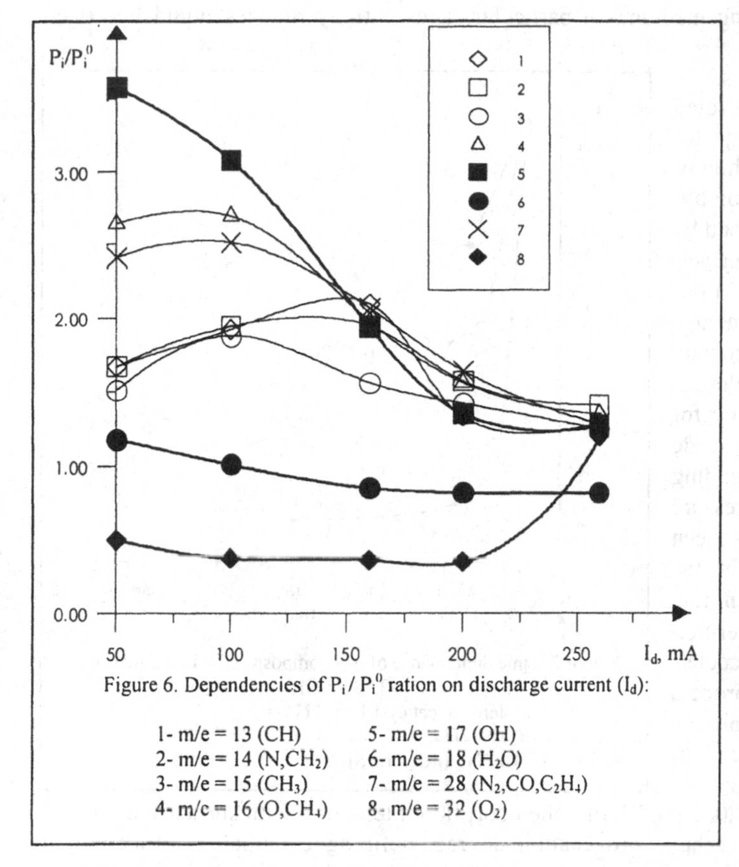

Figure 6. Dependencies of P_i / P_i^0 ration on discharge current (I_d):

1- m/e = 13 (CH)
2- m/e = 14 (N, CH_2)
3- m/e = 15 (CH_3)
4- m/e = 16 (O, CH_4)
5- m/e = 17 (OH)
6- m/e = 18 (H_2O)
7- m/e = 28 (N_2, CO, C_2H_4)
8- m/e = 32 (O_2)

intensities of hydrogen ions peaks (H^+, H_2^+, H_3^+) remain close both in the discharge with metal-hydride cathode, and in reference one. In impurities composition the significant modifications in a range of values of discharge current of 50–200 mA are observed. At further increase of discharge current the value P_i/P_i^o tends to unity, that means an approaching the concentrations of impurity components to values, characteristic for reference discharge. As it has been shown above, hydrogen thermo-desorption from a metal hydride is accompanied by emission of vibrationally-excited H_2 molecules in amounts noticeably exceeding equilibrium ones. The availability of such particles in near-cathode region of the discharge stimulates various plasma-chemical reactions including hydrogen and results in a modification of component composition of a gas phase. However, in accordance with increase of discharge current and corresponding pressure increase, it will increase a rate of vibrational-translation (V–T) and rotational-translation (R–T) relaxations of the excited molecular particles. Therefore, reactions initiated by the excited particles, are inhibited and the component composition of a gas phase comes nearer to the one which is characteristic for the discharge in a medium of cylinder hydrogen at the absence of a metal hydride.

5. Investigation of metal hydrides influence on spin-isomeric conversion of hydrogen molecules

As it is known, energy difference of ortho- (o-) and para- (p-) states of hydrogen molecules makes at low temperatures significant value (about 340 cal per mole of o-H_2). Therefore, o – p-transformation is accompanied by heat release, which plays a noticeable role in thermal balances in a number of power and technological processes using hydrogen as a working medium. In particular, it essentially reduces liquid hydrogen storage time.

The researches on the given line are devoted to study of rate change for spin-isomeric homo-molecular exchange of hydrogen molecules caused by hydride-forming intermetallic compounds of rare-earth and transition metals. The original experimental technique was developed. The catalytic efficiency for o–p-transfer by hydride forming alloys depending on temperature, pressure and sample state has been determined [27]. For the first time it was shown, that hydride forming intermetallides effectively accelerate the mentioned process (Fig. 7). On our opinion, these results correlate with energy parameters of hydrogen molecules excitation of during their sorption interaction with surfaces of various metal hydrides. Thus, the hydrogenation degree, defining essential modification of magnetic properties of a sorbent, determines a level of catalytic effect in hydrogen o–p-conversion reaction at low temperatures. The fact is also interesting, that the spin-isomeric conversion of hydrogen molecules of in a presence of hydride-forming sample goes according to low-temperature mechanism and depends on a magnetic state of a metal hydride working as the paramagnetic catalyst.

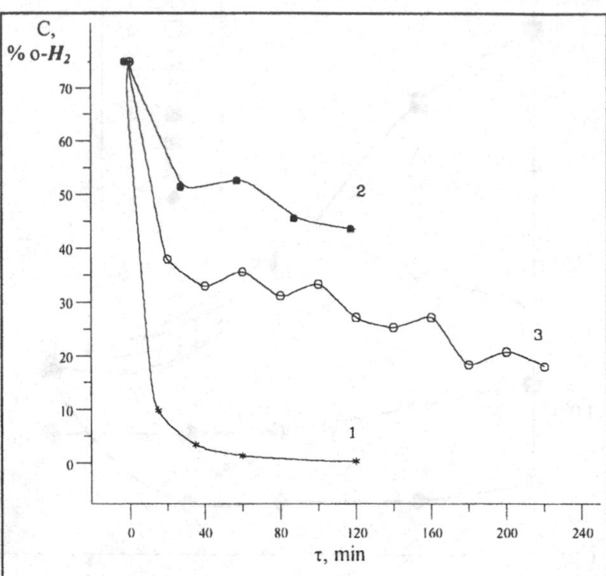

Figure 7. Time dependence of o-p-composition in liquid hydrogen (T=20.4 K):
1 – reference catalyst **$Fe(OH)_3$**;
2 –**$LaNi_5H_x$**;
3 – **$LaNi_5$** (dehydrided)

Apart for only physical interest, these results have also practical significance, if to take into account, that metal-hydride thermo-sorption compressors receive more and more broad applications, especially in a cycles of hydrogen liquefaction.

6. Application-oriented researches

The effect of hydrogen sorption activation by metal hydrides opens new possibilities in increase of parameters of an overall performance of various devices and systems of technical physics (such, as ion sources in structure of accelerating complexes, injectors of intensive beams of neutral atoms, neutron generators, hydrogen masers and lasers, etc.), as well as electrochemical energy converters and chemical current sources.

The most perspective scheme of construction of gas feed systems for electro-physical installations permitting to make leak-in of activated hydrogen immediately into working zone of the vacuum chamber, is the scheme of internal leak-in [28,29]. In difference from a traditional system of a gas feed providing supply of plasma-forming gas (hydrogen isotopes) into the vacuum chamber from an external source through the impulse valve or leak device, in the scheme of internal leak-in hydrogen source is the metal-hydride element included into a structure of electrodes of plasma-forming stage of the system – consumer. Thus the metal-hydride element simultaneously executes source functions of plasma-forming gas, electrode and getter facility for evacuation. Using this circumstance, it is possible to unload considerably or generally to disconnect an evacuation system. The advantages of the internal scheme of gas supply, constructed on the basis of multifunctional metal-hydride elements, are brightly expressed in the case of vacuum-plasma installations using tritium as working gas.

The mentioned scheme was realised in a number of experimental sources of positive and negative hydrogen ions [24,30], where the metal-hydride element, on a basis of hydrogen-gettering alloys Zr–V–Fe and Zr–V–Fe + B_2O_3, was included into a structure of the cathode or anode of plasma-forming stage. The carried out tests of a plasma source with the metal-hydride anode [30] have shown, that the use of sorption activation effect in such sources allows to increase the yield of ions in H^+ generation mode by 30–50%, and in H^- generation mode – by 15–20%. Thus the noticeable reduction of ignition voltage of the discharge (more than 1.7 times) and hydrogen pressure in the discharge chamber (about the order of magnitude), as well as stable burning, necessary for maintaining of the discharge and required beam current were observed.

The current density of a beam of negative ions obtained in a plasma source with the metal-hydride cathode, reached the values of 5–10 mA/cm^2 what corresponds to practically full extraction of H^- ions generated in near-cathode region of the discharge [24]. The current efficiency of a source of hydrogen negative ions based on the Penning discharge with metal-hydride cathode, made 40–50%.

Applied to hydrogen masers, the hydrogen accumulators of and systems for its stabilised leak-in on the basis of metal hydrides [28] were developed. Apart from effective realisation of auxiliary functions (compact hydrogen storage, it deep purification and leak-in), use of these systems has allowed to reduce a power inputs for obtaining a working medium (monatomic hydrogen) in the high-frequency discharge and to increase stability of its burning due to the presence in supplied gas the particles in energetically non-equilibrium state.

7. Conclusion

In this work the attempt of the state-of-the-art review of being available data concerning the effect of hydrogen sorption activation by metal hydrides was undertaken and the aspects of its practical and methodical applications are considered. Though limited volume of the report has allowed to affect the considered problems only fluently, we hope, that this physical phenomenon represents significant interest both in fundamental, and in the applied plan. The further research doubtlessly will be fruitful, as in the plan of more deep understanding of its nature, and concerning the extension of perspectives of its practical use.

8. References

1. Baikov Yu.M., e.a.: Isotopic exchange and chemical reaction in Zr–Ni–H system, *Kinetika i kataliz (Kinetics and Catalysis)* **26**, No.3 (1986). 753–756 (in Russian).
2. Semenenko K.N., Petrova L.A. and Burnasheva V.V.: Synthesis and some properties of hydride phases on the basis of $LaNi_{5-x}T_x$, compounds where T – Al, Cr, Fe, Cu, *J.N.Kh. (Russ. J. Inorg. Chem.)* **28**, No.1 (1983). 195–201 (in Russian).
3. Leonova G.I.: *Catalytic reduction of aromatic nitro-compounds on hydrides of intermetallides on the basis of $LaNi_5$, modified with cerium and copper (Ph.D. Thesis Abstract)*, Moscow, 1989 (in Russian).
4. Carstens D.H.W. and Farr G.R.: The formation of niobium dihydride from niobium catalysed by $LaNi_5$, *J. Inorg. Nucl. Chem.*, **36**, No.2 (1974). 461–462.
5. Suga S. and Uchida M.: Mixing effects of different types of hydrides, *Hydrogen Energy Systems. Proc. 2-nd World Hydrogen Conf. Zurich, 1978*, Oxford e.a., 1979, **vol.3**, p.1561–1573.
6. Oesterreicher H., Ensslen K. and Bucker E.: Water formation on oxygen exposure of some hydrides of intermetallic compounds, *Appl. Phys.*, **22**, No.3 (1980). 303–306.
7. Lobashina N.E.: *Formation and transfer of monatomic hydrogen from metal-activator onto carrier surface (spill-over effect) (Ph.D. Thesis Abstract)*, Moscow, 1984 (in Russian).
8. Lobashina N.E., Savvin N.N. and Myasnikov I.A.: Investigation of the mechanism of H_2 spill-over on sprayed metallic catalysts, *Kinetika i kataliz (Kinetics and Catalysis)*, **25**, No.2 (1984). 503–504 (in Russian).
9. Bachman C.H. and Silbergt P.A.: Thermionic ions from hydrogen palladium, *J. Appl. Phys.*, **29**, No.8 (1958). 1266.
10. Shapovalov V.I.: About form of hydrogen existence in iron and possibility of proton emission from metals, *Izv. vuzov (Proc. Colleges, Ser. Ferrous Metallurgy)*, **No.8** (1978). 56–59 (in Russian).
11. Svoren' I.I.: On problem of hydrogen saturation of solids, *F.Kh.M.M. (Physical-Chemical Mechanics of Materials)*, **16**, No.2 (1980). 7–12 (in Russian).
12. Vanek V., Dixon D.P. and Gekelman W.: Production of negative ions in alkali hydrides arc, *J. Appl. Phys.* **50**, No.11 (1979). 7237.
13. Strokach A.P.: *Perspectives of application of ionic hydrides in the sources of negative hydrogen ions (Pre-print of Research Inst. of Electro-physical Equipment, No. K-0479)*, Leningrad, 1980 (in Russian).
14. Podgorny A.N. e.a.: Hydrogen activation in "hydrogen – intermetallic hydride" systems, *V.A.N.T. (Probl. in Nuclear Science and Engineering, Ser. Nuclear-Hydrogen Power Engineering and Technology)*, **No.1** (1987) 68–72 (in Russian).
15. Galchanskaya S.A. e.a.: Investigation of hydrogen activation process by metal hydrides. I. Mass-spectrographic analysis of glow discharge plasma, *Ibid., Ser. Nuclear Engineering and Technology*, **No.1** (1989) 55–58 (in Russian).
16. Valuiskaya S.B. e.a.: Investigation of hydrogen activation process by metal hydrides. II. Mass-spectrometric determination of potential and cross-section of hydrogen ionisation, *Ibid.*, 58–61 (in Russian).
17. Tyulin V.I.: *Vibrational and Rotational Spectra of Multi-Atomic Molecules: (Introduction into a Theory)*, Moscow, Publ. of Moscow State Univ., 1987 (in Russian).
18. Shmal'ko Yu.F., Klochko Ye.V. and Lototsky M.V.: Influence of isotopic effect on the shift on the ionisation potentials of hydrogen desorbed from metal hydride surface, *Int. J. Hydrogen Energy*, **21**, No. 11/12. (1996). 1057–1059.
19. Huber K.P. and Hertzberg G.: *Constants of Diatomic Molecules (in 2 volumes)*, Moscow, translated into Russian by "Mir" Publ., 1984, **vol.1.**
20. Shmal'ko Yu.F., Lototsky M.V., Klochko Ye.V. and Solovey V.V.: The formation of excited H species using metal hydrides, *J. Alloys and Compounds*, **231** (1995). 856–859.
21. Klochko Ye.V. e.a.: Sorption and elctro-transfer characteristics of hydrogen-gettering material in contact with a hydrogen plasma, *Ibid.*, **261** (1997). 259–262.
22. Klochko Ye.V. e.a.: Investigation of plasma interaction with metal hydride, *Int. J. Hydrogen Energy*, **24** (1999). 169–174.
23. Yartys' V.A. e.a.: Oxygen-, boron- and nitrogen-containing zirconium-vanadium alloys hydrogen getters with enhanced properties, *Z. Phys. Chem.* **183** (1994). 485–489.
24. Borisko V.N., Klochko Ye.V., Lototsky M.V. and Shmal'ko Yu.F.: Technological plasma source of negative ions, *V.A.N.T. (Probl. in Nuclear Science and Engineering, Ser. Physics of Radiation Damages and Radiation Material Science)*, **Nos.3(69), 4(70)** (1998). 179–182 (in Russian).

25. Allan M. and Wong S.F.: Effect of vibrational excitation on dissociative attachment in hydrogen, *Phys. Rev. Letters*, **41**, No.26 (1978). 1791–1794.
26. Lototsky M.V. e.a.: An investigation of the hydrogen ion source with metal-hydride emitter, *Hydrogen Energy Progress XI. Proc. of the 11-th World Hydrogen Energy Conf. (Stuttgart, Germany, 23 – 28 June, 1996)*, Ed. by T.N.Veziroglu, C.-J.Winter, J.P.Baselt and G.Kreysa. – Int. Association for Hydrogen Energy, 1996, **vol. 2**, p.2039–2044.
27. Shmal'ko Yu.F., Milenko Yu.Ya. and Karnatsevich L.V.: Influence of surface metal-hydride on the spin conversion of hydrogen molecules, *J. Alloys and Compounds*, **231** (1997). 364–367.
28. Shmal'ko Yu.F., Solovey V.V., Lototsky M.V. and Klochko Ye.V.: Metal-hydride vacuum technologies for physical-energy systems, *V.A.N.T. (Probl. in Nuclear Science and Engineering, Ser. Nuclear-Physical Researches /Theory and Experiment/)* **No.1(27)** (1994). 13–19 (in Russian).
29. Shmal'ko Yu.F., Solovey V.V. and Lototsky M.V.: Use of hydrides in systems for supplying vacuum physical-energy installations, *Hydrogen Energy Progress X. Proc. of the 10-th Word Hydrogen Energy Conf. (Cocoa Beach, Florida, U.S.A., 20 – 24 June, 1994)*, Ed. by D.L.Block and T.N.Veziroglu. – Int. Association for Hydrogen Energy, 1994, **vol. 2**, p.1311–1319.
30. Shmal'ko Yu.F. e.a.: Application of metal hydrides in hydrogen ion sources, *Z. Phys. Chem.*, **183** (1994). 479–483.

THE MODULATED PERMEATION TECHNIQUE USED AT THE OPEN UNIVERSITY

N St J BRAITHWAITE
Dept Materials Engineering
The Open University
Milton Keynes
MK7 6AA
UK

1 Introduction

The permeability of hydrogen isotopes in first wall materials is a crucial research subject in fusion reactor technology [1]. Preventing the loss of tritium is required to increase the economic viability of the fusion reactors and also permeation of tritium to the surroundings poses a radiological hazard. Recent strategies to reduce hydrogen permeation combine low diffusivity/permeability coatings on a mechanical substrate. The coatings can be put into two categories; one is the compact oxidation films formed by oxidation of metal construction materials, and the other is the thin films created by chemical vapor deposition, ion sputtering, ion implanting, plasma assisted deposition etc. TZM, a molybdenum alloy with small additions of Ti and Zr, owing to its very high tensile strength and high thermal and electrical conductivity, has been proposed as a candidate divertor material for future fusion reactors. A rather large amount of permeation data, showing wide scatter, particularly in solubility and diffusivity of hydrogen in molybdenum, suggest that the permeation process may be convolved with processes other than pure lattice transport [2–4]. Recent reports on hydrogen permeation in TZM show that the permeation data are not so different from those of pure molybdenum. It is claimed that slow surface processes and/or trapping of hydrogen is important in these materials [5–7].

In order to reduce hydrogen permeation further, chemical vapour deposition (CVD) coating of TZM with TiC, which has very close thermal expansion coefficient to TZM, is in attractive option. Several attempts have been made to deduce the permeation parameters, (solubility and diffusivity) and even surface reaction rates of hydrogen in TiC and it is suggested that TiC could be a reasonable permeation barrier in fusion technology applications [8, 9].

C.H. Wu (ed.), Hydrogen Recycling at Plasma Facing Materials, 109–117.

In this paper we present modulated permeation measurements on bare TZM and CVD coated TiC on a TZM substrate to reveal the possibility of identifying permeation characteristics of hydrogen in substrate and coating materials. This involves the measurements and analysis of the frequency dependence of phase lag and modulation amplitude ratio of hydrogen flux in modulated permeation experiments.

2 Experiments

Experiments were performed in a double ultra high vacuum system in which a foil specimen separates two chambers termed input and output. The modulation of the input pressure was provided by a large stainless steel bellows fitted to the input chamber. The input chamber pressure was monitored using capacitance monometers. The output chamber, with a fixed volume, was continuously pumped by a magnetically levitated turbo–molecular pump via a calibrated leak. The output chamber pressure was monitored using an ion gauge. Hydrogen was introduced into the input chamber to the desired equilibrium pressure via a palladium thimble which ensured high purity. Temperature was measured using calibrated chrome/nickel thermocouples with a common fixed reference. The temperature stability achieved was 0.1 K. With the exception of the absolute pressure setting, control of the equipment was fully automated.

Steady state conditions were first established between input and output chambers before commencing sinusoidal modulation of the input pressure, p_s, with an amplitude of $0.1p_s$. After sufficient time had elapsed to allow transients to decay, measurements were made of the pressure waveform in the input and output chambers for cycle times ranging from 20 s to 8000 s. Further experimental details and methods of data analysis have been reported previously [10, 11].

Materials investigated in this work were 0.1 mm foils of bare TZM (a Mo alloy with 0.5% Ti and 0.08% Zr) and 1 µm TiC + TZM prepared by CVD and supplied by Metallwerk Plansee, Austria. Measurements were made within the temperature range of 573–723 K and over a pressure range of 1.3 kPa to 133 kPa.

3 Analysis

Modulated experiments are analysed in terms of two characteristic independent variables (the phase difference and the ratio of relative modulation amplitude of the input and output pressure signals) as functions of the modulation frequency [12–14]. The permeation process is treated as a sequential combination of surface and bulk processes which give rise to characteristic behaviour of the phase and the amplitude variations with the modulation frequency, ω. The modelling of modulated permeation can be extended to laminated composites. Although this approach could readily be incorporated into the overall modulated permeation model, it is of limited applicability since it presumes implicitly that the interface between coating and substrate is in dynamic equilibrium. It was found that a more thorough strategy was required to allow for the possibility that reaction kinetics may limit the flow. It is also possible that local disorder in the interface region, being not unlike the disorder associated with heavily work hardened material, could lead to trapping and accumulation.

The model is conveniently handled in a matrix form in which each step of permeation path (surface processes, bulk diffusions in the coating and the substrate and the interface in between them) is described by matrices linking concentrations and fluxes. Time variation is incorporated through a linearized harmonic analysis [12-14].

The method treats the permeation process as a sequential combination of surface and bulk processes. The surface reaction rate constants, the bulk diffusivity, D, solubility, K_{sm}, the mean input and output pressures p_s and p_s', the pump speed , S, the output chamber volume ,V, and the specimen thickness, l, enter the equations as parameters, the first three of which are determined when the model is fitted to experimental data. The degree to which surface processes influence the overall permeation can be recognised from the characteristics behaviours of the phase and the amplitude variations with the modulation frequency, ω, (Table 1). When the bulk transport limits the permeation one can pick out the bulk diffusivity from the phase lag and the amplitude response independently and the permeability can be obtained from the modulation amplitude ratio. For the bulk diffusion limited flow the resulting equations for phase lag, ϕ, and amplitude ratio, $|R|^{-1}$ to determine the bulk diffusivity and permeability are as follows :

$$\varphi = \arctan\left[\frac{\tan\varsigma - \tanh\varsigma}{\tan\varsigma + \tanh\varsigma}\right] \tag{1}$$

where ζ is the normalised frequency factor :

$$\varsigma = l\left(\frac{\omega}{2D}\right)^{1/2} \tag{2}$$

This functional relationship is of potential experimental importance, since it is independent of solubility K_{sm} and so allows direct experimental evaluation of a diffusion coefficient.

TABLE 1. Characteristic behaviour of phase and amplitude data

Limiting process	Phase	Characteristic	Amplitude Characteristic	
	$\phi(\omega = 0)$ high ω limit	remarks $\phi(\omega)$	$\Lambda(\omega = 0)$ exponent n	remarks $\Lambda(\omega)$
bulk diffusion	$-\pi/4$	solely depends on D	1/2	solely depends on D
adsorption	$\leq+\pi/4$	pressure dependent	$1/2 < n < 1$	pressure dependent
dissociation	$\leq+\pi/4$	pressure dependent	$1/2 < n < 1$	pressure dependent
accumulation (molecules)	$\leq+3\pi/4$	pressure independent	1/2	pressure independent
accumulation (atoms)	$\leq+3\pi/4$	pressure dependent	1/2	pressure dependent
surface penetration	$\leq+\pi/4$	pressure independent	1/2	pressure independent

$$|R|^{-1} = \frac{|p'|}{|p|} = \frac{DK_{\rm sm}\sigma}{p_s^{1/2} Sl} = \frac{\varsigma}{(\cosh 2\varsigma - \cos 2\varsigma)^{1/2}} \tag{3}$$

This results is important in two ways. First, it provides a separate way of evaluating the diffusion coefficient and, hence, by comparison with the phase data, gives an internal consistency check . Second , it allows evaluation of the permeability, independent of the outgassing effects which can modify the precision of estimates based on the steady-state pressures p_s and P_s . Results are conveniently displayed in a normalised form for the amplitude ratio:

$$\Lambda = \frac{|p'|P_s}{p_s'|p|} = \frac{\varsigma}{(\cosh 2\varsigma - \cos 2\varsigma)^{1/2}} \tag{4}$$

Data will be presented later in figures 1 and 2 which show typical plots of ϕ and Λ against $\omega^{1/2}$. Analysis is done by non-linear least squares fitting curves for ϕ and $|R|^{-1}$. Different expressions for ϕ, Λ and $|R|^{-1}$ are found for surface limited flow[13]. If D, the bulk diffusion coefficient, is a function of temperature only, that is, it is independent of the concentration and the degree of cold work and so on, then the process involved is purely lattice diffusion; otherwise , the lattice diffusion is convolved with impurity effects such as trapping.

4 Results and discussion

Figs. 1 and 2 show the variation of phase lag, ϕ, and relative modulation amplitude ratio, Λ, with root angular frequency, $\omega^{1/2}$, for 0.1 mm TZM at 623 K for three different base pressures. Similar responses are obtained at other temperatures. These data show behaviour which the model identified as strongly characteristic of bulk diffusion limited flow together with diffusivity impeded by trapping. The diffusion limited flow is strongly associated with the zero frequency extrapolation of the phase variation at high frequencies together with the value of 0.5 for the relative modulation amplitude ratio at zero frequency; note how all phase lag curves on extrapolation pass through $-\pi/4$. The pressure dependence indicates that bulk diffusivity increased when higher pressures (higher internal concentrations) give rise to reduced phase shift at a given frequency as traps are becoming more filled and less effective as delaying sites. This pressure effect also indicates that the traps are saturable and hydrogen is trapped as atoms. If the nature of the lattice–trap equilibrium (Sievert's law) were that of molecular trapping, the trapped concentration should be an increasing fraction of the total hydrogen concentration, as the lattice concentration increases. This would give the opposite effect of traps becoming more effective as delaying sites and would increase the phase shift at a given frequency [15].

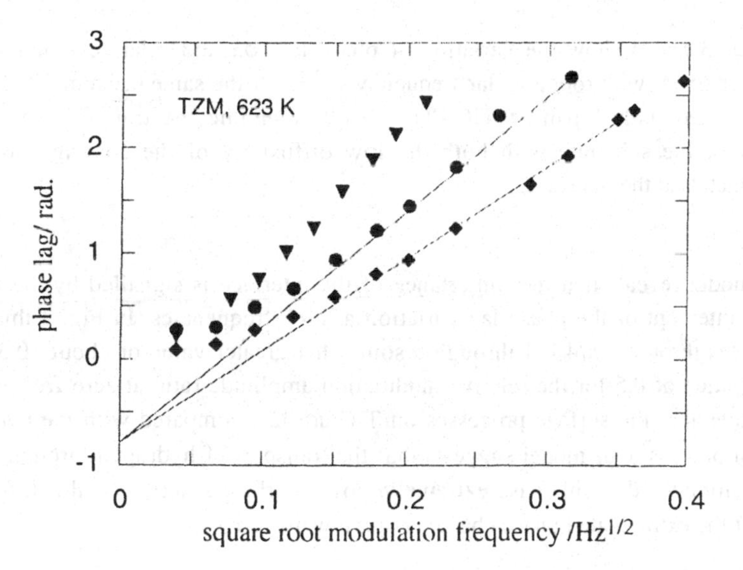

Figure 1. Measured variation of phase lag, Φ, with root frequency, $\omega^{1/2}$, for TZM at 623 K. ◆ 133 kPa; ● 13.3 kPa; ▼ 1.3 kPa

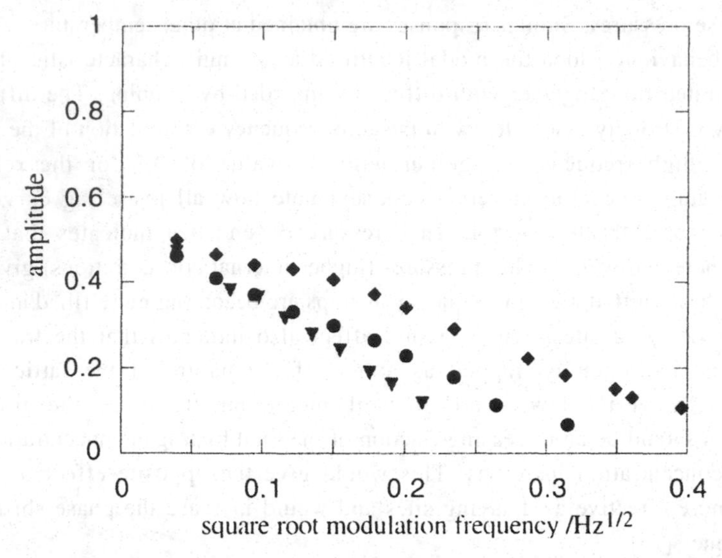

Figure 2. Measured variation of relative amplitude modulation ratio, Λ, with root frequency $\omega^{1/2}$ for TZM at 623 K. ♦ 133 kPa; ● 13.3 kPa; ▼ 1.3 kPa

Figures 3 and 4 show the variation of phase lag, ϕ,, and relative modulation amplitude ratio, Λ, with root angular frequency, $\omega^{1/2}$, for the same 0.1 mm TZM foil but now coated with 1 μm of TiC. The results combine the trapping impeded diffusivity of the substrate with both the low diffusivity of the coating and the reaction kinetics at the interface.

The model reveals that the importance of the interface is signalled by the zero frequency intercept of the phase lag variation at high frequencies. In Fig. 3 this no longer passes through $-\pi/4$ but through a somewhat greater value of about –0.3. In Fig. 4 the value of 0.5 for the relative modulation amplitude ratio at zero frequency also indicates that the surface processes on TiC are fast compared with the overall permeation process. Our model suggests that the transport of hydrogen through TiC is mainly atomic and in this case extremely slow interface kinetics would shift the intercept of the extrapolated phase characteristic towards zero.

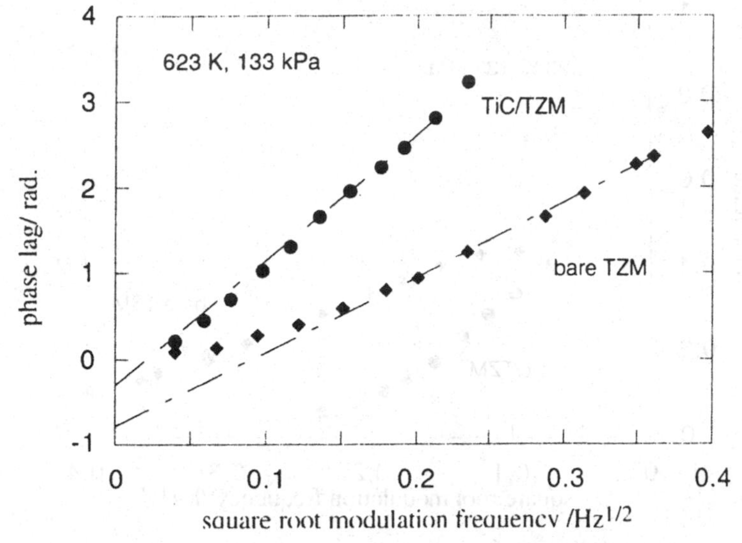

Figure 3. Measured variation of phase lag Φ with root frequency $\omega^{1/2}$ for TiC/TZM at 623 K at a base pressure of 133 kPa. ● TiC/TZM; ♦ TZM

♦ TZM

5 Conclusions

Modulated permeation experiments have been performed on foils of TZM, bare and coated with the TiC. Results reveal that hydrogen is reversibly trapped within the bulk during permeation in TZM. It has been found that diffusion through a TiC coating is atomic. Both diffusion through the coating and slow interface kinetics reduce the permeation through a TiC coated TZM foil.

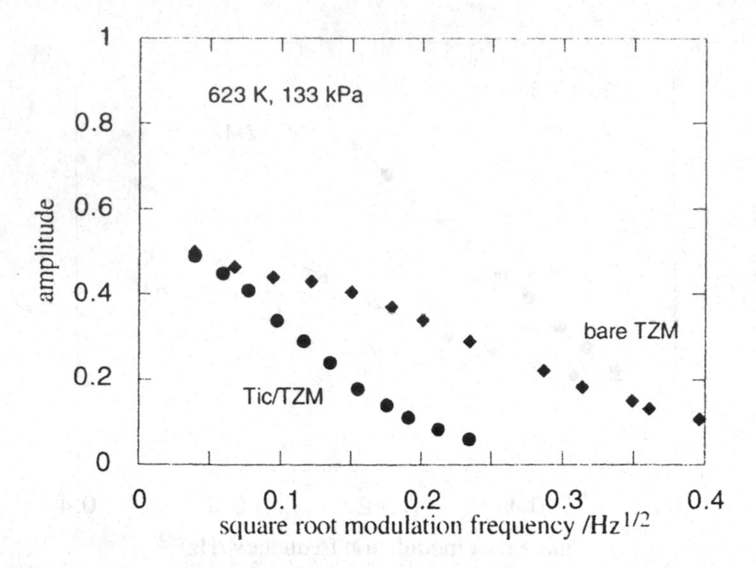

Figure 4. Measured variation of relative amplitude modulation ratio, Λ, with root frequency $\omega^{1/2}$ for TiC/TZM at 623 K at a base pressure of 133 kPa. ● TiC/TZM;

Acknowledgments

This work forms part of a programme supported by the UK Engineering and Physical Sciences Research Council (EPSRC) and was presented at the 1996, Int. Symp. Metal Hydrogen Systems, Les Dialberets, Switzerland (Journal of Alloys and Compounds **231** (1997), 302-306).

References

1. A.D. Le claire, Diffusion and Defect Data, **34** (1983) 1.

2. H.Katsuta *et al.*, J. Phys. Chem. of Solids, **43-6** (1981) 533.

3. G.R. Gaskey *et al.*, J. Nucl. Mat. **55** (1975) 279.

4. T. Tanabe, Y. Yamanashi and S. Sumoto, J. of Nucl. Mat, **191-194** (1992) 439.

5. K.S Forcey , A. Perujo, F. Reiter and P.L. Lolli-Ceroni, J. of Nucl. Mater. **203** (1993) 36.

6. S. Tominetti, M. Caorlin , J. Campolsivan , A. Perujo and F. Reiter, J. of Nucl. Mater. **176-77** (1990) 672.

7. M. Caorlin , J. Campolsivan , and F. Reiter, J. of Nucl.Mat **176-77** (1990) 672.

8. K.S Forcey , A. Perujo, F. Reiter and P.L. Lolli-Ceroni, J. of Nucl. Mater. **200** (1993) 417.

9. M. Caorlin, PhD thesis (1992), The Open University, UK.

10. A.K. Altunoglu, D.A. Blackburn, N.St.J. Braithwaite, D.M. Grant, J. Less Com. Met.**172-174**, (1991), 718.

11. A.K. Altunoglu, N.St.J. Braithwaite, D.M. Grant, Z. Phys. Chem. **181** (1993) 133.

12. D.L. Cummings, R.L. Reuben and D.A. Blackburn, Metall. Trans. A, **15A** (1984) 639.

13. D.L.Cummings and D.A. Blackburn, Metall. Trans. A, **16A** (1985) 1013..

14. A.K. Altunoglu, N.St.J. Braithwaite, J. of Nucl. Mater. **224** (1995) 273-287.

15. A.K. Altunoglu, N.St.J. Braithwaite, Journal of Alloys and Compounds **231** (1995), 302-306.

A MODEL FOR CALCULATION OF TRITIUM ACCUMULATION AND LEAKAGE IN PLASMA FACING SANDWICH STRUCTURES

A.A. PISAREV, M.A. CHUDAEVA
Moscow State Engineering and Physics Institute,
Moscow 115409, Russia.

1. Introduction

Tritium accumulation and leakage in thermonuclear reactor during plasma-surface interaction is usually calculated by using computer codes. The most developed one is TMAP-4 [1]. For steady state regime, which is of interest for reactor application, there is no need to use time-consuming codes, and one may solve steady state equations directly. This offers an opportunity of fast estimations. This is especially useful when it is necessary to repeat calculations many times to understand possible tendencies in tritium accumulation and leakage.

A simple steady state model for plasma-driven implantation was described in [2]. A delta-function ion source, constant temperature, and second order thermodesorption were considered in that work. In the present work we use about the same approach as in [2], but consider a sandwich structure. Example calculations are performed for oxidised beryllium on a copper heat sink for conditions typical for the buffle of the ITER divertor.

Model

If energetic particles of a high flux come from plasma and enter the plasma-facing material, a steady state concentration profile establishes rather soon. This profile $u(x)$ is described by the following equations:

$$\frac{\partial J_d(x)}{\partial x} = I_0 \varphi(x)$$

$$\frac{\partial u(x)}{\partial x} = -\frac{J_d(x)}{D(x)}$$

here I_0 and $\varphi(x)$ are the flux and the stopping profile of the incoming particles, J_d and D are the diffusion flux and the diffusion coefficient. If the PFC is a sandwich consisting of several layers having different characteristics,

C.H. Wu (ed.), Hydrogen Recycling at Plasma Facing Materials, 119–124.

these two equations are to be written for every layer. The concentrations in the adjacent layers are supposed to be proportional to the solubilities in the respective materials:

$$\frac{u_{ij}}{u_{ji}} = \frac{S_i}{S_j}$$

Here u_{ij} – is the concentration in the layer i at the interface with the layer j. The boundary conditions are the balances of diffusion and desorption fluxes at the front and the back sides of the sandwich:

$$-J_d(0) = K_0 u_0^2$$

$$J_d(L) = K_L u_L^2$$

Here K_0, u_0 and K_L, u_L are the recombination coefficients and concentrations on the front and the back boundaries of the sandwich.

The recombination coefficient can be taken as proposed by Pick [3]:

$$K = \frac{2.635 \cdot 10^{20}}{\sqrt{MT} \cdot S_0{}^2} \exp \frac{2(Q_S - E_C)}{kT}, \text{ cm}^4\text{s}^{-1},$$

where T is the temperature in K, S_0 -pre-exponent in the expression for the solubility (taken in units of atoms·cm^{-3}·Pa$^{-1/2}$), Q_S - the heat of solubility, E_c - the activation energy for chemisorption, and M - the molecular mass of the hydrogen isotope (M=6 for tritium).

2. Model calculations

Let us consider tritium accumulation in and permeation through BeO-Be-Cu sandwich in applications to the ITER divertor baffle. The, the data for solubility (S) and diffusivity (D) in BeO and Be are taken as recommended by Longhurst [4]. The diffusivity in the ion stopping region x=0-2R_p was taken to be different from that in the rest of the BeO layer. The data for Cu were taken from the review of Federichi [5]. The diffusivities and solubilities accepted are summarised in Table 1. Other parameters used were in ranges given in [5]: the thicknesses of BeO, Be and Cu layers L_1=0-10^{-3}, L_2=0.5, and L_3=0-0.5cm respectively, the ion range R_p=10^{-7}cm, the ion flux I_0=10^{19}at·cm^{-2}·s^{-1}, the temperature 550-1100K, and the activation energy for chemisorption E_c=0-0.5eV.

Figure 1 shows an example of calculations of the temperature dependencies of the net accumulation of tritium in various sandwiches. Sandwiches considered are both with BeO and without BeO; with a Cu substrate and without it. The contributions of the layers to the net accumulation are shown in figure 2 for two thicknesses of the BeO layer.

TABLE 1 Data for model calculations

Layer	D_0, cm^2s^{-1}	E_d, eV	S_0, $at\cdot cm^{-3}\cdot Pa^{-1/2}$	E_s, eV	Reference
Be			22×10^{15}	0.17	Shapovalov [6]
Be	6.7×10^{-5}	0.294			Abramov [9]
BeO			9.4×10^{11}	-0.8	Macaulay-Newcomb[7]
BeO, rad.induced	1.33×10^{-11}	0.025			Longhurst [4]
BeO, normal	1.31×10^{-5}	1.335			Longhurst [4]
Cu			3.13×10^{18}	0.572	McLellan [8]
Cu	6.6×10^{-3}	0.387			Katz [10]

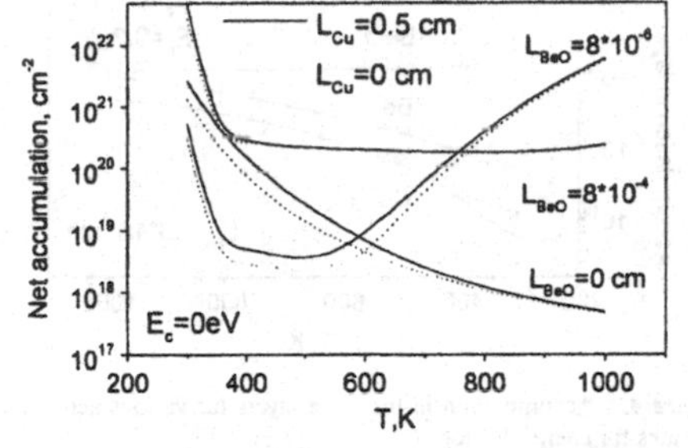

Figure 1. Net accumulation of tritium in the sandwich for various thicknesses of the Beo and Cu layers as a function of temperature

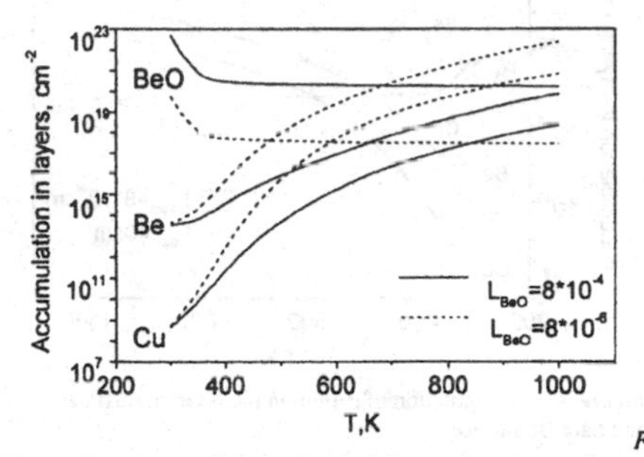

Figure 2. Accumulation of tritium in various layers for two values of the BeO thickness as a function of temperature

The first observation from figures 1,2 is that the contribution of the Cu substrate is always negligible. The second one is that the tritium accumulation in BeO and Be layers behave in different ways as a function of temperature. This is the primary reason of why the temperature dependence of the net accumulation in the sandwich is strongly influenced by the oxide layer thickness. Indeed, if the oxide is thick, $L_{BeO}=8\cdot10^{-4}$cm here, practically all the tritium is accumulated in the oxide layer, which determines the temperature dependence of the net accumulation. For a thin oxide, $L_{BeO}=8\cdot10^{-6}$cm, the net accumulation has a minimum at some temperature: at low temperatures the accumulation is determined by the oxide layer while at high temperatures - by beryllium metal.

The third significant feature of accumulation in the oxidized Be, that follows from fig.2, is that the accumulations in BeO and in the underlying Be metal behave in different way with variation of the oxide thickness. Indeed, the decrease of L_{BeO} gives a decrease of accumulation in BeO, and this is quite obvious. But at the same time the accumulation in Be surprisingly increases. Therefore, at low temperatures where the accumulation is determined by BeO, the net tritium content in fig.1 is higher if the oxide layer is thicker. But at high temperatures, the accumulation is determined by Be layer, and therefore the net tritium content increases if the oxide gets thinner. The peculiar thing here is that the concentration of tritium in Be under BeO is higher if the BeO thickness is less. Analyses of the concentration profiles show that the reason is as follows: the concentration in BeO diminishes at $x>R_p$, so both concentrations u_{12} and u_{23} at the BeO/Be interface diminish with L_{BeO} increase.

The plasma facing surface can be free of oxides due to ion cleaning by high plasma fluxes. In this case (L_{BeO}=0) the accumulation in the sandwich Be-Cu has a qualitatively different temperature dependence if to compare with BeO covered sandwich (fig.1). This is because both the diffusion coefficients and the recombination coefficients in BeO and Be are very different. Figure 3 demonstrate the difference between the two cases. Important to note that no oxide does not mean the smallest accumulation. Accumulation in Be for the BeO free sandwich can be very high. Therefore the net accumulation for the BeO free surface can be both less and higher than the net accumulation in the oxidized Be as we have already seen from fig.1.

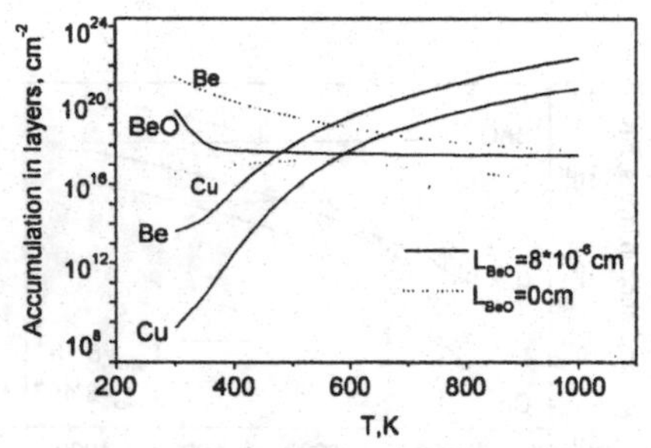

Figure 3. Accumulation of tritium in the layers for oxidized and bare Be surfaces

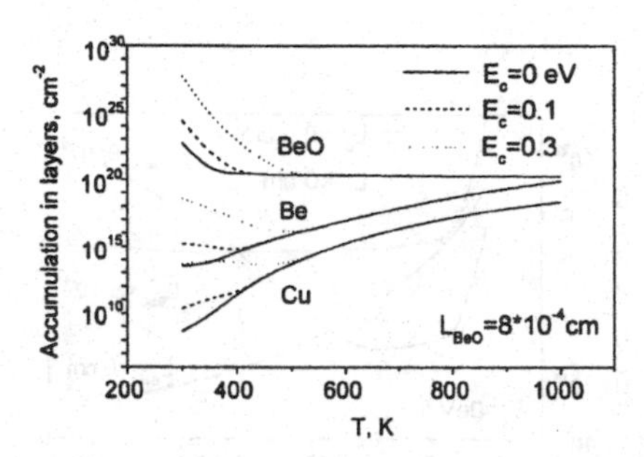

Figure 4. Accumulation in the three layers for various activation energies for chemisorption

Analyses shows that the accumulation of tritium in the sandwich is insensitive to the surface conditions on the backside. The characteristics of the surface in this model is the recombination coefficient. The reason of this independence is that the concentration on the back side is always very low in this system, and therefore, from the point of view of accumulation, there is no difference if this concentration is a little bit less or higher. In contrast, the recombination coefficient on the front side, which determines the concentration at $x=0$, can influence the concentration profile and the integral accumulation. Figure 4 shows the temperature dependences of the accumulation for three activation energies for chemisorption E_c, which determine the recombination coefficient on the surface. One can see that for these particular conditions, the surface parameters influence only at low temperatures. There is no E_c dependence at high T.

Analyses of the concentration profile explains the E_c dependence. At low temperatures, increase in E_c leads to increase in u_0 and hence to increase in the accumulation while at high temperatures the front surface concentration is always low, and the accumulation does not depend on E_c.

The diffusion coefficient is to be an important factor in the diffusion-limited regime. We have accepted in calculations that the diffusion coefficient in the ion-stopping region near the front surface is much higher than that in the bulk of the material because of some ion induced processes. If we take the diffusion coefficient in the near surface region of the same low value as in the bulk of the material outside the ion-stopping region, the accumulation increases drastically at low temperatures. The levels of the concentrations obtained at $D_{BeO,r}=D_{BeO}$ at low temperatures are enormously high and physically unreal. This can be considered as a support of the choice of $D_{BeO,r}>>D_{BeO}$ made in [4].

Features of tritium permeation through the sandwich can be easily understood from consideration of the features of the accumulation and profile variations. Figure 5 shows, as an example, variation in the permeation rate with the parameters of modelling. One can see that the permeation is sensitive to the thickness of the oxide layer and diffusivity in the ion stopping region as we have seen for the net accumulation. Copper layer and the back side conditions do not influence permeation seriously as it was the case also for the accumulation.

From the later one can expect that an additional steel tube on the heat sink side will be of no influence on tritium accumulation and permeation.

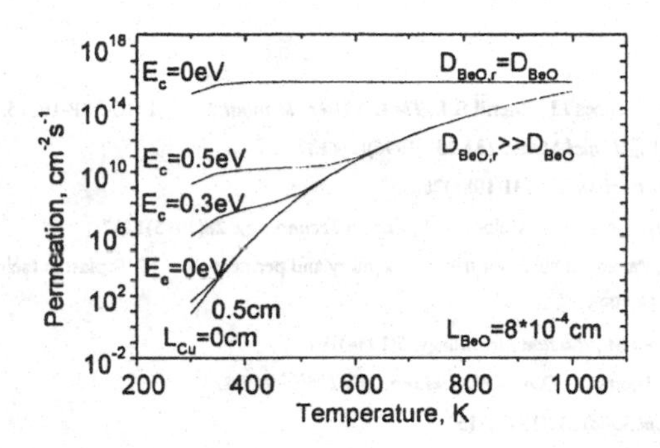

Figure 5. Tritium permeation rate through the sandwich: influence of radiation enhanced diffusion in the stopping region, activation energy for chemisorption on the front side and the thickness of the BeO layer.

3. Conclusion

A model for steady state tritium accumulation in and leakage through sandwich structures of plasma facing materials of the thermonuclear rector is described. Steady state condition is useful for fast estimations and qualitative analyses of tendencies of accumulation and leakage variation with the parameters of the plasma – PFC interaction expected.

Model calculations for the ITER divertor baffle show the folowing features of accumulation and permeation.

The recombination coefficient on the back surface has no influence both on the permeation and accumulation. The recombination coefficient on the front surface is of significance at low temperatures only. Decrease of the Be oxide layer thickness can both increase and decrease the net accumulation in different conditions. The permeation rate decreases with BeO thickness increase. The temperature dependences of the accumulation with and without the BeO layer are qualitatively different. Accumulation and permeation without BeO can be both less and higher than that with BeO on the surface. The accumulation and permeation both strongly depend on the diffusion coefficient in the region, where the incoming ions slow down. To give physically reasonable values of tritium concentrations, it must be accepted that the mass transport rate in this region is much higher than that outside it. If an additional layer of steel between Cu and coolant is added, it has no effect on the permeation rate.

4. Acknowledgement

The work was performed as a part of the Contract 98-3-021-344 with the Ministry of the Atomic Energy of the Russian Federation.

5. References

1. Longhurst G.R., Holland D.F., Jones J.L., Merill B.J., *TMAP4 Users Manual*, INEL, EGG-FSP-10315, 1992
2. Pisarev A.A., Grachova M.L., *J.Nucl.Mater.* 233-237(1996)[2]1137
3. Pick P.M, Sonnenberg K., *J.Nucl.Mater.* 131(1985)208
4. Longhurst G.R., Anderl R.A., Dolan T.J., Mulock M.J., *Fusion Technology.* 28(1995)1217
5. Federichi G., Holland D.F., Parameter study for tritium inventory and permeation in ITER plasma facing components, ITER Internal Report, Jan 18,1995
6. Shapovalov V.I., Dukelski Yu.M., *Russian Metallurgy.* 5(1984)201
7. Macaulay-Newcomb R.G., Thompson D.A., *J.Nucl.Mater.* 212-215(1994)942
8. McLellan R.B., *J.Phys.Chem.Solids.* 34(1973)1137
9. Abramov E., Riehm M.P., Thompson D.A., Smeltzer W.W., *J.Nucl.Mater.* 175(1990)90
10. Katz L., M.Guinan,.Borg R.J, *Physical Review B.* 4,2(1971)330

TRANSPORT OF HYDROGEN THROUGH AMORPHOUS ALLOY

I.E. GABIS, E.A. EVARD, N.I. SIDOROV*
Institute of Physics, St.Petersburg State University, Ulianovskaya 1, 198904, St.Petersburg, Russia
**Institute of Metallurgy, Urals Division of Russian Academy of Sciences, 620016, Ekaterinburg, Russia*

Hydrogen permeation through amorphous and recrystallized foils of Fe-based alloys was investigated. Only the ionization of the gas in glow discharge near the upstream side of foils resulted in noticeable permeation fluxes. The kinetics of flux relaxation to the steady-state values revealed reversible trapping in amorphous alloy. Permeation flux demonstrated Arrhenius linear dependence for recrystallized samples and non-monotonous for amorphous ones. The mechanism of hydrogen transport is proposed taking into account re-emission and diffusion processes. Activation energies of thermal desorption and diffusion are evaluated.

1. Introduction.

The outlook for the utilization of hydrogen as an energy carrier may suggests its refinement. Permeation through palladium-based alloys is one of possible ways to obtain hydrogen of highest purity. It is well known that amorphous alloys, unlike ordinary metals, do not have ordered atomic structure, possess the excess of free volume that could affect the diffusion of interstitial impurities. Unfortunately the information about diffusion in disordered materials is very poor due to narrow temperature range available for experimental studies [1-3]. The purpose of this investigation was to clear out the question whether the peculiarities of structure of amorphous metal influence the permeation of hydrogen.

2. Experiment.

In this study hydrogen permeability through membranes of amorphous and crystallized Fe-based alloy (Fe-77.353 Ni-1.117 Si-7.697 B-13.622 C-0.202 P-0.009) was investi-

C.H. Wu (ed.), Hydrogen Recycling at Plasma Facing Materials, 125–131.

gated. Samples were discs 15 mm diameter cut from the foil 25 μm thick. Each sample was mounted as a membrane that separated the upstream-side chamber, where hydrogen could be admitted, and the downstream-side one, including mass-spectrometer for registration of permeation flux. Preliminary outgassing was carried out in vacuum 10^{-6} Pa at $300\,^0$ C. Long annealing under the higher temperature resulted in the formation of polycrystalline structure with dimension of grains about 1 μm.

Permeation flux from the molecular phase on the upstream side of membrane was very low even at maximum allowed temperatures ($300\,^0$ C for amorphous sample and $400\,^0$ C for recrystallized one): at hydrogen pressure 10 Torr (1330 Pa) the flux was equal to $3.8{\cdot}10^{16}$ $m^{-2}s^{-1}$. It is known that sticking probability of hydrogen atoms is several orders greater than that of molecules [4]. However the dissociation on the heated tungsten filament located at 10 mm from sample also did not allow to obtain noticeable permeation flux.

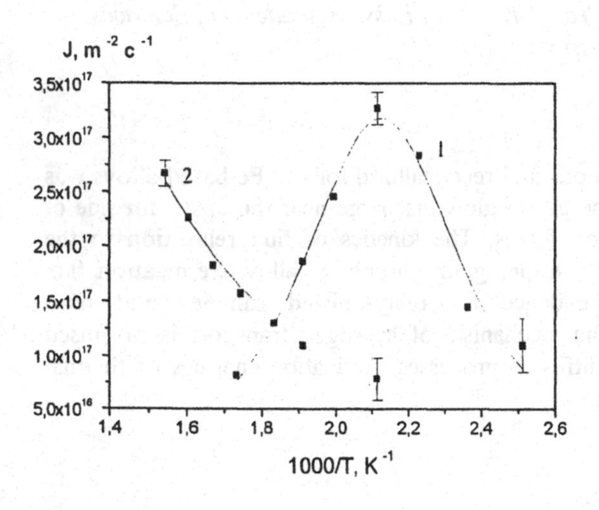

Figure 1. Temperature dependence of permeation flux: 1 – amorphous specimen. 2 – crystallized one.

It is known [5] that Fe-based alloys doped by Si and B tends to enrich surface layers with these elements at the depth about 50 Å that results in inhibition of adsorption and permeation of hydrogen. After layerwise etching by argon ions the content of impurity decreased down to normal level, but after subsequent heating it increased again.

To overcome the passivation layer the glow discharge in hydrogen atmosphere was used. Ions of hydrogen generated in discharge area easily permeate into the bulk of metal [4]. Under this conditions we observed the noticeable permeating fluxes. All experiments were carried out at input pressure of hydrogen 2 Torr (266 Pa) at which the glow discharge was stable.

Temperature dependencies of steady-state flux for amorphous and recrystallized specimens are presented in *Figure 1.* Lower bound of temperature range was defined by reliability of flux detection and was around $120\,^0$C for amorphous and $200\,^0$C for crystallized samples. The most important distinction of amorphous alloy is the non-monotonous behavior of permeation flux with rising temperature. It increased within temperature range 125..200 °C, achieved maximum value $3.3 \cdot 10^{17}$ $m^{-2}s^{-1}$ at 200 °C and then showed abnormal decrease at further heating. The flux through crystallized membrane obeyed linear Arrhenius dependency with activation energy 17.9 kJ/mole and reached value $2.7 \cdot 10^{17}$ $m^{-2}s^{-1}$ at 375 °C. It should be noted that flux through amorphous sample is several times higher compared to flux through crystallized one at 200 °C.

The permeation relaxation to the steady state for these types of samples also differs. Rapid phase with characteristic time about 30-60 s was revealed for both types in all temperature range. In the case of amorphous membrane below 225 °C a very slow increase of the permeation flux was observed after the rapid phase (*Figure 2*). A typical time of this stage was about 6000 s. Raising of the temperature resulted in vanishing of the slow phase and in general decrease of penetrating flux. Thus the temperature dependence of the flux is non-monotonous and reveals maximum value near 200 °C.

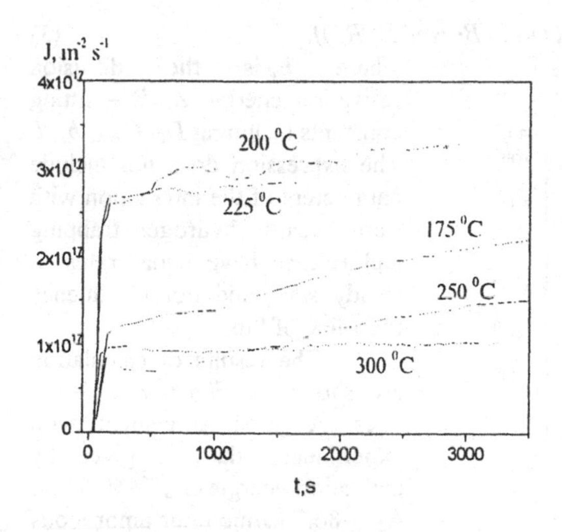

Figure 2. Kinetics of flux relaxation through amorphous alloy.

3. Discussion.

The overextended time of flux relaxation is most likely associated with reversible trapping of diffusant [6]. We suppose that the rate of detrapping of hydrogen atoms increases more rapidly than the trapping rate with rising temperature. One can see (*Figure 2*) that the slow stage of flux increases in the magnitude below 200 °C. Above this temperature the slow stage of flux doesn't increase. This behavior is typical for traps with activation energies of release and trapping $E_{rel} > E_{trap}$.

The decrease of the flux through amorphous membrane in temperature range 200..300 °C can be explained by surface processes under ion irradiation. Relatively low-energy hydrogen ions (~200 eV) are absorbed in the surface layer 30..70 Å thick [7]. Thus the penetration depth of ions approximately corresponds to the thickness of passivation layer. Since the permeation flux is less by three orders of magnitude than the falling flux (ν_f~10^{20} $m^{-2}s^{-1}$), the balance of fluxes can be written as

$$v_j = v_T + v_r \tag{1}$$

where

$$v_r = v_j \cdot C_1 / C_{max} \quad \text{and} \quad v_T = b_1 \cdot exp(-E_1/RT) \cdot C_1 \tag{2}$$

are ion-induced reemission and thermal desorption fluxes at the upstream surface. Here C_1 is the hydrogen concentration in non-disturbed membrane area near the input surface, E_1 is activation energy and b_1 is pre-exponential factor of thermal desorption on the upstream surface.

According to Fick's law the value of steady-state flux is determined by the expression $J=D\,(C_1 - C_2)/l$, where $D = D_0 \exp(-E_d/RT)$ is hydrogen diffusivity and l is the membrane thickness. For the case of negligible desorption at 300 K the maximum available surface concentration C_{max} was evaluated in paper [8] as $10^{24}\ m^{-3}$. Assuming that the concentration near the downstream side C_2 is much less compared to the upstream one and using equations (2) we receive the expression for the steady-state permeation flux:

$$J = A \cdot \exp(-E_d/RT)/(1 + B \cdot \exp(-E_1/RT)), \tag{3}$$

where E_d is the diffusion activation energy, A, B – fitting constants including D_0, C_{max}, b_1, l. The expression does not include parameters of the interaction with traps because hydrogen trapping and release have equal rates in steady state and don't influence the value of flux.

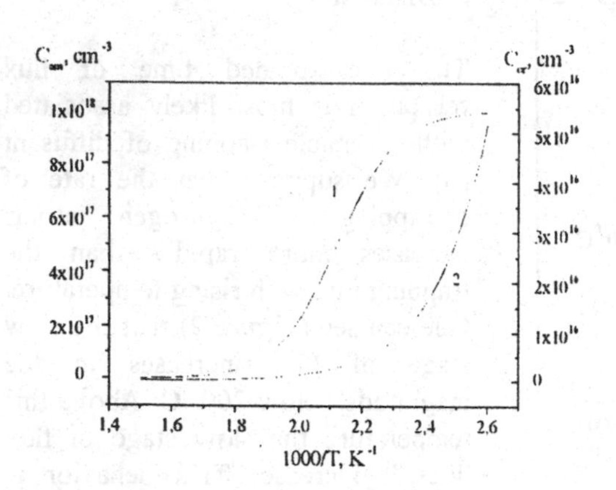

Figure 3. Calculated temperature dependence of hydrogen concentration near upstream side of membranes.

The results of calculation are shown in *Figure 1* by solid lines. A good agreement with experimental data is given by activation energies E_d^{am}=40.8 and E_1^{am}=86.7 kJ/mole for amorphous sample and E_d^{cr}= 71.2 and E_1^{cr}=51.7 kJ/mole for crystallized one. Simulation of the surface concentration C_1 was performed using equations (1)–(2) and values of activation energies pointed above. The parameter C_{max} accounted in calculations does not affect activation energies but only pre-exponential factors. In *Figure 3* temperature dependence of the surface concentrations for both types of samples is presented.

So, the surface processes and the relation of E_d and E_1 define temperature dependence of steady-state flux. In the case of amorphous alloy the concentration C_1 has value C_{max} up to temperature 170 ^{0}C and permeation is determined by diffusion only. Steady-state flux increases at these temperatures. At further rising of temperature the concentration C_l is falling down exponentially, and the relation E_d^{am} E_l^{am} results in general decreasing of the flux. As to recrystallized alloy the input concentration decreases in all temperature range (*Figure 3*, line 2). Under this condition relation $E_d^{cr} \cdot E_l^{cr}$ results in Arrhenius dependency

$$J \sim exp\ (-E_a / RT) \tag{4}$$

where $E_a = E_d^{cr} - E_l^{cr} \sim 19.6$ kJ/mole according to our calculations, which is closed to E_a=17.9 kJ/mole derived from the experimental data.

The difference in activation energies of diffusion may be undoubtedly associated with interior structure of the samples. Some excess of free volume in amorphous alloy provides lower energetic expenditure at migration of hydrogen atoms in the bulk. Besides facilitated ways are possible for hydrogen migration in the form of Bernal' chains [2].

Activation energy of thermal desorption is lower for the amorphous alloy compared to the crystallized one: E_l^{cr} E_l^{am}. This fact can be explained by the surface reconstruction and the change of influence of passivation layer on hydrogen desorption after annealing.

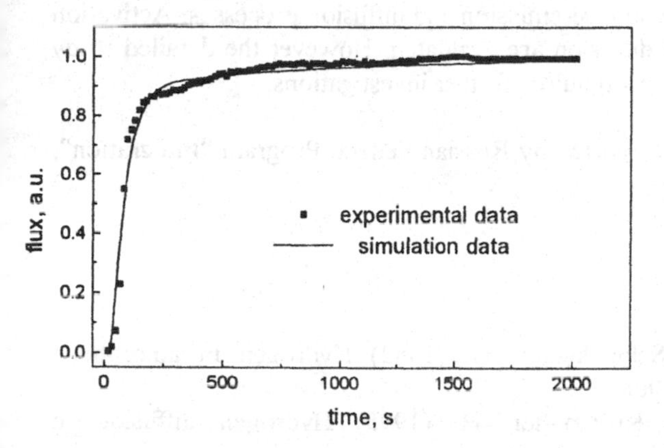

Figure 4. Simulation of steady-state flux relaxation in amorphous alloy at 225 ^{0}C.

In order to evaluate kinetics parameters of interaction of hydrogen with traps an approximation of experimental flux relaxation was performed. We used simple model of Fick's diffusion and reversible trapping [9]. As the glow discharge became stable in a fraction of a second we might neglect the time of establishing of hydrogen concentration C_l. The result of simulation for the amorphous alloy at 225 ^{0}C is shown in *Figure 4*. At the temperature about 200 ^{0}C, as it was mentioned above, competing processes of trapping and detrapping reach approximately equal rates. In our simulation we received values of rate constants of trapping $k_t = (1.1 \pm 0.4) \cdot 10^{-3}$ s^{-1} and detrapping $k_d = (1.2 \pm 0.6) \cdot 10^{-3}$ s^{-1}. The diffusivity was $D = (1.1 \pm 0.3) \cdot 10^{-8}$ cm^2/s. Pre-exponential factor was $D_0 \sim 2 \cdot 10^{-4}$ cm^2/s. The large uncertainties was due to comparatively high noise of experimental data.

4. Conclusion.

Hydrogen permeation through amorphous and recrystallized foils of Fe-based alloys was investigated by means of permeability method within relatively wide temperature range. It was founded that surface layers of both samples are enriched by metalloids. These layers prevented permeation of hydrogen from molecular and atomic phases. Only the ionization of the gas in glow discharge near the upstream side of foils resulted in noticeable permeation fluxes.

The kinetics of flux relaxation to the steady-state values revealed reversible trapping in amorphous alloy. Detrapping activation energy is found to be larger than trapping activation energy. Temperature dependence of the steady-state flux essentially differs for two types of structure of studied alloy. Permeation flux demonstrates linear Arrhenius dependence for recrystallized samples and non-monotonous for amorphous ones. Activation energy of hydrogen diffusion in amorphous sample is about two times lower compared to the recrystallized one.

Values of steady-state flux through amorphous foil were several times higher than through ordered metal at temperatures below 200 ^{0} C. The mechanism of hydrogen transport is proposed taking into account re-emission and diffusion processes. Activation energies for thermal desorption and diffusion are evaluated. However the detailed study of hydrogen transport in these materials requires further investigations.

Acknowledgement. This work was supported by Russian Federal Program "Integration", project № 326.25.

References.

1. Kirchheim, R., Sommer, F., Schluckebier, G. (1982) Hydrogen in amorphous metals-I, *Akta Metal.* **30**, 1959-1068.
2. Hirscher, M., Mossinger, J., Kronmuller, H. (1995) Hydrogen diffusion in nanocrystalline, mesoscopic and microcristalline heterogeneous alloys, *J. Alloys and Compounds* **231**, 267-273.
3. Kim, J.J., Stevenson, D.A. (1998) Hydrogen permeation studies of amorphous and crystallized Ni-Ti alloys, *J. Non-Cryst. Solids* **101,** 187-197.
4. Livshits, A.I. (1979) Superpermeability of solid membranes and gas evacuation. Part I, *Vacuum* **29**, 103-112.
5. Walter, J.L., Bacon, F., Luborsky, F.E. (1976) The ductile-brittle transition of some amorphous alloys, *Mater. Sci. and Eng.* **24(2)**, 239-245.
6. Shu, W.M., Hayashi, Y., Tahara, A. (1991) Hydrogen permeation and diffusion in Fe-Ti alloys, *J. Less-Common Metals* **172-174,** 740-747.
7. Ziegler, J.F., Biersack, J.P., Littmark, U.L. (1985) *The stopping and range of ions in*

solids. Pergamon Press, London.

8. Sokolov, Yu.A., Gorodetsky, A.E., Grashin, S.A., Sharapov, V.M., Zakharov, A.P. (1984) Interaction of hydrogen with the material of discharge chamber of tocamak TM-4, *J. Nucl. Mater* **125**, 25-32.
9. Herst D.G. (1962) Diffusion of fusion gaz. Calculated diffusion from sphere taking into account trapping and return from the traps, *in CRRP-1124. Atomic Energy of Canada: Ist. Conf. Oct.-Nov., Balk River*, 129-135.

AN INTERACTION OF HYDROGEN ISOTOPES WITH AUSTENITIC Cr - Ni STEELS WITHOUT AND DURING REACTOR IRRADIATION

Hydrogen Recycle At Plasma Facing Materials

B.G. POLOSUKHIN , L.I. MENKIN.
Sverdlovsk branch of Research and Development Institute of Power Engineering, Zarechny, 624051, Russia.

Experimental data on an interaction of hydrogen isotopes (protium and deuterium) with austenitic Cr - Ni steels are reported. The data were obtained with the special equipment designed and installed at the research reactor IVV -2M in the conditions without and during irradiation by fast neutrons of a flux density up to 1.8 x $10^{18} n/m^2s$ (E > 0.1MeV), fluence up to $10^{25} n/m^2$ and a gamma - absorbed dose rate< 3.6 W/g, a hydrogen pressure being up to 2.4 MPa.

Steel samples were tested out - of - pile in the protium medium under static and dynamic modes, reaching the pressure of 50 MPa and rupture of the samples .The experimental data showed that the radiation effects increased with a decreasing temperature during irradiation. Thus, at T < 673K a relative increase of both hydrogen permeability and solubility values can reach several orders of magnitude. Besides, there were observed the substantial deviations of isotopic effects, changing of physico - mechanical properties and other radiation effects.

1. Introduction.

In the program of fusion reactor (FR) development great attention is paid to the investigation of an interaction of hydrogen isotopes and hydrogen-bearing media with candidate structural materials (SMs) subjected to various kinds of ionizing irradiation. Unfortunately, works on this issue are not numerous and reported data are contradictory [1-7]. The experimental data on both joint and separate gamma-neutron reactor irradiation are especially required. It is aimed at keeping the most important physico-mechanical characteristics of SMs, reaching low permeability of hydrogen isotopes (first of all tritium) through the SMs of the plasma-facing wall, heat-exchange surfaces and blanket; and ensuring ecological and radiation safety under FR operation conditions. At the same time the literature data though very few indicate a possibility of initiation of hydrogen isotopes transfer in SMs under neutron, gamma and other irradiation. By the present time measurements of hydrogen isotopes permeability and diffusion through SMs under gamma-neutron irradiation were conducted only at three nuclear reactors.

C.H. Wu (ed.), Hydrogen Recycling at Plasma Facing Materials, 133–138.

Thus, for the first time protium and deuterium permeability (P_h and P_d) and diffusion (D_h and D_d) in steel of the 304SS type were measured at the "Astra" reactor. The experiments showed that P_h increased 3 times due to its solubility [3] at T=843K and neutron flux densities $f_{fast}=1x10^{17}$ n/m^2 s and $f_{therm} \approx 4.2x10^{17}$ n/m^2 s and absorbed gamma-dose rate G=0.5 W/g compared to an unirradiated specimen.

Later, special equipment was created at the IVV-2M reactor to measure P_h, P_d and D_h, D_d in austenitic Cr-Ni steels and high nickel alloys during reactor irradiation at $f_{fast}<1.8x10^{18}$ n/m^{2}s and $f_{therm}<3x10^{18}$ n/m^2 s, by fluence F <1x10^{25} n/m^2, and G<4 W/g, and T=473-1200K [4-6]. It was found that each SM has its peculiarities in addition to its general relationships. And the irradiation effect increased with a temperature decrease. For example, at T>673K P_h, P_d and D_h, D_d increased 10 times and more with respect to its value. At first approximation Sieverts law was true and a radiation effect on permeability materialized owing to radiation-stimulated diffusion of hydrogen isotopes. At T<673K temperature dependencies of P_h and P_d were observed to be weak and permeability could reach several orders of magnitude by their value including the increase due to a substantial growth of solubility. The Sievert's law did not hold here and essential deviations of isotopic effects from their initial values were observed.

There was also the slow growth of an oxide layer and the simultaneous decrease of P_n by ~ 1000 times in steel 0.12 C - 18 Cr - 10 Ni - Ti exposed to a fluence of 7.8 x 10 ^{24}n/m^2 (E > 0.1 MeV) in the protium medium with ~ 0.01 vol. % of oxygen added. The following irradiation of the steel up to a fluence of 9.5. ≅ 10 ^{24}n/m^2 in diffusion - refined protium slowly increased the coefficients P_h and D_h by 10 times at lower T and decreased them at higher T, correspondently. Besides, the physico - mechanical properties of the steel were changing essentially.

At the third reactor, IVG-1M, P_h and P_d of two Cr-Ni steels were measured during reactor irradiation at $f_{fast}=(5x10^{16} - 2x10^{17})$ n/m^2 s; $f_{therm}<1x10^{18}$ n/m^2 s and fluence F<4x10^{21} n/m^2 and T=600-1000K [7]. A maximum increase of P_h was 1.7 times and that of D_h was 5 times at T=773K.

This work mainly presents the experimental data on the interaction of protium and deuterium with the steels 0.05 C -16Cr - 15Ni - 3Mo –0.5Ti and 0.07C - 25Cr - 16 Ni - 6Mn – 0.4N – 0.4V tested without and during reactor irradiation . These data are compared with the data of the earlier investigated alloys of the type 304SS and 316L and out - of - pile test results on austenitic Cr - Ni steels tested in protium medium up to high pressures.

2. Experimental Equipment

Special experimental equipment was created at the water-cooled water-moderated research reactor IVV-2M of the rated power 15 MW with relevant maximum values of $f_{fast}=2x10^{18}$ n/m^2 s, (E>0.1 MeV), and $f_{therm}=5x10^{18}$ n/m^2 s. The equipment provides studying of transfer parameters (permeability, diffusion, solubility, sorption and desorption) of hydrogen isotopes and hydrogen-bearing media and helium in different materials unirradiated, post-irradiated and during reactor irradiation [6] at T=370-1300 K and more and medium pressure up to 20 MPa.

The equipment consists of a device for hydrogen diffusion purificaton and experimental test-beds for in-pile and out-of-pile investigations. The test-beds comprise replaceable channel devices, where the test specimens are placed to. The test-beds include the systems of vacuum-pumping, gas feed, calibration, measuring and a collecting of information. Measurements were performed by an accumulation method (Diness) and a dynamic one with a constant pumping-off and a usage of mass spectrometer. In order to expand the experimental capabilities of FR SMs investigation a possibility of creating gas-discharged plasma of E<1 keV and ion flux density up to 10^{23} ion/m^2 s simultaneously with neutron irradiation is provided at the IVV-2M.

The specimens can be transported into the hot-cell laboratory for complex metal-physics investigations after the tests in hydrogen-bearing media under reactor irradiation.

3. Results and Discussion.

Out- of - pile tests of all austenitic Cr- Ni steels were performed with disk specimens, dia. 30 mm, and cylinder specimens, external dia. 14mm. Their effective length was 83 mm. For measurements during reactor irradiation only cylinder specimens were used. The effective thickness values of all the specimens were 1 - 1.5mm. Besides, diffusion - refined hydrogen isotopes with impurity content not more than 10^{-5} vol. % were used there, excluding specific measurements. After their preparation the specimens were expected to obey Fick's law and the velocities of all the processes to be high compared to the slowest one, i.e. a diffusion over a volume.

In real ITER operational conditions the cyclic loads are anticipated to be on the structural materials of the plasma - facing wall and other elements. Besides, after a long - term neutron irradiation structural materials accumulate a great number of radiation defects , which can lead to a high concentration of hydrogen isotopes (especially at lower temperatures). All that will effect the physico - mechanical properties of the structural materials.

Therefore the authors in collaboration with the VNII «NEFTECHIM» Saint - Petersburg, performed out -of -pile tests of the structural materials in static and dynamic modes in protium medium up to (high) P = 50 MPa and sample rupture . In the tests only cylinder specimens of the above dimensions and the wall thickness of 1 to 3 mm were used. The following Cr - Ni steels: 0.08C -18 Cr -10 Ni - 0.65Ti, 0.07C - 25Cr - 16 Ni - 6Mn - 0.4V - 0.4N were tested in different modes in protium medium. At the first stage permeability P_h of diffusion - refined protium through these steels of the thickness δ = 1 to 3mm was measured at T = 473 - 973K and P = 1to 25 MPa.

VNII NEFTECHIM provided the experimental installation which comprises a chromatographer calibrated against protium and a thermal conductivity detector and employs a process gas argon. Then the similar results were obtained for steels in technical hydrogen at T = 573 to 1073K and P = 1 to 50 MPa.. In this case the P_h values decreased two times. It is found that the isobars of P_h are straight lines in the coordinates log P_h - $1/_T$, i.e. P_h increased with T by the exponential law. A linear dependence of P_h on $\sqrt{P}$ holds true as well, which is typical for the atomic mechanism of protium permeability through steel is inversely proportional to δ. Here the activation energy of P_h is ~ 64 to 72 kJ/ mol and a difference in P_h values is related to the content and presence of easily oxidizing elements in steels. It is also found that P_h values keep stable in long -term static tests up to 400 hours at T = 823K and P = 25MPa. Each steel was subjected to 5 loading cycles. The working areas of diffusion cells with the samples under study keep their integrity under cyclic loads up to P = 25 MPa at T = 523K.

At the end of this cycle two samples of 0.08C - 18Cr - 10 Ni - 0.65Ti and 0.07C-25Cr-16Ni-6Mn-0.4V-0.4N were tested for rupture by internal pressure of protium as well as the cylinder samples of the wall thickness of 1mm made. The samples of these structural materials were damaged at the stresses close to the calculated ones (at p = 68 and 100 MPa , respectfully).

At the next investigation stage the similar cylinder samples with diffusion cells were fabricated and protium and duterium transfer parameters in the austenitic steels 0.05C – 16Cr –15Ni –3 Mo –0.5Ti and 0.07 –25 Cr –16Ni- 6Mn –0.4V – 0.4N of the wall thickness 1.5 and 1mm, correspondently, were measured out –of –pile. Thus, in the case without irradiation the coefficients P_h P_d and D_h ,D_d for protium and deuterium in the first steel are well described by the Arrhenius dependence in the temperature range T = 573 – 1073K and P< 1MPa. The experimental data showed that the activation energy of permeability and diffusion of protium and deuterium are close by the corresponding values and their numeric values for protium were 1.3 – 1.5 times more than the similar values for deuterium. Hence, under these conditions, the isotopic effects of permeability and diffusion are approximately described by the “mass law” $(Md / Mn)^{0.5}$. Besides, the Sieverts law holds true for the permeability and solubility of hydrogen isotopes. The experimental results for this structural material are close to the earlier obtained data for steel 0.04C –16Cr - 11Ni - 3 Mo – 0.3Ti (steels of the type 316L) [4].

The out of pile tests followed were performed on the steel 0.07C-25Cr –16N – 6Mn –0.4V –0.4N, which contained a substantial portion of nitrogen. The choice was determined by the data available on beneficial effect of nitrogen on the steels of austenitic grade. For example, with an increasing nitrogen content their strength increase, plasticity remains stable, corrosion decreases, resistance to radiation form – changing increases under fast neutron irradiation, as well as the resistance against hydrogen embrittlement. All those factors are very important for ITER operation.

This austenitic steel has some peculiarities in the investigated range T = 600 – 1100 K when compared to the earlier studied 0.08C – 18 Cr – 10 Ni – 0.65Ti (being of the type 304SS). Thus, in the case without irradiation (Fig.1) the values of permeability and diffusion of protium and deuterium were less that those of the steel 0.08 C – 18 Cr – 10 Ni – 0.65 Ti (e.g. P_h and P_d are ~ 3 times less at T < 873K). However, when the steel, containing nitrogen, was placed into the hydrogen medium for several hours at T > 873K the values P_h P_d and D_h ,D_d increased substantially with a simultaneous increase of activation energies of these processes (E_p incresed from 74.7 to 92.5kJ/ mol.). The single transition of the first rank occured, and permeability and and diffusion polymers showed dramatic abnormalties (ramp changes) in both conditions of the well ordered sample heating and cooling followed. It means the sample sensitivity to temperature – time conditions of the experiment was observed as well as to the admixtures presence in the steel , which impacted other physical properties of structural materials.

In the process of reactor irradiation ($f = 1.6 \times 10^{18} n/m^2 s$, E> 0.1 MeV , the absorbed gamma dose rate G ~ 3w /g) of the steel containing nitrogen, the values of the coefficients (being indicated by the process increase substantially and their activation energies decreased at the same time, thus, E_p from 92.5 to 64 kJ /mol at T > 773 K and from 74.7 to 55.2 kJ /mol at the lower T). In these irradiation conditions the increase of P_h and P_d reached ~ 10 times at T = 873K P_d and became higher than that of the steel 0.08C – 18Cr – 10 Ni – 0.65Ti . At the same time the numeric values of P_h and D_h were 1.3 – 1.5 times more than P_d and D_d. Therefore the isotopic effects were approximately described by the " mass law" $(M_d/M_h)^{0.5}$. The Sieverts Law holds true in the investigated range P< 1 MPa.

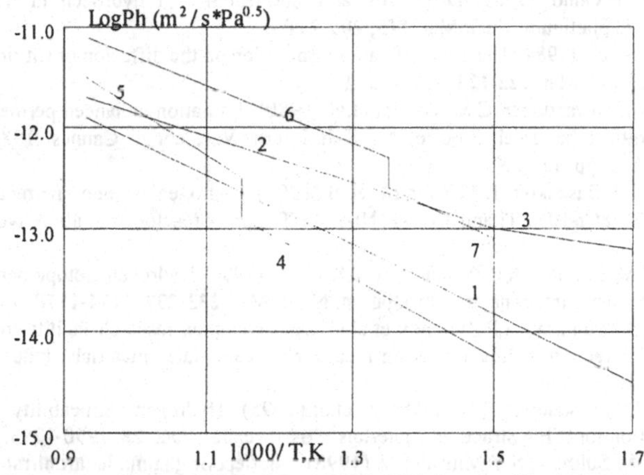

Figure 1. P_hT dependencies for 0.08C-18Cr-10Ni-0.65Ti (1-unirradiated; 2,3-during irradiation) and 0.07C-25Cr-16Ni-6Mn-0.4V-0.4N (4,5-unirradiated; 6,7-during irradiation)

If necessary parameter measuring of hydrogen isotope transfer in SM can be followed by materials science study of these irradiated specimens. Thus, materials science investigation of steel specimens showed that austenite solid solution of this materials were unstable under irradiation in hydrogen media and began to degrade forming carbides and ferrite components. Then an intreraction of steels with gaseous media in the condition studied isncrease of its strengthening, chemical activity of grain boundaries, a decrease of plasticity, etc.

4. Conclusion

The experimental results show that protium and deuterium permeability solubility and diffusion coefficients in austenitic Cr-Ni steels greatly increase during neutron irradiation at $f<1.8 \cdot 10^{18}$ n/m^2s. Each of these SMs has its own peculiarities in addition to common laws. Thus, irradiation effect increases with decreasing temperature. At temperatures above 673 K, Sievert's law holds to a first approximation, and irradiation effect is realized through radiation-stimulated diffusion. Very weak temperature dependencies of hydrogen isotope permeability were shown at T < 673 K. That is why a relative increase of hydrogen isotope permeability can reach values several orders of magnitude higher within this temperature range. In this case, Sievert's law does not hold (the dependence on pressure is more strong), substantial deviations of isotopic transfer effects from their initial values were observed and hydrogen solubility can increase dramatically, resulting in changes in the SMs mechanical properties.

Materials science studies of SMs showed property changes under irradiation in protium. Understanding of mechanisms of hydrogen isotope transfer requires continuation of investigations.

References

1. H.Katsuta, T.Iwai, H.Ohno. (1983) Diffusivity and permeability of hydrogen in neutron irradiated molybdenum and plattinum. Nucl. Mat. **115**, 206-210.
2. R.A.Causey, L.M.Steck. (1984) The effect of gamma radiation on the diffusion of tritium in 304 stainless steel. Nucl. Mat. **122-123,** 1518-1522.
3. R.Dobrozemsky, G.Schwarzinger, C.Stratowaet et.al.(1980) Radiation-enhanced permeation of hydrogen through austenitic tubes steel. Proc. of the 8-th Intern. Vac. congr. Cannes. 1980. (Paris; Society Francaise du Vide, pp. 15-18.
4. B.G.Polosukhin, E.P.Baskakov, E.M.Sulimov et.al.(1992) Hydrogen isotope permeability through austenitic steels 18Cr-10Ni-Ti and 16Cr-11Ni-3Mo-Ti during reactor irradiation. Nucl. Mat. **191-194,** 219-220.
5. B.G.Polosukhin, E.M.Sulimov, A.P.Zyrianov, A.V.Kozlov. (1996) Hydrogen isotope permeability through austenitic Cr-Ni steels under neutron irradiation. Nucl. Mat. **233-237**, 1174-1178.
6. B.G.Polosukhin, E.M.Sulimov, A.P.Zyrianov et.al.(1996) Complex research facility to study radiation effect on the interaction hydrogen-bearing media with candidate materials under irradiation. Nucl. Mat. **233-237**, 1573-1576.
7. I.L.Tazhibaeva, V.P.Shestakov, E.V.Chikhray et.al.(1995) Hydrogen permeability technique in sity reactor irradiation for ITER Structural Materials. Fus. Technol. Oct. 28, 1290-1293.
8. B.G.Polosukhin, E.M.Sulimov, S.Y.Mitrofanov. (1998) Aneffect of gamma-neutronirradiation and oxygen admixtures on interaction of hydrogen with austenitic 18Cr- 10Ni-Ti steel. Fus.Engin. and Design **41,** 135-141.

DIFFUSION OF TRITIUM IN V, Nb AND Ta UNDER CONCENTRATION, TEMPERATURE AND ELECTRIC POTENTIAL GRADIENTS

M. SUGISAKI, K. HASHIZUME and K. SAKAMOTO
Department of Advanced Energy Engineering Science,
Interdisciplinary Graduate School of Engineering Sciences,
Kyushu University
Hakozaki, Fukuoka, 812-8581, Japan

1.Introduction

The diffusion data of hydrogen isotopes, especially tritium, are essential to evaluate the hydrogen recycling rate, hydrogen retention and tritium inventory in the reactor core materials. Such data, however, have not necessarily been provided for candidate materials of plasma facing components. In the cases in which the tritium data are not provided, the data of protium or deuterium is usually used in place of tritium, or certain kind of extrapolation from the data of protium and deuterium is made. Therefore, it is important to know a limit of the validity of such treatment. In addition, it should be noted that the severe thermal gradient and the electric potential gradient may be imposed to the plasma facing components. Then, it is also important to evaluate the influence of such potential gradients on the diffusion flux. From such viewpoints, we have accumulated the diffusion data of hydrogen isotopes under concentration, temperature and electric potential gradient in some materials which have a potential for the plasma facing component.

In the present paper, the data base for hydrogen isotopes H, D and T in Va metals (V, Nb and Ta), which have been established by our recent work [1-11], will be introduced, and their impacts upon fusion technology will be discussed in terms of the evaluation of tritium permeation through reactor core materials. The reasons why these materials have been selected in our study are as follows: These

C.H. Wu (ed.), Hydrogen Recycling at Plasma Facing Materials, 139–145.

metals are well-known for the large mass dependence of diffusion coefficients of hydrogen and some alloys of V and Nb have a potential to be used for core materials of fusion reactors.

2. Diffusion equation under concentration, temperature and electric potential gradient

The diffusion flux J_H of hydrogen isotopes under concentration, temperature and electric potential gradient is represented by the equation,

$$J_H = -D\left(\frac{dC}{dx} + \left(\frac{Q^*C}{RT^2}\right)\frac{dT}{dx} + \left(\frac{Z^*C}{RT}\right)\frac{dV}{dx}\right), \tag{1}$$

where D is a diffusion coefficient of hydrogen isotopes, Q* represents a heat of transport and Z* stands for a effective charge. The notations C, T and V represent hydrogen concentration, temperature and electric potential, respectively. The first term is the ordinary diffusion flux due to the concentration gradient; the second term is usually called the thermomigration and its magnitude and direction are represented by the magnitude and sign of Q*. The third term is usually called electromigration and its magnitude and direction depends upon the magnitude and sign of Z*. Therefore, the data of D, Q* and Z* are necessary to evaluate the diffusion behavior of hydrogen isotopes under concentration, temperature and electric potential gradients.

3. Measurement of Q* and Z*

When a specimen in which hydrogen isotopes are uniformly dissolved is subjected to a thermal gradient, diffusion flux is brought about by the thermal gradient [the second term in eq.(1)] in spite of the fact that the concentration gradient does not exist initially. Then, the concentration gradient is gradually built up in the direction either down or up the thermal gradient depending on the sign of Q*; simultaneously the ordinary diffusion [the first term in eq.(1)] starts in the direction against the induced concentration gradient; finally those two diffusion fluxes are counterbalanced and the stationary concentration distribution is established. In the stationary state, Q* is related to the concentration distribution by the equation,

$$\ln C(x) = \frac{Q^*}{RT} + A(const.) \quad (2)$$

Then, Q* can be determined by measuring the built-up concentration distribution, which is determined by a sectioning method based on a vacuum extraction method [12]. The similar experimental procedures are also adopted in the case of the electromigration.

In the experiment, rod-shaped specimens were used, whose dimension was 2mm in diameter and 50 mm in length. They were loaded with H, D and tritium with a gas absorption method. The atomic ratio of hydrogen isotopes to metal was adjusted to be 0.005~0.01. The details of the experiment is given in our papers [1-5].

4. The experimental data

4.1. HEAT OF TRANSPORT Q*

The data of heat of transport Q* of hydrogen isotopes in vanadium, niobium and tantalum measured by the present authors [1-5] are summarized in Fig.1, in which the data of protium and deuterium in vanadium are cited from Peterson and Smith [13]. These authors also reported the data of protium and deuterium in niobium and tantalum, though those data are not cited to avoid redundancy because they are almost identical with our data. Common features of these data are; (1)the sign is all positive; (2) the magnitude of Q* increases in the order of vanadium, niobium and tantalum for each isotope, Q*(Ta)>Q*(Nb)>Q*(V); and (3) the magnitude of Q* increases in the order of protium, deuterium and tritium in each metal, $Q^*_T > Q^*_D > Q^*_H$.

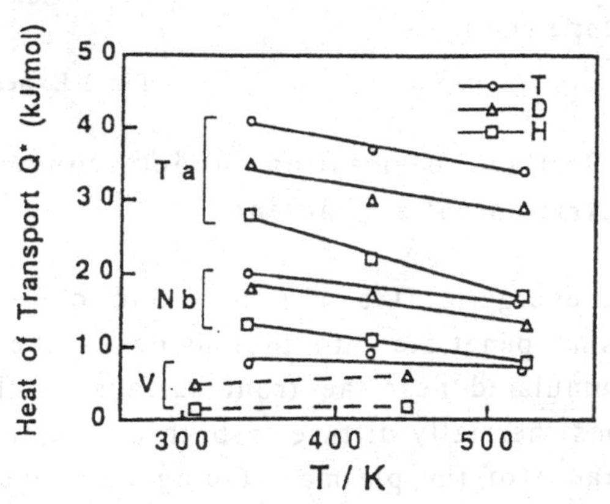

Fig.1 Experimental data of heat of transport Q*

The mass dependence of the heat of transport Q* is compared with that of the activation energy of diffusion, which has been reported by Qi et al. [14]

in Fig.2. As is seen in this figure, the strong correlation of mass dependence can be observed though the value of Q* is not exactly in agreement with that of the activation energy of diffusion.

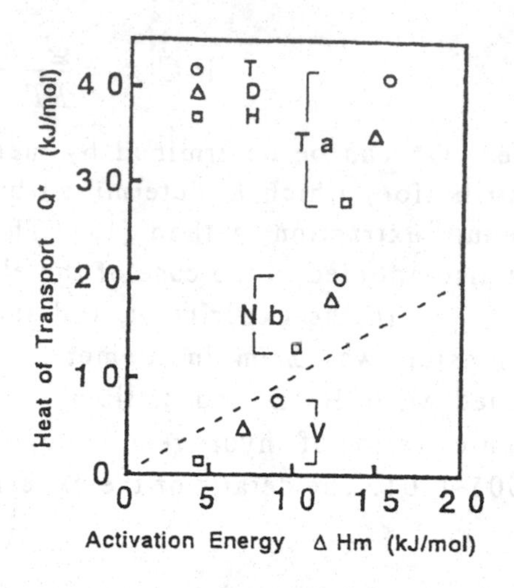

Fig.2 Comparison of heat of transport with activation energy of diffusion

4.2.EFFECTIVE CHARGE Z*

The experimental data of the effective charge of hydrogen isotopes in vanadium, niobium and tantalum measured by the present authors [6-10] are summarized in Fig.3. The characteristics of the present data can be summarized as follows: (1)Z* is positive and dependent upon the isotope mass and they are not necessarily +1. (2)Z* is dependent upon temperature and approaches to +1 with increasing temperature. (3)Deviation of Z* from +1 tends to increase with the isotope mass.

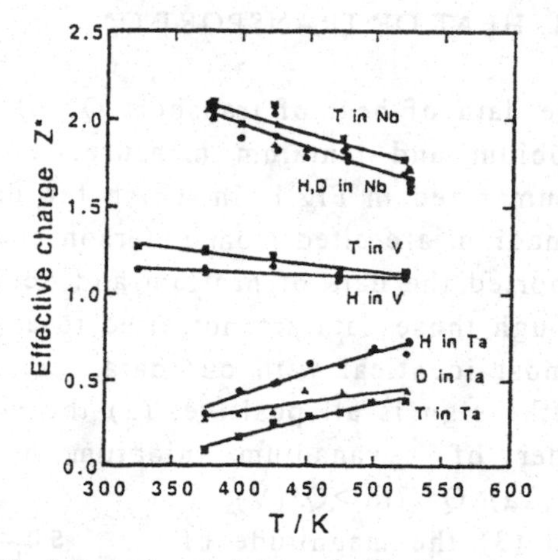

Fig.3 Experimental data of effective charge Z*

5. Tritium permeation under concentration, temperature and electric potential gradients

The energetic D and T ions and neutral atoms escaping from the plasma penetrate into the inside of the plasma facing material and accumulated near the front surface of the material. These D and T atoms thermally diffuse in both directions towards the front and back surfaces of the plasma facing component. When the plasma facing component is subjected to the thermal and electric potential gradient, these diffusion fluxes are inevitably influenced by those potential gradients. These potential gradients, however, may not have an appreciable direct influence upon the diffusion towards the front surface of the plasma facing material, because the length of the diffusion path from the range of the incident ions to the front surface is extremely short; e.g. this length is calculated to be several hundred angstroms under the usual plasma conditions. On the other hand, the diffusion towards the back surface may be appreciably influenced by these potential gradients because the length of the diffusion path is considered to be of the order of 0.1-1.0cm in the case of the future commercial reactors. Therefore, the influence of these potential gradients is considered to be important especially for permeation.

As is shown in Fig.4, the ordinary diffusion flux due to the concentration gradient is enhanced by the thermal gradient in the case of Q*>0 and suppressed in the case of Q*<0 under the usual reactor conditions. The influence of the electromigration is rather complicated because the direction of the electric potential gradient depends upon the plasma conditions.

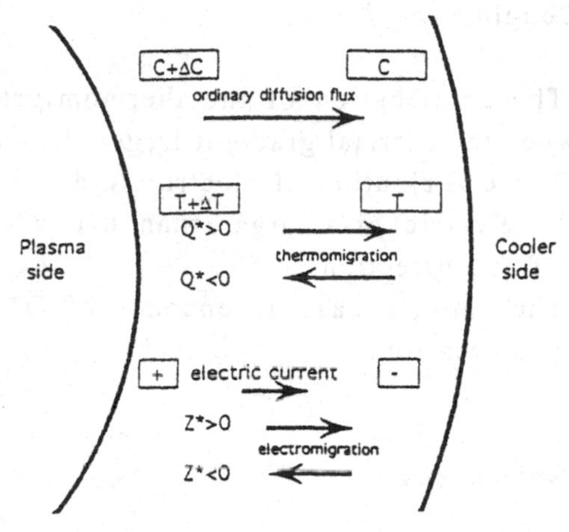

Fig.4 Schematic description of three kinds of diffusion flux

In this figure, the inside of the plasma facing component is assumed to be positively biased. The influence of the electromigration depends on the sign of Z* as is described in the figure. In order to evaluate the influence of the thermomigation and electromigration, we shall assume that the thickness of the plasma facing component is 1 cm and the hydrogen concentration at the inside surface is 1 at% and the concentration gradient is linear. In addition, we shall assume that the heat of transport Q* is about 30 kJ/mol; the effective charge is +1; and the average temperature is 700K and the electric resistively of material is about 50μ Ω cm. Furthermore, we shall suppose two cases of the potential gradient; one is that the thermal gradient is 10 K/cm and the electric potential gradient is about 0.01V/cm(i.e. the electric current is 100A/cm^2). The other is that the thermal gradient is 100 K/cm and the electric potential gradient is about 0.1V/cm(i.e. the electric current is

$1000 A/cm^2$.In these cases, the ratio of contributions of the ordinary diffusion, thermomigration and electromigration is calculated as 1: 0.04 : 0.04 in the first case, and 1 : 0.4 : 0.4 in the latter case. Therefore, it can be concluded that the influence of the diffusion flux due to the temperature and electric potential gradient cannot be neglected in these conditions of the plasma facing components, in particular the mass effect is very important; e.g. the contribution of the thermomigration is largest in the case of tritium permeation because of the fact that the heat of transport Q* is largest for tritium.

6.Conclusions

(1)The contribution of the thermomigration should not be neglected when the thermal gradient larger than 100 K/cm exists.
(2)The contribution of electromigration should not be neglected when the electric bias larger than 0.1 V/cm exists continuously in the definite direction.
(3)The isotope mass dependence of Q* and Z* should be taken into consideration.

7. References

[1] Sugisaki, M., Idemitsu, K., Mukai, S. and Furuya, H. (1981) Thermal diffusion of tritium in Nb metal, *J. Nucl. Mater.*,103&104,1493-1498

[2] Sugisaki, M., Mukai, S., Idemitsu, K. and Furuya, H. (1983) Isotope effect in heat of transport of H, D and T in Nb, *J. Nucl. Mater.*,115,91-94

[3] Sugisaki, M., Furuya, H. and Mukai, S (1985) Temperature dependence of Q* for protium, deuterium and tritium in niobium, *J. Less-Common Metals*,107,79-88

[4] Sugisaki, M., Furuya, H. and Narai, T (1985) Isotope and temperature dependence of heat of transport of hydrogen isotopes in Ta, *Z. Physik. Chemie Neue Folge*, 145,251-260

[5] Sugisaki, M., Furuya, H., Ichigi, N., Koori, I. and Kumabe, I. (1991) Tritium release behavior of a Al-Mg-Li alloy irradiated by thermal neutrotns, *J. Nucl. Mater.*, 179-181,312-315

[6] Hashizume. K., Kawabata, Y., Matsumoto, K., Tsutusmi, N. and Sugisaki, M. (1993) Effective valence of protium and deuterium in α phase of niobium, *Defect and diffusion forum*, 95-98, 329-334

[7] Hashizume, K., Kawabata, Y., Fujii, K. and Sugisaki, M. (1994) Isotope

dependence of the effective valence of protium and deuterium in α phase of tantalum, *J. Alloys and Compounds*, 215,71-76

[8] Hashizume, K., Fujii, K. and Sugisaki, M. (1995) Effective valence of tritium in the alpha phase of niobium, *Fusion Technology*, 28,1179-1181

[9] Hashizume, K., Fujii, K. and Sugisaki, M. (1997) Mass effect of effective charge of hydrogen isotopes in Ta, *Defect and Diffusion forum*, 143-147,1689-1692

[10] Fujii, K., Hashizume, K., Hatano, Y. and Sugisaki, M. (1999) Hydrogen isotope electromigration and impurity effect in vanadium, *J. Alloys and Compounds*, 282,38-43

[11] Sugisaki, M., Hashizume, K. and Fujii, K. (1997) Correlation of mass dependence between heat of transport and effective charge of hydrogen isotopes in V, Nb and Ta, *J. Alloys and Compounds*, 253-254,401-405

[12] Sugisaki, M., Idemitsu, K. and Furuya, H. (1982) A new technique for liquid scintillation counting of tritium present in metals, *Radiochem. Radioanal. Letters*, 51,5,293-300

[13] Peterson, D. T. and Smith, M. F. (1983) Thermotransport of hydrogen and deuterium in vanadium-niobium, vanadium-titanium and vanadium-chronium alloys, *Mettal. Trans. A*, 14A,871-874

DEUTERIUM SUPERPERMEATION THROUGH NIOBIUM MEMBRANE

M. BACAL, A.-M. BRUNETEAU and M.E. NOTKIN *
Laboratoire de Physique des Milieux Ionisés, U.M.R. 7648 du C.N.R.S., École Polytechnique, 91128 Palaiseau, France
**Permanent address: Bonch-Bruyevich University of Telecommunications, 61 Moika, St. Petersburg 191186, Russia*

1. Introduction

The wide investigations of superpermeation phenomenon [1, 2] undertaken during recent years [3, 4, 5] demonstrated that the superpermeable membranes can be successfully used for pumping of hydrogen as well as for its purification and separation from impurities in various fusion and plasma devices.

The present work is the continuation of these studies and was undertaken to investigate the membrane interaction with hydrogen isotope, deuterium which is widely used in various applications as well as to compare the deuterium and hydrogen superpermeation. Although the basic thermodynamic characteristics of hydrogen – niobium and deuterium – niobium systems are very close to each other, the sputtering efficiency by deuterium exceeds that of hydrogen. As a result, the bombardment of the input membrane surface responsible for superpermeation by fast neutral and ionized particles may lead to the change of its state and to the destruction of the superpermeable regime of membrane operation. One can also expect other isotopic effects too.

2. Experimental

The present experiment was carried out on the plasma - membrane test stand presented in fig. 1 [5, 6]. Two tubular membranes of niobium and vanadium (1 cm in diameter, 18 cm long, and 0.01 cm wall thickness) hermetically separate the input and output vacuum chambers.

The membranes are immersed into a uniform hydrogen (deuterium) plasma filling the input chamber of 44 cm diameter and 45 cm height. The plasma is generated by the electric discharge between a set of sixteen hot tantalum cathode filaments (0.5 mm diameter, of a total area 30 cm^2) located close to the chamber walls in a multicusp magnetic field and the chamber walls serving as an anode.

Switching on of filament heating or of the plasma discharge leads to the generation of energetic hydrogen particles. Some of these particles are absorbed by the membrane, penetrate through it, and are released at its output side in molecular form. That results in an increase of the output pressure. A quadrupole mass spectrometer and a set of vacuum

C.H. Wu (ed.), Hydrogen Recycling at Plasma Facing Materials, 147–155.

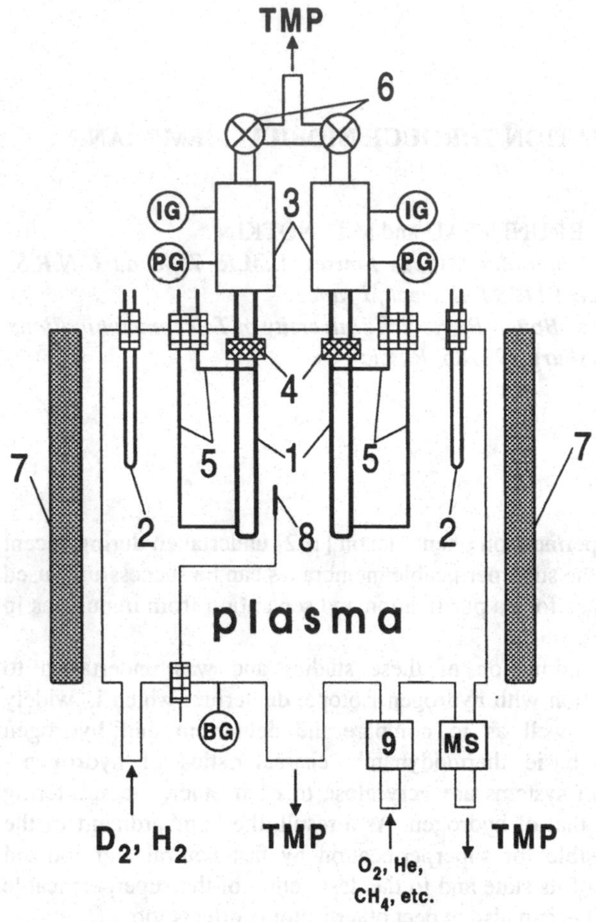

Figure 1. Schematic of the plasma-membrane setup
1 - membranes, 2 - filaments, 3 - output chamber,
4 - ceramic breaks, 5 - heavy current leads, 6 - valves,
7 - magneticsystem, 8 - probe, 9 - gas admittance system

gauges serve to register the hydrogen flux permeating through the membrane.

To provide the possibility to bias the membranes with respect to the plasma, they are electrically insulated with ceramic breaks from the metallic vessel. The membranes can be Joule-heated over a wide temperature range up to 2000 °C, and their temperature is measured with thermocouples. The input and output chambers are pumped with turbomolecular pumps. Various gases such as O_2, CH_4, He, etc. can be introduced into the input vacuum chamber using the gas admittance system.

The typical plasma parameters observed under usual experimental conditions (discharge current 25-40 A, discharge voltage 60-70 V, and hydrogen pressure 1-3 mTorr) were: plasma density of 10^{10} – 10^{11} cm^{-3}, and the electron temperature of 0.6 eV [6].

The detailed study and comparison of the hydrogen and deuterium plasma was fulfilled in Ref. 7. As an example of the typical results of these investigations, the dependence of electron density on discharge current is presented in fig. 2. As it is seen, the parameters of deuterium and hydrogen plasma are very close to each other. Thus, one can suppose that the densities of the deuterium and hydrogen atoms generated in the plasma are very close too. Therefore, in the case of deuterium plasma driven experiment one can expect a large permeability (up to superpermeability) of deuterium atoms.

Two types of experiments were performed in the course of the present investigation. First, experiments in which the membrane was under a floating potential of about 3 V. Second, a bias voltage was applied to the membrane. The bias voltage varied over the range of 20 - -250 V, and, correspondingly, the membrane inlet surface was acted upon by accelerated hydrogen and deuterium ions of an energy up to 250 eV.

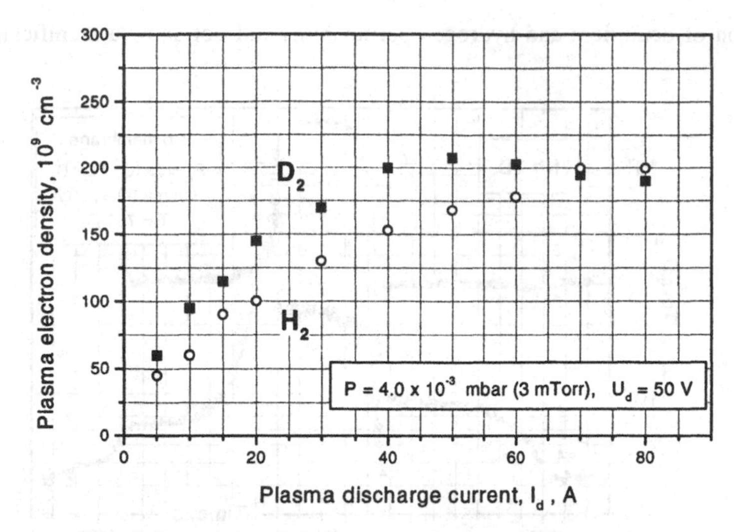

Figure 2. Dependence of plasma electron density on discharge current.

3. Deuterium plasma driven permeation at membrane floating potential

The thermal atom and plasma driven permeation of deuterium through the superpermeable niobium membrane was investigated over a wide range of membrane temperatures, deuterium particle fluxes and plasma parameters. The temperature independence of the deuterium permeation through the membrane was observed both in the case of thermal deuterium atoms and for energetic particles generated by plasma. So, it was demonstrated that the superpermeable to energetic hydrogen particles niobium membrane is superpermeable to energetic deuterium particles.

The direct comparison of the deuterium and hydrogen plasma driven permeation can be made from the data presented in fig. 3 (deuterium) and in fig. 4 (hydrogen). These results were obtained in the same experimental run under the same temperature, vacuum, etc. conditions. As it is seen, in the first approximation the data for hydrogen and deuterium permeation are very close to each other.

To understand if some isotope effect in hydrogen isotope permeation takes really place one has to take into account the following. First, the pumping speed of the output vacuum chamber is smaller for deuterium than that for hydrogen by a factor of $\sqrt{2}$. So, the equality of hydrogen and deuterium output pressures (fig. 3 and fig. 4) means that the permeating flux of deuterium particles is smaller than that for hydrogen. On the other hand, the incident flux of energetic deuterium particles is smaller than that for hydrogen by the factor of $\sqrt{2}$ due to smaller value of deuterium particle velocity.

Taking also into account that the parameters of hydrogen and deuterium plasma are very close (fig.2), one conclude that the observed membrane permeability to deuterium energetic particles is very close or even equals that for hydrogen. Thus, the direct comparison of deuterium and hydrogen permeability did not show a significant isotope effect.

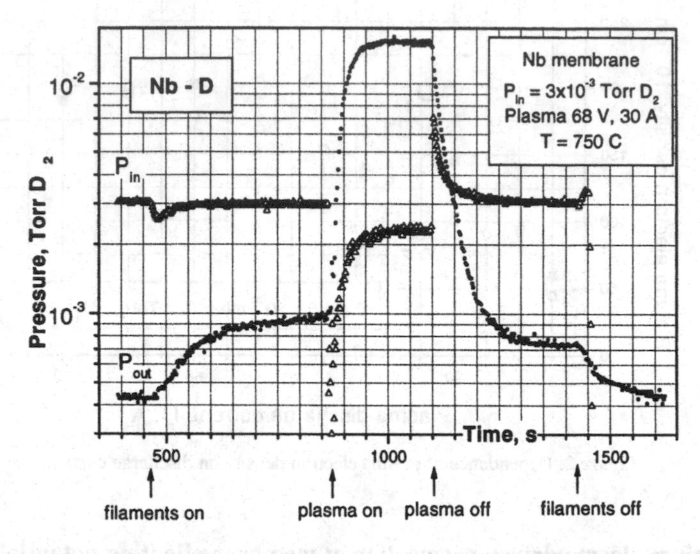

Figure 3. Permeation of energetic deuterium particles through the niobium membrane

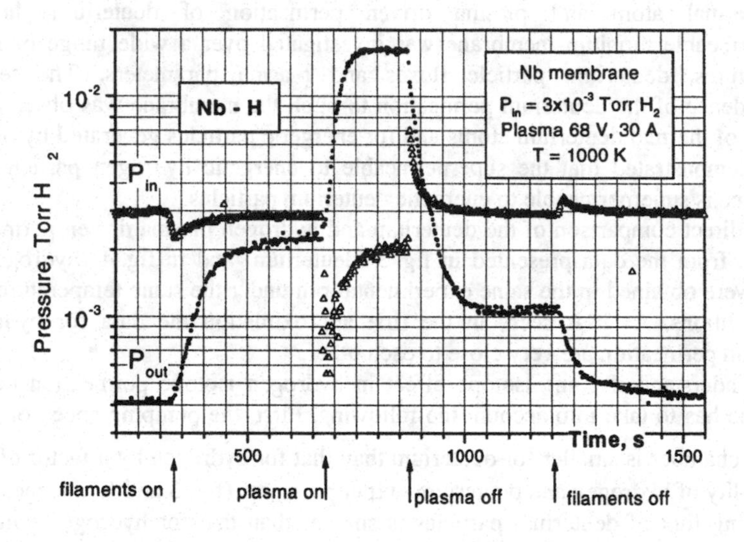

Figure 4. Permeation of energetic hydrogen particles through the niobium membrane

3.1. CHARACTERISTIC TRANSIENT TIMES

Along with the measurements of membrane permeability, the study of the characteristic transient times of the establishment of the steady-state level of permeation flux was carried out. After switching on of thermal atomization or of plasma discharge, the steady-state level of the hydrogen (deuterium) particle permeation through the membrane (output pressure in figs. 3 and 4) establishes with the characteristic transient times, $t_{0.5}$. Correspondingly, after switching off the energetic particle generation the steady-state level of the permeation flux decays with the characteristic transient times, t_{des}. These characteristic times depend on membrane temperature, permeation flux value, state of the membrane surfaces, the diffusion rate of absorbed particles over the membrane bulk, etc. and substantially exceed the average diffusion time of single crossing of the membrane by diffusing hydrogen atom, t_D: $t_{0.5}$, $t_{des} >> t_D$. The diffusion time depends on the kind of absorbed particle. As a result, characteristic times are different for hydrogen and deuterium. The temperature dependence of the transient characteristic times of the establishment and of the decay of the permeation flux steady-state level for hydrogen and deuterium permeation is shown in fig. 5.

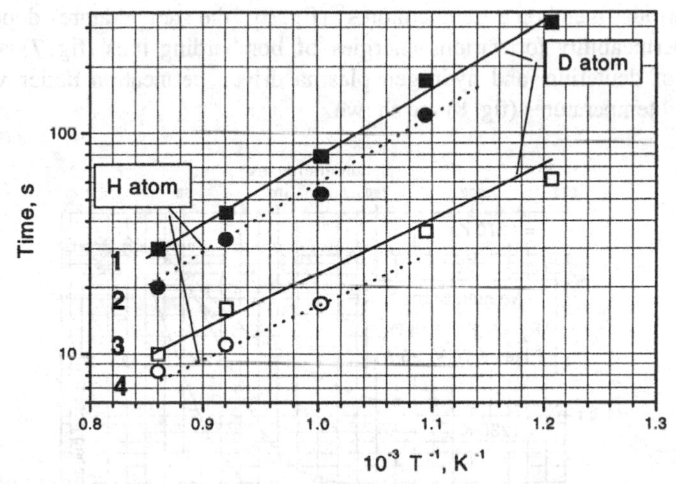

Figure 5. Temperature dependence of transient characteristic times, $t_{0.5}$ and t_{des}
1, 2 - $t_{0.5}$ for deuterium and hydrogen, respectively,
3, 4 - t_{des} for deuterium and hydrogen, respectively.

As one can see in fig. 5, there is a distinct difference in hydrogen and deuterium transient characteristic times. The values of characteristic transient time of permeation are longer in the case of deuterium than those for hydrogen by a factor of about $\sqrt{2}$.

4. Bias experiments with deuterium plasma

Under conditions of the actual plasma and fusion devices, the membranes are affected by fluxes of energetic particles which bombard the membrane surface and sputter non-metallic impurity films responsible for superpermeability. The bias experiment was undertaken to model these processes.

In comparison with niobium membrane interaction with hydrogen ions, the threshold energy of sputtering by deuterium ions is lower [8, 9] and the sputtering efficiency is higher than that for hydrogen ions [8, 9]. So, one can expect that the range over which the superpermeable regime of membrane operation remains undisturbed under deuterium ion bombardment will be narrower than that in the case of hydrogen.

4.1. EXPERIMENTAL RESULTS

The next procedure was used during bias experiment. After switching on of energetic particle generation and establishing of a steady state level of hydrogen permeation, the potential was applied to the membrane, and its effect was investigated over a wide range of membrane temperatures and potentials. The results of this investigation are presented in figs. 6 – 8, where the dependence of membrane permeability on energy of hydrogen ions for various membrane temperatures (fig. 6), the temperature dependence of membrane permeability for various energies of bombarding ions (fig. 7) and a direct comparison of deuterium and hydrogen plasma driven permeation under various bias potentials and temperatures (fig. 8) are shown.

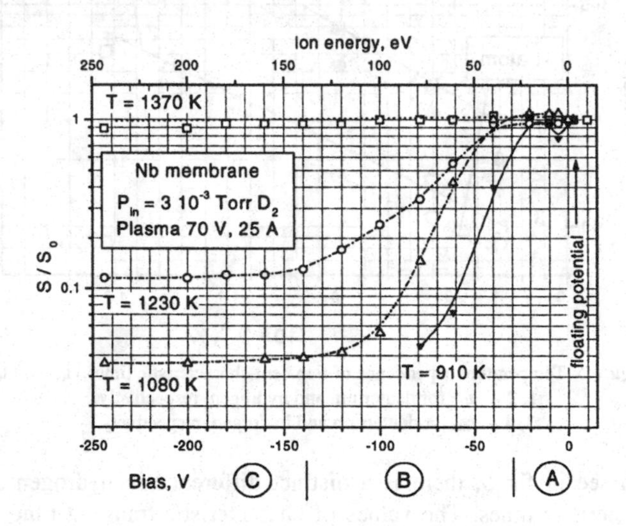

Figure 6. Dependence of plasma driven permeation on membrane bias (energy of deuterium ions)
S - membrane permeation under ion bombardment,
S_0 - membrane permeation at membrane floating potential (without ion bombardment).

As it is seen in figs. 6 and 7, the effect of deuterium ion bombardment on plasma driven superpermeability is very similar to that observed during our experiments with hydrogen (see Notkin et al in these Proceedings). Membrane permeability strongly depends on energy of deuterium ions, and there are three zones of ion energy over which the influence of membrane bombardment on membrane permeability is very different and varies from the total independence of the membrane permeability of ion energy at ion energies lower than the threshold energy for sputtering (zone **A**) to its very strong dependence in zone **B** where a small increase, of several tens of eV in ion energy, results in significant permeability decrease.

As it was expected, the range over which the superpermeable regime of membrane operation remains unperturbed under ion bombardment (zone **A**) is narrower than it was in the case of hydrogen, and is limited by the value of threshold energy of about 30 eV [9].

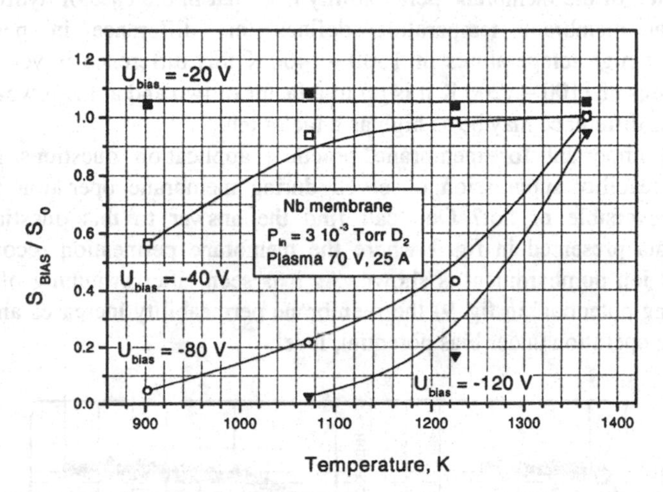

Figure 7. Dependence of deuterium plasma driven on membrane temperature under various bias potential S_{bias} and S_0-are the membrane pumping speed with and without ion bombardment, correspondingly.

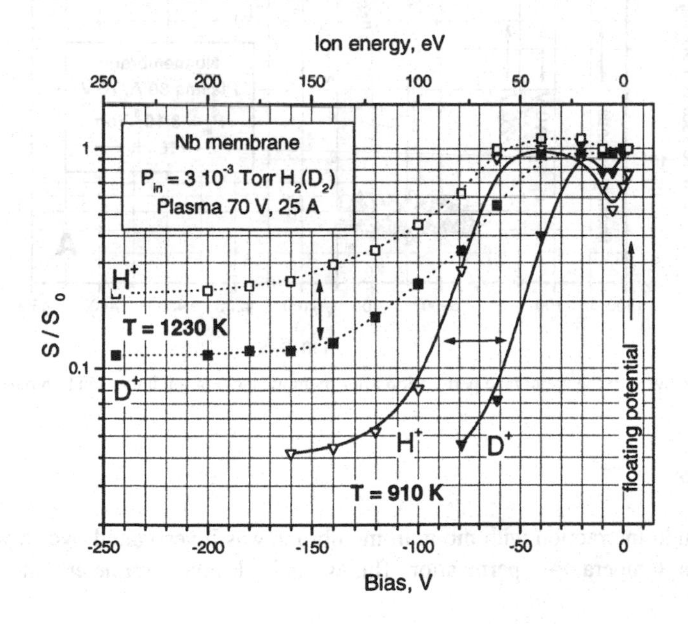

Figure 8. Comparison of deuterium and hydrogen plasma driven permeation under various bias potentials

As for membrane temperature it has a very strong influence on the effect of the membrane bombardment by ions - figs. 6 and 7. As it is seen in fig. 7, the membrane permeability does not depend on ion energy at highest membrane temperatures and decreases significantly at low temperatures.

A direct comparison of the effect of ion bombardment on deuterium and hydrogen plasma driven permeation under various bias potentials is presented in fig. 8 for two membrane temperatures. The higher sputtering efficiency by deuterium ions results in higher suppression of the membrane permeability than that in the case of hydrogen ions. Note also, that membrane temperature defines this difference in permeability suppression: if at high temperatures of 1300 – 1400 K this difference is very small, at lower temperatures of 1100 – 1200 K it is up to a factor of two and at the lowest working temperatures this difference may be as high as a factor ten.

One of the important for membrane practical application questions is: is the membrane permeability suppression observed during membrane operation under ion bombardment reversible or not? One can find the answer to this question in the experimental data presented in fig. 9 where the membrane permeation recovery after operation under ion bombardment is shown. As it is seen, after switching off the bias voltage ("floating potential" in fig. 9) the membrane permeability increases and reaches its initial, before operation under bias potential, level.

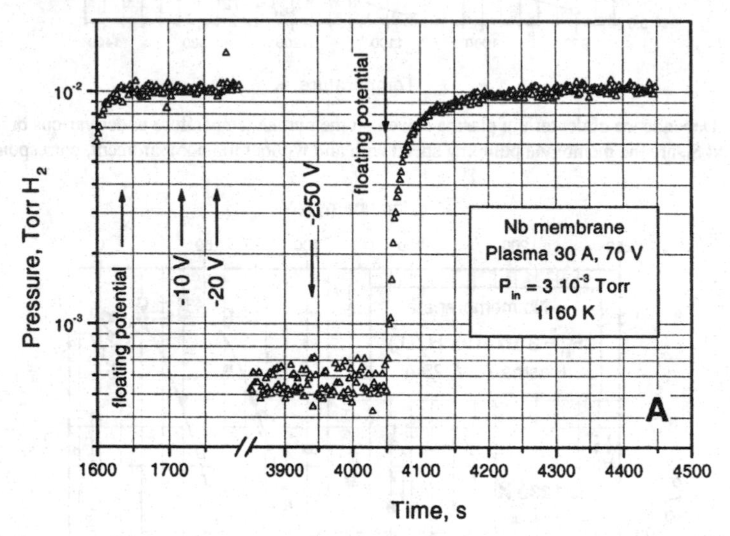

Figure 9. Recovery of the membrane permeation after membrane operation under ion bombardment.

5. Conclusion

1. Deuterium interaction with niobium membrane was investigated over a wide range of membrane temperature, permeation fluxes and plasma parameters at membrane

floating potential, and plasma driven superpermeation of deuterium through the niobium membrane was demonstrated for the first time.

2. Direct comparison of deuterium and hydrogen permeation did not show a significant isotope effect at membrane floating potential. It was found that the value of characteristic transient times of permeation of deuterium was higher than that of hydrogen by a factor of $\sqrt{2}$.

3. The effect of ion bombardment of the input membrane surface on the deuterium plasma driven superpermeation was investigated over the range of ion energy of 0 – 250 eV and of membrane temperature of 910 – 1420 K.

4. It was demonstrated that the range of "spontaneous" superpermeability under ion bombardment depends both on the membrane temperature and on the energy of bombarding ions and is possible as over the range of ion energies below the threshold energy at all membrane temperatures investigated and over the range of the membrane temperatures of 1420 K and higher at all bias voltages investigated.

5. It was found that the effect of deuterium ion bombardment on membrane permeability is very similar to that observed for hydrogen ions. Still, bombardment by deuterium ions having higher sputtering efficiency resulted in the narrowing of the range of working parameters, over which the "spontaneous" superpermeability is possible.

Acknowledgement. This work was supported by the Association Euratom – CEA under Contract CEA/V.3094.002 with EURATOM.

6. References

1. Livshits, A.I. (1979) Superpermeability of solid membranes and gas evacuation ,*Vacuum* **29**, 103 –113.
2. Livshits A.I., Notkin M.E. and Samartsev A.A. (1990) Physico-chemical origin of superpermeability – large-scale effects of surface chemistry on "hot" hydrogen permeation and absorption in metals, *J. Nucl. Mater.* **170**, 74-94.
3. Livshits A.I., Notkin M.E., Samartsev A.A. and Solovyov A.A. (1996) Interaction of low energy hydrogen ions with niobium: effects of non-metallic overlayers on reemission, retention and permeation, *J.Nucl.Mater.* **233–237**, 1113-1117.
4. Livshits A., Ohyabu N., Notkin M., Alimov V., Suzuki H., Samartsev A., Solovyov M., Grigoriadi I., Glebovsky A., Busnyuk A., Doroshin A. and Komatsu K. (1997) Applications of superpermeable membranes in fusion: the flux density problem and experimental progress, *J.Nucl.Mater.* **241–243**, 1203–1209.
5. Livshits A., Sube F., Soloviev M., Notkin M. and Bacal M. (1998) Plasma driven superpermeation of hydrogen through group Va metals, *J. Appl. Phys.* **84**, 2558-2564.
6. Courteille C., Bruneteau A.M. and Bacal M. (1995) Investigation of a large volume negative hydrogen ion source, *Rev. Sci. Instrum.* **66**, 2533 - 2540.
7. Courteille C. (1993) Etude d'une drande source multipolaire hybride d'ions négatifs d'hydrogène et de deuterium. Développement des techniques de mesures par photodétachement laser, *These de doctorat de l'university de caen,* PMI 2795,.
8. Behrisch R. (1981) *Sputtering by particle bombardment*, Springer, Berlin-Heidelberg-New York, 194-280.
9. Eckstein W., Garcia-Rosales C., Roth J., Ottenberger W. (1993) *Sputtering data*, Max-Plank-Institut fur Plasmaphysik, Report IPP 9/82.

EFFECTS OF HELIUM ON THE SUPERPERMEATION OF THE GROUP Va METALS

M.E. NOTKIN *, A.I. LIVSHITS *, A.-M. BRUNETEAU, M. BACAL
Laboratoire de Physique des Milieux Ionisés, U.M.R. 7648 du C.N.R.S., École Polytechnique, 91128 Palaiseau, France
**Permanent address: Bonch-Bruyevich University of Telecommunications, 61 Moika, St. Petersburg 191186, Russia*

1. Introduction

A metal membrane of macroscopic thickness may be superpermeable to hydrogen particles whose energy (kinetic, internal or chemical) exceeds ~ 1 eV. Such a membrane can let through the energetic hydrogen almost like an opening of the same area, compress the permeating gas, and isolate it from a gas mixture: the superpermeability [1, 2]. The phenomenon is controlled by catalytic processes at the membrane surfaces and is conditioned by the state of the upstream surface, in particular, by the upstream potential barrier created by mono-atomic films of non-metallic elements such as O, C, S, etc.

During the last years, superpermeability was investigated for a number of metal – hydrogen systems, and a long-term stable membrane operation was demonstrated under various operational conditions. Still, there are a few subjects important for the practical applications of membranes requiring a further detailed study. The problem of the helium effect on the superpermeable membrane operation is among them.

As it is known, helium is expected to be one of the main gas impurities in various fusion devices. Although thermal helium atoms can hardly cause the change of the state of the membrane surface, its bombardment by fast neutral and ionized helium particles may result in the destruction of the surface film and lead to the decrease of the membrane permeation. Moreover, such an undesirable effect as blistering formation cannot be excluded too. For this reason study of the effects of helium on membrane operation appears to be rather interesting and important from the practical point of view.

2. Experimental

The present investigation was carried out with a tubular membrane of niobium (1 cm in diameter, 18 cm long, and 0.01 cm wall thickness) in the plasma - membrane test stand described in detail in Ref. 3 (see also M. Bacal, A.-M. Bruneteau and M.E. Notkin in these Proceedings).

The membrane is immersed into a uniform hydrogen plasma filling the input chamber of 44 cm diameter and 45 cm height. The plasma is generated by electric discharge

C.H. Wu (ed.), Hydrogen Recycling at Plasma Facing Materials, 157–165.

between a set of sixteen hot tantalum cathode filaments (0.5 mm diameter, of a total area 30 cm^2) located close to the chamber walls in a multicusp magnetic field and the chamber walls serving as an anode.

To provide the possibility to bias the membrane with respect to the plasma, it is electrically insulated with a ceramic break from the metallic vessel.

2.1. EXPERIMENTAL PROCEDURE

Two types of experiments were performed in the course of the present investigation. First, the experiments in which the membrane was under floating potential of about 3 V. Second, a bias voltage was applied to the membrane. The bias voltage varied over the range of 20 - -250 V, and, correspondingly, the membrane inlet surface was acted upon by accelerated hydrogen and helium ions of an energy up to 250 eV.

Switching on of filament heating or of the plasma discharge leads to the generation of energetic hydrogen particles. Those particles which are absorbed by the membrane, penetrate through it, and are released at its output side in molecular form. That results in an increase of the output pressure. After establishment of a steady state level of hydrogen permeation, helium is admitted into the input chamber of experimental setup, and the effects of helium are investigated over a wide range of membrane temperature, plasma parameters and helium content in the hydrogen / helium mixture.

3. Experiments with membrane at a floating potential in the presence of helium

3.1. THERMAL ATOM DRIVEN PERMEATION

The results of the experiment on the effects of helium on thermal hydrogen atom driven permeation through the superpermeable niobium membrane are presented in fig. 1.

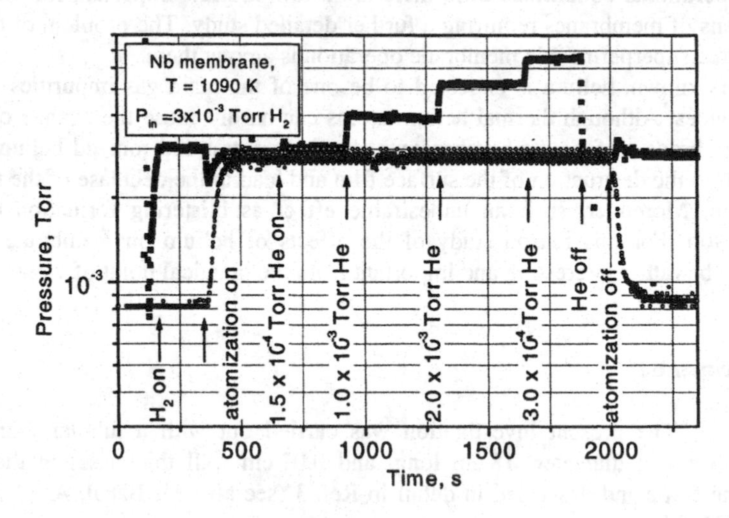

Figure 1. Effects of helium on thermal atom driven permeation

After switching on the hydrogen atomization ("filaments on" in fig. 1) and the establishment of a steady state level of hydrogen atom permeation, helium was admitted, and its pressure was continuously increased up to a value of about 3×10^{-3} Torr which corresponds to a 50% content of helium in the hydrogen-helium mixture.

As one can see, helium had no effect on the membrane pumping of hydrogen, and the output pressure of hydrogen remained unchanged over the whole investigated range of helium pressures.

3.2. PLASMA DRIVEN PERMEATION

Next, the interaction of Nb membrane with the plasma produced in hydrogen-helium mixture was investigated. The experiment was performed in a way similar to that with hydrogen atoms. After switching on of hydrogen plasma and establishing of a steady state level of hydrogen permeation, helium was admitted into the input chamber of plasma set-up. Helium pressure was varied from zero up to 4×10^{-3} Torr - fig. 2.

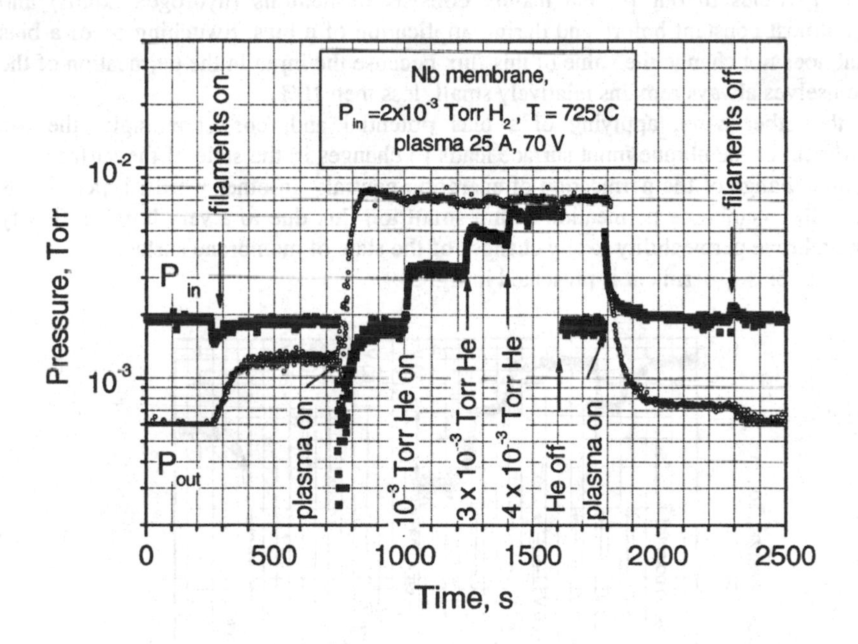

Figure 2. Plasma driven permeation of hydrogen through the Nb membrane under helium additive condition

As one can see, the helium additive had no effect on the membrane pumping of the atomic hydrogen generated by plasma, as it was already observed during experiments with thermal atoms. Moreover, the membrane pumping speed remained unchanged even when the ratio of helium / hydrogen pressures was increased up to a value of 2.

So, the investigation of thermal atom and plasma driven permeation showed that both neutral and ionized helium particles had no effect on hydrogen permeation through the superpermeable niobium membrane at floating potential.

One can explain these results by the fact that, in the case of the operation with thermal atoms, helium hardly can cause any change of the membrane permeation: chemical interaction of helium with the membrane is improbable. As for the ionized helium particles, their energy (of about 2.7 V) is lower than the threshold energy of sputtering and destruction of the non-metallic films on the membrane surface at the membrane floating potential.

4. Bias experiments

Under conditions of the actual plasma and fusion devices, the membranes are affected by fluxes of energetic particles which bombard the membrane input surface and sputter non-metallic impurity films responsible for superpermeability. The bias experiment was undertaken to model these processes.

The idea of the bias experiment was based on the fact that the incident flux of energetic particles in our plasma mainly consists of neutrals (hydrogen atoms) and remains almost constant before and during application of a bias. Switching on of a bias potential does not change the value of this flux because the input in the permeation of the ions themselves always remains relatively small, less then 10%.

On the other hand, applying of a bias potential and, correspondingly, the ion bombardment of membrane input surface leads to changes in the state of the surface and results in a change of the permeation of energetic neutrals. In other words, it permits us to control the membrane permeation with a small ion flux due to a very high sensitivity of the membrane permeability to any changes of the state of membrane surface.

A typical bias experiment is presented in fig. 3.

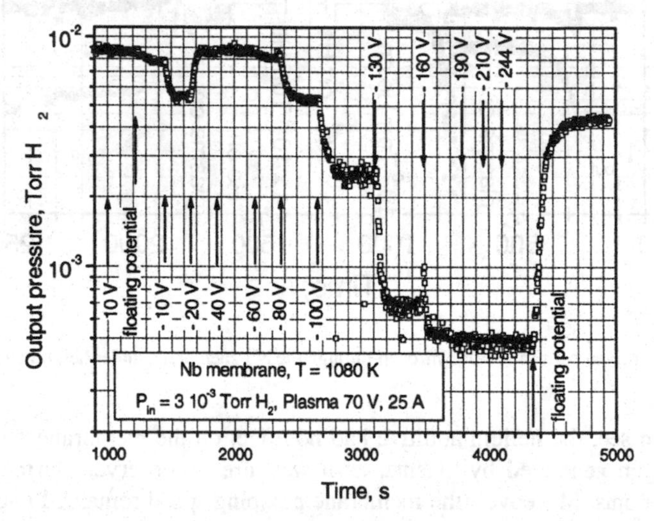

Figure 3. Experiment on plasma driven permeation of hydrogen through the niobium membrane under membrane bombardment by hydrogen ions

After switching on of the plasma discharge and establishing of the steady state level of hydrogen permeation, various bias potentials were applied to the membrane, and hydrogen permeation was measured at each bias voltage. The dependence of plasma driven permeation on membrane bias is given in fig. 4. As one can see, the membrane permeability strongly depends on the energy of hydrogen ions, and one can single out at least three ranges of ion energy where the membrane permeability behaves differently.

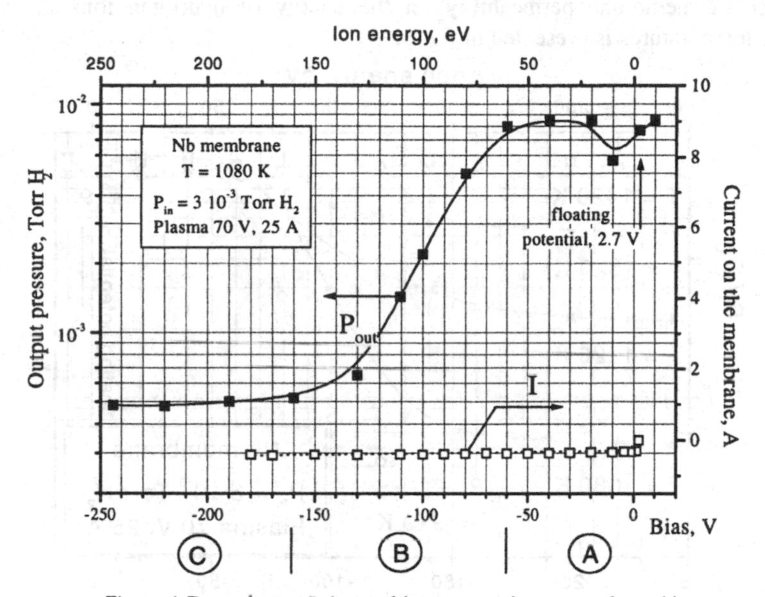

Figure 4. Dependence of plasma driven permeation on membrane bias

Putting aside the peculiarity of the membrane permeability behavior at an ion energy of 10 eV which is probably related to chemical sputtering of surface non-metallic film, these ranges are:

- **A** - ion energy of (0 - 60) eV. Membrane permeability (superpermeability) remains almost constant over this range;
- **B** - ion energy of (60 - 160) eV. Membrane permeability drops more than 15 times;
- **C** - ion energy of (160 - 250) eV. Membrane permeability remains almost constant at its lowest level.

The independence of the membrane permeability of ion energy over the range A is most probably connected with the fact that the ion energy here is lower than threshold energy of the physical sputtering of non-metallic surface film composed of such light impurities as oxygen and carbon. For instance, the value of threshold energy for the sputtering of carbon by hydrogen ions equals about 30 eV [4].

At a higher ion energy (range **B**), a strong permeation decrease is observed. Most probably, it is connected with physical sputtering of the surface film which results in degradation of the superpermeation regime.

Further, over the range **C**, this rapid decrease is replaced with a weak dependence of permeability on ion energy. That may be explained by the weak dependence of the sputtering coefficient that is almost constant over this energy. For example, the value of carbon sputtering coefficient rapidly increases with ion energy and reaches a maximum at an ion energy of about 200 - 400 eV [4].

The bias effect was investigated over a wide range of membrane temperature, and the dependence of membrane permeability on the energy of hydrogen ions for various membrane temperatures is presented in fig. 5.

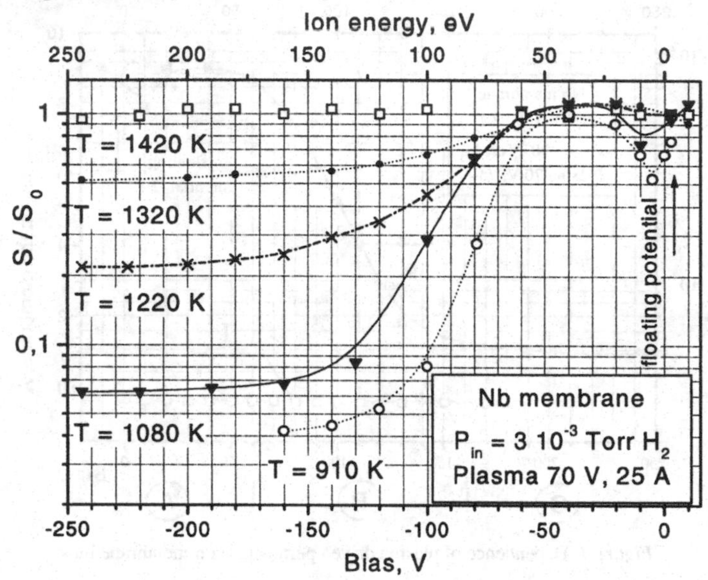

Figure 5. Dependence of plasma driven permeation on membrane bias and membrane temperature
S - membrane pumping speed under ion bombardment,
S_0 - membrane pumping speed at membrane floating potential (without ion bombardment).

One can see that the membrane permeability strongly depends both on membrane temperature and on the energy of bombarding ions. This dependence varies from the total independence of the membrane permeability on ion energy at the highest membrane temperatures to its more than 20-fold decrease at the lowest temperatures.

One can understand the data observed by taking into account that the state of the input membrane surface is in a permanent dynamic equilibrium during the bias experiment. Two factors affecting the membrane surface: the ion bombardment and membrane temperature, have a quite different influence on the state of the surface film.

Ion bombardment leads to surface film sputtering, with the increase of ion energy resulting in a quicker and more effective degradation of membrane permeability.

In contrast to that, membrane temperature governs the process of the segregation of non-metallic impurities from bulk metal resulting in the restoration of the surface film. The higher the membrane temperature, the quicker the membrane permeability recovery process. Finally, the magnitude of membrane permeability at given ion energy and membrane temperature values is defined by the correlation of the rates of surface film destruction by ion sputtering and surface film recovery by impurity segregation.

5. Bias experiment with helium additive

One can suggest that under bias the bombardment of membrane surface by relatively fast helium ions (up to 250 eV) may result in an additional destruction of the surface film and of the potential barrier at the membrane inlet side. Although the threshold energies of sputtering of light impurity by hydrogen and helium ions are very close to each another, the sputtering efficiency (the value of sputtering coefficient) for helium ions is more than ten times higher than that for hydrogen ions [4].

After switching on of plasma and establishing of a steady state level of hydrogen permeation at a given bias potential and a given membrane temperature, helium was admitted into the input chamber of plasma setup. That resulted in an additional sputtering of the input surface film by helium ions and led to a decrease of the plasma driven permeation and to the establishment of its new level. Then a new value of bias potential was applied. As a result, the effect of helium additive was measured by means of direct comparison of membrane permeability with and without helium additive at each bias potential. The results of these experiments are presented in fig. 6 where the dependence of hydrogen plasma driven permeation on energy of bombarding ions is given for a helium additive of 10% for different membrane temperatures.

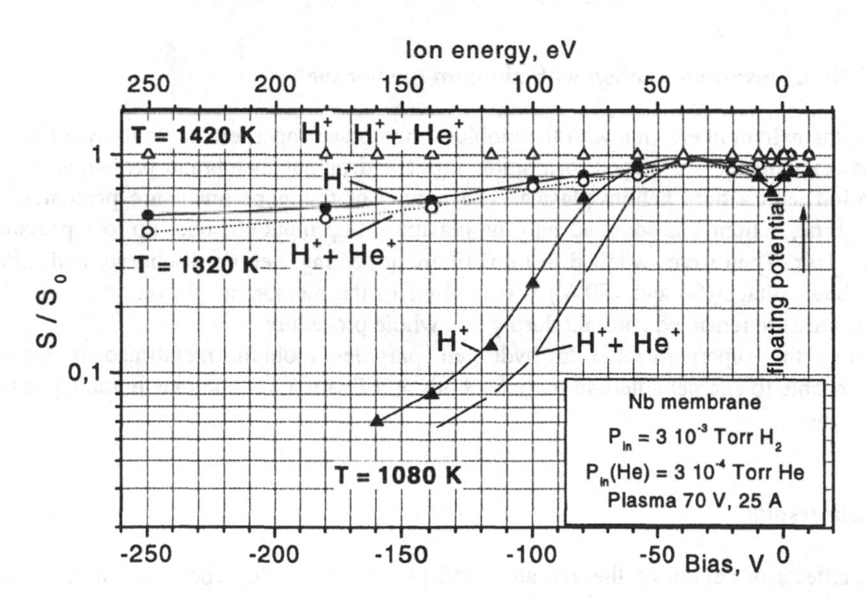

Figure 6. Dependence of hydrogen plasma driven permeation on energy of bombarding ions under condition of 10% helium additive

S - membrane pumping speed under ion bombardment,

S_0 - membrane pumping speed without ion bombardment

In contrast to the case of the membrane operation under floating potential, the helium additive caused, as expected, an additional membrane permeability decrease.

On the whole, helium additive did not change qualitatively the dependence of plasma driven permeation on ion energy. There are three various zones of ion energy characterized by different influence on membrane permeability, as it was observed earlier for hydrogen. Particularly remarkable, the helium additive had no effect on membrane permeability at ion energies below the level of threshold energy.

The high efficiency of helium ion sputtering resulted in an additional suppression of membrane permeability, as it is seen in fig. 6 from a direct comparison of plasma driven permeation with and without helium additive.

As it was observed at the membrane bombardment by hydrogen ions, the effect of helium depends on membrane temperature. It varies from the total independence of membrane permeability of helium additive at the highest membrane temperature to a very strong influence of helium ions at lower membrane temperatures. As before, the higher the ion energy and the lower the membrane temperature, the higher the additional suppression of membrane permeability due to helium additive. For instance, at membrane temperature of about 1000 K, the 10 % helium additive resulted in an additional decrease of the membrane permeability by a factor of 2.

Further increase of the helium content led to a dramatic decrease of membrane permeability at low membrane temperatures. For instance, at a helium content of 30°%, the membrane permeability decreases more that 15 times compared with the case of membrane operation with a pure hydrogen plasma.

6. Helium plasma interaction with niobium membrane

As for the helium interaction with the niobium membrane superpermeable to hydrogen, a special experiment was staged to monitor input and output membrane pressures at the operation with a pure helium plasma. The results of this experiment are presented in fig. 7. First, helium was admitted into the plasma set-up input chamber up to a pressure 3×10^{-3} Torr. Then were switched in turn: filament heating, helium discharge, and, after all, a bias voltage (50 and 250 V) was applied to the membrane. As one can see, the output pressure remained constant during this whole procedure.

Thus, the superpermeable to hydrogen particles niobium membrane is totally impermeable to energetic helium particles both at a floating and the bias potential at the membrane.

7. Main results

1. Effects of helium on thermal atom and plasma driven superpermeation through a niobium membrane were investigated at a floating and bias membrane potential.

2. It was found that both the neutral and ionized helium particles had no effect on the plasma driven permeation of hydrogen through the superpermeable niobium membrane when the membrane was operating at a floating potential.

3. In contrast to the above, the helium additive caused a significant additional

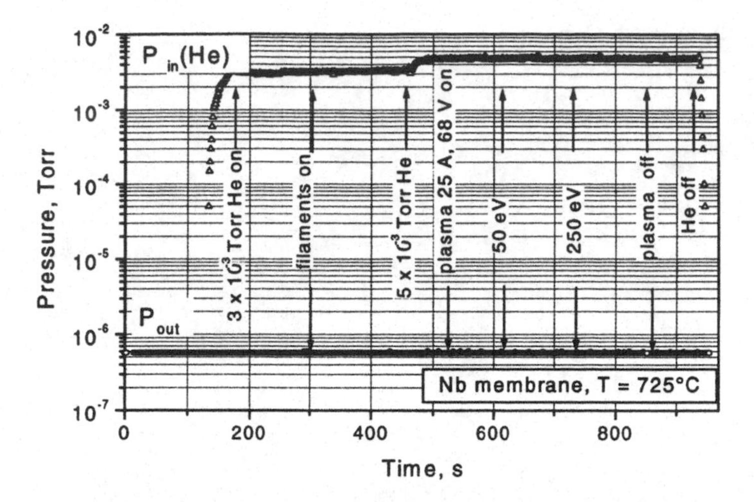

Figure 7. Interaction of pure helium plasma with the superpermeable niobium membrane

suppression of membrane permeability at membrane biasing, strongly depending on helium content in hydrogen / helium mixture, on membrane temperature, and on ion energy. In particular:

- the higher the membrane temperature, the smaller the effect of helium additive, coming up to the permeability independence of helium additive at membrane temperatures of 1420 K and higher,
- the lower the ion energy, the smaller the effect of helium additive, with the permeability not depending on helium additive at ion energies lower than the threshold energy of surface film sputtering,
- the higher the helium content in hydrogen / helium mixture, the stronger the membrane permeability suppression.

4. It was demonstrated that the niobium membrane superpermeable to hydrogen energetic particles was totally impermeable to any helium particles.

Acknowledgement. This work was supported by the Association Euratom – CEA under Contract CEA/V.3094.002 with EURATOM.

8. References

1. Livshits, A.I. (1979) Superpermeability of solid membranes and gas evacuation ,*Vacuum* **29**, 103 –113.
2. Livshits A.I., Notkin M.E. and Samartsev A.A. (1990) Physico-chemical origin of superpermeability – large-scale effects of surface chemistry on "hot" hydrogen permeation and absorption in metals, *J. Nucl. Mater.* **170**, 74-94.
3. Courteille C., Bruneteau A.M. and Bacal M. (1995) Investigation of a large volume negative hydrogen ion source, *Rev. Sci. Instrum.* **66**, 2533 - 2540.
4. Roth J., Bohdansky J., Ottenberger W. (1979) *Data on Low Energy Light Ion Sputtering*, Max-Plank-Institut fur Plasmaphysik, Report IPP 9/26 .

MEMBRANE BIAS EFFECTS ON PLASMA DRIVEN PERMEATION OF HYDROGEN THROUGH NIOBIUM MEMBRANE

Y. NAKAMURA[a], N. OHYABU[a], H. SUZUKI[a],
V. ALIMOV[b], A. BUSNYUK[b], A. LIVSHITS[b]

[a] *National Institute for Fusion Science, Oroshi, Toki 509-52, Japan*
[b] *Bonch-Bruevich University, 61 Moika, St.Petersburg 191065, Russia*

Niobium membrane was immersed in hydrogen plasma in order to study its interaction with hydrogen ions in the energy range from a few eV to 200 eV, controlled by changing the membrane bias. Degradation of plasma driven permeation (PDP) and enhancement of permeation driven by thermal molecules of cold gas (GDP) were observed when bias voltage was applied to the membrane exposed at plasma because of sputtering of non-metal impurity layer at upstream surface. Both dissolution of oxygen and increase of membrane temperature lead to decreasing of membrane bias effect due to rise of surface layer recovering rate.

1. Introduction

1.1. ROLE OF NONMETAL IMPURITIES

Nonmetal monolayer on the upstream membrane surface impedes dissolution of molecular hydrogen in membrane. In the same time, *suprathermal* hydrogen particles (whose energy exceeds ≈1eV) can pass this layer and be absorbed by the metal with very high probability (0.1 - 1) [1]. The non-metallic layer impedes, however, the back release of absorbed hydrogen particles through inhibiting their thermal recombination at the surface since the retained particles "cool down" quickly inside the metal lattice. The impediment to release can be made so great that the probability for a retained H atom to reach the opposite surface by interstitial diffusion becomes larger than the one to be desorbed into the gas at the upstream side. Absorbed hydrogen particles many times pass membrane before desorption and "forget" they initial position. As a result distribution of hydrogen over the membrane becomes almost uniform and absorbed hydrogen particles release to the both membrane sides proportionally to reemission constants. Permeation flux can be expressed in this case in terms of an incident flux J_i, absorption probability ξ of suprathermal particles, and rate constants of recombinative release at the membrane upstream and downstream sides ($k_{r\,u}$ and $k_{r\,d}$ correspondingly).

C.H. Wu (ed.), Hydrogen Recycling at Plasma Facing Materials, 167–175.

$$J_p = J_i \xi \frac{k_{rd}}{k_{ru} + k_{rd}} \tag{1}$$

Permeation probability χ_H in the steady state regime equals to the ratio of J_p and J_i. Taking in consideration that in the case of thermal desorption k_r is proportional to sticking coefficient into the state of dissociative adsorption of thermal H_2 molecules α, we can express χ_H as

$$\chi_H = \xi \frac{\alpha_d}{\alpha_u + \alpha_d} \tag{2}$$

If a membrane has symmetrical surfaces ($\alpha_u = \alpha_d$) half of absorbed particles release to the downstream side. If the desorption at the downstream surface is easier ($\alpha_u << \alpha_d$), then virtually all the particles retained will be released at the downstream side, i.e. $\chi \approx \xi$. Hence, the membrane becomes "transparent" for suprathermal particles. This phenomenon was called "*Superpermeability*" [2].

1.2. POSSIBLE APPLICATIONS IN FUSION AND UPSTREAM MEMBRANE SURFACE PROBLEMS

Due to these characteristics, superpermeable membranes could be employed in fusion machines for D/T mixture pumping and separation from impurities [3-5]. One of the possible applications of the superpermeable membranes was a suggestion to employ such membranes for pumping of the suprathermal hydrogen particles existing in the divertor region [5,6]. As it is expected it may result in a significant improvement of core plasma performance due to an effective evacuation of hydrogen from the divertor chamber. One of the critical points in application of superpermeable membranes in the divertor region is the possible modification of membrane upstream surface by energetic hydrogen particles that can lead to degradation of superpermeability.

Membrane experiments with atom beam [1,7] and hydrogen atomization on incandescent atomizer [3–5,8] show very stable superpermeation. This indicates that thermal atomic hydrogen has no effect on the membrane surface properties. In the same time there are many evidences that energetic hydrogen particles are able to modify membrane surface in such a way that permeation degrades. That is, first of all, the "permeation spike" phenomenon [9–11] when permeation flux drops drastically after its initial growth because of increase of hydrogen reemission coefficient in the course of sputtering of nonmetal impurity layer from the membrane upstream surface by hydrogen ions that has energy higher than threshold of impurity sputtering. Another kind of surface modification under bombardment with H_2^+ ions of 70 eV per proton energy in the presence of carbon impurity that leads to permeation flux degradation due to a decrease of hydrogen ions absorption probability ξ was observed in [12].

2. Experimental

2.1. EXPERIMENTAL DEVICE

A Plasma Membrane Test Device (PMTD) [13,14] was designed to model operational regimes of the membrane pumping systems in the divertor of plasma fusion machines. Main part of PMTD is a ultrahigh-vacuum chamber where a membrane and a plasma generator are installed. Niobium membrane manufactured in the form of a tube of 30 mm diameter, 150 mm length and 0.3 mm wall thickness is located in the center of vacuum chamber and separates upstream and downstream vacuum volumes. Both vacuum chambers are continuously pumped with turbo molecular pumps through the calibrated diaphragms that provides possibility to measure permeation flux and absorption/desorption fluxes at the both membrane sides. The membrane can be resistively heated with electric current.

The axis-symmetric hybrid-type plasma generator consists of an auxiliary hot-cathode duopigatrone plasma generator and a Penning cell placed inside transverse electric and magnetic fields. It produces plasma of cylindrical shape coaxial with the membrane with electron temperature of few eV, and electron density approximately 10^{10} cm^{-3}. Ion charge exchange process at cold hydrogen gas generates hydrogen suprathermal neutrals that may impinge on the membrane. One also can extract ions from the plasma directly onto the membrane by applying a corresponding potential. All these energetic hydrogen particles may be absorbed and pumped by the membrane. The permeation flux is measured by gas flow changes in the upstream and downstream chambers. The ion energy may be varied with membrane potential starting with few eV, and the ion flux density with plasma discharge current.

2.2 EXPERIMENTAL PROCEDURES

A typical permeation experiment is shone on the fig. 1. Admittance of 22 mTorr of H_2 into the upstream chamber results in the appearance of hydrogen in the downstream chamber due to GDP. Switching on the plasma discharge leads to additional increase of hydrogen pressure in downstream chamber due to PDP. Almost all permeation flux is pumped by TMP and only negligible part of it returns back to upstream side. Hence we can calculate the permeation flux J_p from the value of downstream pressure P_d as $J_p(H_2/s) = P_d(\text{Torr})\,\sigma_d(\text{cm}^3/\text{s})\,3.3\times10^{16}$, where σ_d–downstream volume pumping speed (26000 cm^3/s in this experiment). We observe quick drop of permeation flux to a

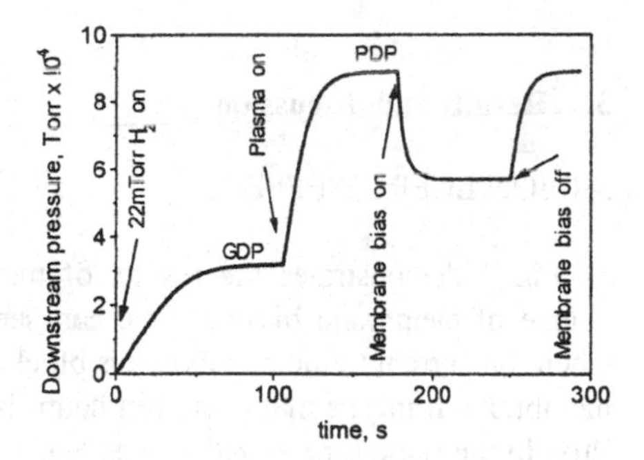

Fig. 1. Effects of switching on plasma and membrane bias on hydrogen pressure in downstream chamber. $T_M = 710^0$ C. Plasma discharge parameters: 3 A, 100 V. Membrane bias voltage is –304 V, ion current density is 0.7 mA/cm^2 during the membrane biasing.

lower steady state level at switching on of the membrane bias and return to previous level at switching off.

Total permeation flux can be expressed as a sum of two permeation fluxes – permeatin flux driven by suprathermal particles of plasma (plasma-driven permeation flux J_{PDP}) and permeatin flux driven by thermal molecules of cold gas (gas-driven permeation flux J_{GDP}). J_{GDP} makes up a significant part of total permeation flux in our experiment. Both GDP and PDP fluxes can be affected by the membrane biasing. Hence we have to measure GDP flux in the course of plasma experiment to extract it from the total permeation flux in order to find PDP flux. The following procedure of self compression of permeating flux in the downstream chamber have been used for measurement of GDP.

After the new value of permeation flux has been established in the course of the membrane biasing the pumping of downstream chamber by TMP was stopped by closing of a corresponding valve. Hydrogen accumulated in the downstream volume until new steady state downstream pressure (P_d^*) established. We can find permeation probability of hydrogen molecules χ_{H_2} from a condition of a balance of direct (J_p) and back (J_b) permeation fluxes ($J_p = J_b$).

$$\chi_{H2} = \frac{\sigma_d \, 3.3\times 10^{16}}{Z A} \frac{P_d}{P_d^*} \tag{3}$$

where A – membrane surface area (cm^2), $Z = (2\pi m k T)^{-1/2}$ is the kinetic theory coefficient (where m is the molecular mass and k is the Boltzmann's constant).

Knowing χ_{H_2} and P_u we can calculate J_{GDP} and hence J_{PDP}:

$$J_{PDP} = J_P - J_{MDP} = P_d \, \sigma_d \, 3.3\times 10^{16}\left(1 - \frac{P_u}{P_d^*}\right). \tag{4}$$

3. Results and discussion

3.1. ION ENERGY EFFECT

Fig. 2 demonstrates the results of measurements of PDP and GDP fluxes in the course of membrane biasing. One can see that there is no change of permeation flux when ion current to the membrane is blocked by applying of positive bias voltages to the membrane. It means that hydrogen neutrals make main contribution to the total incident flux. In the same time negative bias leads to a decrease of J_P. Both fluxes J_{PDP} and J_{GDP} calculated with formula (4) also presented in fig. 2. One can see that J_{PDP} drops significantly while J_{GDP} grows.

Let us consider the J_{PDP} and J_{GDP} fluxes separately in order to understand the physical cause of the phenomenon. One can get an expression for permeation probability of hydrogen molecules χ_{H_2} from the equation (2) taking into consideration that absorption probability ξ for molecules equals to sticking coefficient α_u:

$$\chi_{H_2} = \frac{\alpha_u \alpha_d}{\alpha_u + \alpha_d}. \tag{5}$$

According to equations (2) and (5) growth of α_u due to sputtering of non metal impurity layers from the membrane upstream surface should lead to a decrease of J_{PDP} and an increase of J_{GDP} as that was observed in the course of our experiment. The question was raised as to whether all the change of J_{PDP} can be explained only by the change of α_u.

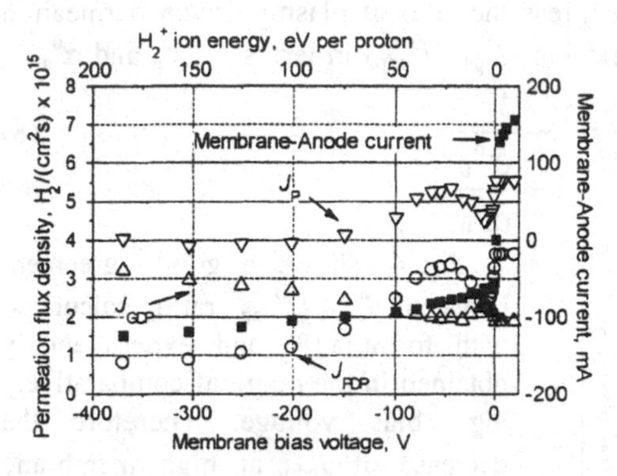

Fig. 2. Dependence of plasma driven permeation flux (J_{PDP}), permeation flux driven by molecule (J_{GDP}) and total permeation flux (J_p) membrane bias voltage. Membrane-anode current is also presented. Membrane temperature is 710^0 C. Plasma discharge current is 3A. Anode-cathode voltage is 100 V. Hydrogen upstream pressure is 22 mTorr.

Special desorption test showed that membrane is symmetric in the absence of biasing, i.e.

$$\alpha^0_u = \alpha^0_d = 2\chi^0_{H_2}, \tag{6}$$

where α^0_u, α^0_d and $\chi^0_{H_2}$ are α_u, α_d and χ_{H_2} in the absence of the membrane biasing.

The fact that incident ions do not influence on the membrane downstream surface gives a possibility to know the change of α_u from the change of χ_{H_2}, in view of Eqs. (5) and (6):

$$\alpha^b_u = \frac{\chi^0_{H_2} \chi^b_{H_2}}{\chi^0_{H_2} - \frac{1}{2}\chi^b_{H_2}}, \tag{7}$$

where α^b_u and $\chi^b_{H_2}$ are α_u and χ_{H_2} in the course of the membrane biasing.

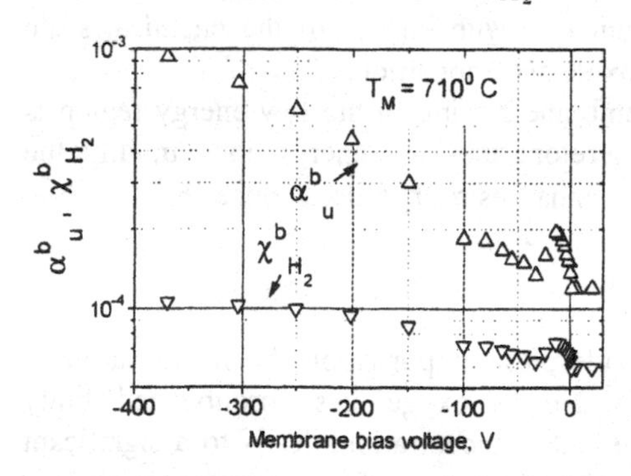

Fig. 3. Dependence of H_2 molecules sticking coefficient of the membrane inlet surface α^b_u and permeation probability $\chi^b_{H_2}$ on membrane bias voltage.

The results of calculations of α^b_u and $\chi^b_{H_2}$ with Eqs. (3) and (7) from measurements of P_d^* and P_d are shown on fig. 3.

Notice, that according to (5), maximum possible rise of $\chi^b_{H_2}$ due to an increase of α^b_u is twice in the case of an initially symmetrical membrane. This case is realized in the our experiment (see fig. 3.) when χ_{H_2} grows almost two times because of significant growth of α_u (almost one order of magnitude) due to sputtering of impurity layers by H_2^+ ions of energy 185 eV per proton.

Increase of α_u leads to redistribution of hydrogen desorption fluxes in the favour of release to the membrane upstream side and consequently to a decrease of J_{PDP}. Assuming that absorption probability of suprathermal hydrogen particles ξ remains constant at membrane biasing, one can express the ratio of plasma driven permeation fluxes with and without the membrane biasing ($J^b{}_{pdp}/J^0{}_{pdp}$) in terms of $\alpha^b{}_u$ and $\alpha^0{}_u$.

$$\frac{J^b{}_{PDP}}{J^0{}_{PDP}} = \frac{2}{1+\dfrac{\alpha^b{}_u}{\alpha^0{}_u}} \tag{8}$$

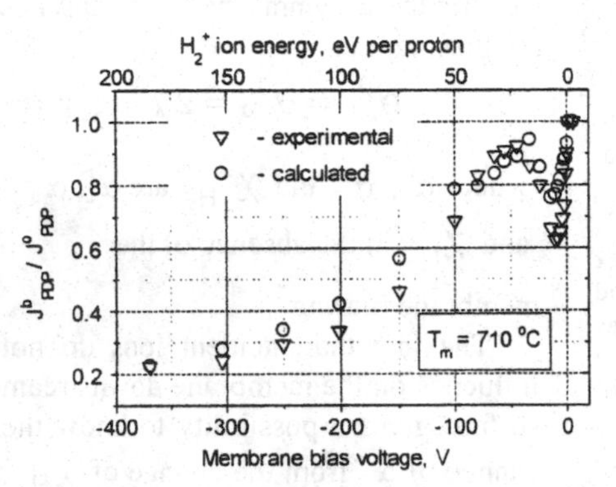

Fig. 4. Effect of membrane bias on PDP flux both experimentally obtained and calculated with formula (8).

Fig. 4. shows a good agreement between $J^b{}_{pdp}/J^0{}_{pdp}$ ratio calculated with formula (8) and experimentally obtained in the range of comparatively high bias voltage. Therefore the decrease of J_{pdp} at high membrane bias voltages where ion energy exceeds a threshold of physical sputtering of nonmetal impurities can be explained in the framework of a simple model given above.

Membrane bias effect on J_{PDP} and J_{GDP} in low voltage range that was observed previously in [15] is well pronounced on the figures 2-4. Physical origin of low energy peak is not clear. One can see that membrane bias effects on J_{PDP} and J_{GDP} start with very small negative voltages that corresponds to ion energy lower than threshold of physical sputtering. It is possible that this is connected with a process of chemical sputtering of oxygen impurity film from the upstream membrane surface. For instance it is believed that the hydrogen ions with energy of the chemical scale ($1 \div 10$ eV) induce the reaction of niobium oxide decomposition.

Notice (see. fig.4), that influence of membrane biasing in the low energy region is stronger that it follows from Eq. (8). Therefore the low energy ions modify the membrane surface in such a way that not only enhances α_u but also reduces ξ.

3.2. EFFECT OF OXYGEN DISSOLUTION

Oxygen along with carbon is a most widespread impurity on the metal surfaces. Dissolved in the bulk of V-a group metals (V, Nb, Ta) oxygen has a pronounced ability to segregate onto surface. Even very few its bulk concentrations leads to a significant surface coverage [16]. Most likely that it is oxygen surface segregation that is responsible for restoring nonmetal layers sputtered by incident ions. That is why the investigation of influence of dissolved oxygen on the PDP flux behavior under the action of the membrane biasing is important.

Dissolution of oxygen have been carried out at 800^0 C membrane temperature and 2×10^{-4} O_2 pressure in the upstream chamber. Amount of dissolved oxygen was proportional to O_2 pressure drop during dissolution (when the membrane heating was switched on), upstream chamber pumping speed and procedure time.

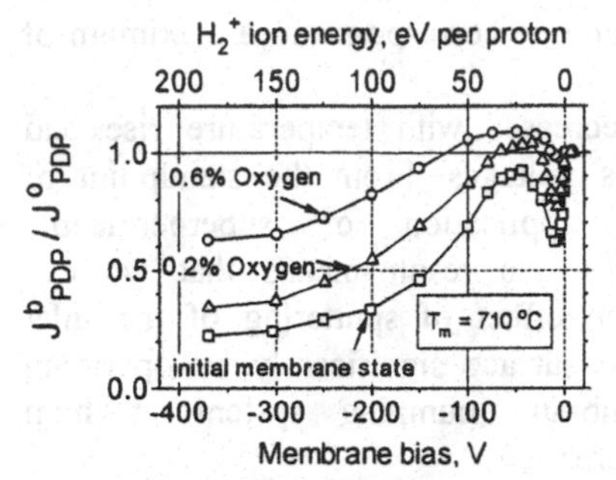

Fig. 5. Dependence of plasma driven permeation flux J^h_{pdp} (normalized to a flux at membrane floating potential J^0_{pdp}) on membrane bias voltage at different contents of dissolved oxygen.

The two portion of oxygen (0.2 %(at.) and 0.4%(at.)) were dissolved in the membrane one after another. One can see on the fig. 5 that the sputtering effect reduces with the increase of the dissolved oxygen content. The possible explanation of the observed effect is following.

The change of the surface oxygen coverage (θ_O) that is responsible for membrane bias effects on GDP and PDP establishes in the course of counterbalancing of the two competing processes – oxygen layers removing from the surface by ion sputtering and its recovering due to segregation onto the surface from the bulk of the membrane.

Surface coverage θ_O in the absence of sputtering can be described by the Langmuir-McLean isotherm [17]

$$\frac{\theta_O}{1-\theta_O} = K_O c_O \exp\left(\frac{\Delta H}{RT}\right), \tag{9}$$

where K_O is a constant, c_O is oxygen bulk concentration and ΔH is the difference between the enthalpy of solution and the enthalpy of adsorption of oxygen. For oxygen in tantalum it was measured in [15] that $K_O = 0.18$ and $\Delta H = 38.9$ kJ mol^{-1} in the temperature range $1100\div1800^0$C. Extrapolating to the membrane operation temperature (700^0 C) formula (9) gives surface coverage 0.81 and 0.93 at c_O 0.2%(at.) and 0.6%(at.) correspondingly. Thus surface coverage is almost saturated at oxygen concentrations and temperatures that we have in our membrane experiment. Taking into consideration that all V-a group metals (V, Nb, Ta) have very similar chemical properties [18] we can suppose that θ_O on our niobium membrane is also close to saturation. Therefore sputtering rate that is proportional to θ_O depends only weakly on the oxygen bulk content. In the same time the surface layer recovering rate is limited by oxygen atoms diffusion in the metal lattice and hence is proportional to c_O. So the resulting sputtering effect must decrease with the increase of the bulk oxygen content, as it was observed in the experiment.

3.3. MEMBRANE TEMPERATURE EFFECTS

The membrane temperature dependence of the membrane bias effect on PDP and GDP was investigated at two bias voltages -200 V and -12 V. The first voltage corresponds to H_2^+ ion energy 100 eV that is higher than threshold of the process of physical sputtering of impurity layer and the second one corresponds to a maximum of low-energy pick.

Fig. 6 demonstrates that sputtering effect decreases with temperature rise and vanishes at 800° C at the both investigated bias voltages. From the standpoint of practical application of superpermeable membranes this result means that we can reduce the effect of sputtering of the inlet membrane surface significantly by operating the membrane pumping system at high temperatures.

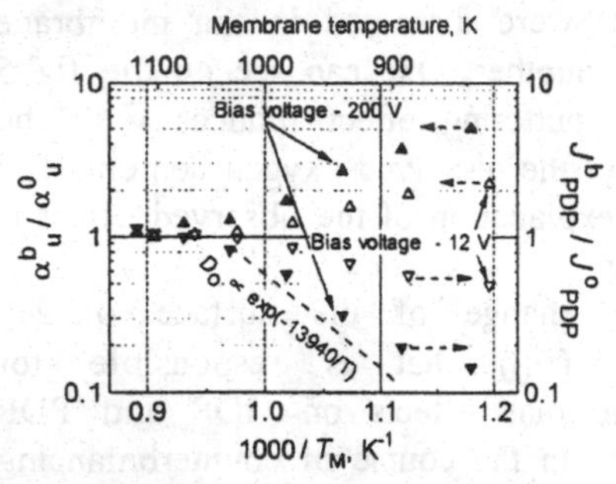

Fig.6. Dependence of relative changes of plasma driven permeation flux J^b_{pdp} (down triangle symbol) and H_2 sticking coefficient at upstream membrane surface (up triangle symbol) at two membrane bias voltages -12 V (open symbol) and -200 V (closed symbol) on the membrane temperature. 0.2%(at.) of oxygen was dissolved in the membrane. Temperature dependence of diffusion coefficient of oxygen atoms in niobium lattice D_O is also presented (dashed line).

Taking into consideration that physical sputtering process does not depend on the surface temperature we can ascribe temperature dependence of bias effect at -200 V on PDP and GDP to the temperature dependence of the surface recovering process. One can see that angle of slope of Arrenius plot for J^b_{PDP}/J^0_{PDP} in this case is very close to the angle of slope of Arrenius plot for a diffusion coefficient of oxygen in niobium lattice (D_O) [18]. It means that oxygen diffusion from the membrane bulk to the surface is a rate-determining factor of the recovering process. Different slope angle in the case of U_{MB} = -12 V probably indicates on temperature dependence of chemical sputtering process.

Conclusions

(1) Applying negative bias to the membrane exposed in plasma leads to an increase of GDP and decrees of PDP fluxes due to rise of a rate constant of thermal absorption/desorption process under the sputtering of impurity layers at the membrane upstream surface by extracted hydrogen ions.

(2) There are two zones were observed in the dependence of membrane bias effect on the bias voltage. High voltage branch starts with several tens volt and corresponds to physical sputtering of nonmetal impurity layer. Low-voltage zone has a maximum at approximately -12 V and probably corresponds to process of chemical sputtering of non-metal impurity layer.

(3) Both oxygen dissolution in the membrane and increase of membrane temperature lead to a decrease of the membrane bias effects due to increase of surface impurity recovering rate.

References

1. A. I. Livshits, M. E. Notkin and A. A. Samartsev, *J. Nucl. Mater.* 170 (1990) 79-94.
2. A. I. Livshits, Sov. Phys. Tech. Phys. 21 (1976) 187.
3. A. I. Livshits, M. E. Notkin, A.A. Samartsev, A.O. Busnyuk, A.Yu. Doroshin and V.I. Pistunovich. *J. Nucl. Mater.* 196-198 (1992) 159-163.
4. A. I Livshits, M. E. Notkin, A. A. Samartsev, M. Bacal and A. O. Busnyuk, *J.Nucl.Mater.* 220–222 (1995) 259.
5. A. I. Livshits, N. Ohyabu, M. E. Notkin, V. N. Alimov, H. Suzuki, A. A. Samartsev, M. N. Solovyov, I. P. Grigoriadi, A. A. Glebovsky, A. O. Busnyuk, A. Yu. Doroshin and K. Komatsu, *J. Nucl. Mater.* 241–243 (1997) 1203.
6. N. Ohyabu, A. Komory, K. Asashi et al, *J. Nucl. Mater.* 220–222 (1995) 298.
7. A. I. Livshits, M. E. Notkin, Yu. M. Pustovoit and A. A. Samartsev, *Vacuum* 29 (1979) 113.
8. A. Yu. Doroshin, A. I. Livshits and A. A. Samartsev, *Phys.Chem.Mech.Surf.* 4 (1987) 2321.
9. M. Yamawaki, K. Yamaguchi, T. Kiyoshi and T. Namba, *J. Nucl. Mater.* 145-147 (1987) 309.
10. M. Yamawaki, T. Namba, *J. Nucl. Mater.* 133-134 (1985) 292.
11. A.I. Livshits, M.E. Notkin, A.A. Samartsev and I.P. Grigoriadi, *J. Nucl. Mater.* 178 (1991) 1.
12. A. I. Livshits, M. E. Notkin, A. A. Samartsev and M. N. Solovyov, *J.Nucl.Mater.* 233-237 (1996) 1113.
13. A. I. Livshits, N. Ohyabu, M. Bacal, Y. Nakamura, A. O. Busnyuk, M. E. Notkin, V. N. Alimov, A. A. Samartsev, H. Suzuki and F. Sube, *J.Nucl.Mater.* 266-269 (1999) 1267-1272.
14. Y. Nakamura, N. Ohyabu, A. I. Livshits, V. N. Alimov. A. O. Busnyuk and I. P. Grigoriadi, *International Workshop on Hydrogen Recycle at Plasma Facing Materials, Tokyo, October 15–16, 1998*; published in the Workshop Proceedings.
15. A. I. Livshits, F. Sube, M. N. Solovyev, M. E. Notkin and M. Bacal, *J.Appl.Phys.* 84 (1998) 2558.
16. G. Horz, H. Kanbach and H. Vetter, *Mat. Sci. and Eng.* 42 (1980) 145.
17. J. M. Blakeley and J. C. Shelton, in J. M. Blakeley (ed.), Surface Physics of Materials, Vol. 1, Academic Press, New York, 1975.
18. E. Fromm and E. Gebhardt, eds., Gase und Kohlenstoff in Metallen (Springer, Berlin, 1976).

PLASMA DRIVEN PERMEATION THROUGH THE NB MEMBRANE AT LOW TEMPERATURE

A.A. SKOVORODA, Yu.M. PUSTOVOIT, V.S. SVISHCHOV, *V.S. KULINAUSKAS

*INF RRC Kurchatov Institute, *INP Moscow University,*
123182, Kurchatov Square 1, Moscow, Russia

The CW plasma driven superpermeation finds out presence of threshold temperature ~400°C. This phenomenon contacts with the percolation phase transformation on crystallite defects and with β→α phase transformation of diluted hydrogen in Nb.

1. Introduction

In Plasma Neutraliser for ITER NB injection system the metallic membrane pumping is planed to use [1]. It is known, that a metallic membrane of macroscopic thickness may be superpermeable to the hydrogen atoms and ions whose energy exceeds ~1eV [2]. The unique opportunities, which were opened by this phenomenon in fusion devices, have determined the large interest to research of the energetic hydrogen permeation through membranes [3-6]. The Nb membranes showed the best results in experiments.
However membrane pumping is not widely used now. In what is the reason of such situation? The existing explanation of the superpermeation phenomena is based on the determinative influence of the surface potential barriers on the absorption and permeation in *uniform* metals. In the case of energetic particles the surface barrier has no effect on the implantation process. But the surface barrier determines the back thermal desorption, that can be described by Sievert's law $J = KC^2$ (here J desorbed molecular flux, C concentration of gas atoms in metal, K recombination coefficient). Entering the dimensionless parameters $\gamma = K_L / K_0$ and $\beta = D / L\sqrt{JK_0}$ (here D diffusion coefficient, L membrane thickness, J total flux, $K_{0,L}$ recombination coefficients on entry and outlet sides of membrane), it is possible to show, that the permeation efficiency η (ratio of the permeation flux to the total flux) is equal

$$\eta = \frac{\gamma}{1+\gamma} \tag{1}$$

at the condition β>>1 [2,7]. For Nb membranes with L<0.1mm and J<10 eq.ma/cm^2 this condition is carried out at temperature T since 25 °C (we use the numerical quantity of required parameters from [8]). The relation (1) shows, that superpermeation (η>0.1)

C.H. Wu (ed.), Hydrogen Recycling at Plasma Facing Materials, 177–183.

should be realised in wide temperature range at $\gamma \geq 1$, i.e. at a more clean outlet surface of a membrane. The UHV (10^{-8}-10^{-10} torr) experiments with atomic hydrogen (energy >1eV) fortified this conclusion [6]. The temperature independence of $\eta \sim 0.3$ was experimentally demonstrated in Pd at T since 25 ^{0}C. But in real vacuum conditions (the residual gas pressure 10^{-6}-10^{-7} torr) the superpermeation of hydrogen atoms is observed usually at high $T \sim 600$-900 ^{0}C. Moreover, the complex uncontrollable gas composition in real plasma devices makes practically impossible the long time control of the membrane surface quality that is necessary for the superpermeability. Nevertheless, the different schemes of atomic hydrogen pumping by membranes are now under R&D [6].
The atomic hydrogen has low energy to change the entry surface properties. The situation is different for hydrogen plasma ions with energies 10-100eV. They can effectively modify the entry surface. The plasma bombardment leads usually to entry surface cleaning, i.e. $\gamma \rightarrow 0$, and the superpermeation should disappear. But plasma driven superpermeation is observed in experiments with Nb membranes [9,10].
The experiments [9,10] were carried out at large hydrogen pressure (10^{-1}-10^{-3} torr), when the atom stream is much higher as the ion stream. In [9] the large permeation was obtained at temperature since 250^{0}C. In [10] the permeation was investigated at T=800^{0}C.
In present work the plasma driven superpermeation in plane NB membrane at the determinative role of ion stream is investigated. The real vacuum conditions are realised.

2. Experimental installation

Two types of the experiments with the plasma flows (1-10 ma/cm^{2}) on the Nb plane membrane (22 µm, disc, S_m = 25 cm^{2}) in H_2 pressure 1÷6 10^{-4} torr at membrane temperature 20-700^{0}C were carried out: (1) "plasma driven permeation" experiment and (2) "Nb membrane properties" experiment. The used experimental set-up is shown on *Figure 1.* The different investigative boxes (I) (see *Figures 1.* and 2.) were applied.

2.1. PLASMA DRIVEN PERMEATION EXPERIMENT

The investigative box (I) for this type of experiments is shown on *Figure 1.* Two metallic flanges 1 with the insulating coatings 2 clamp the Nb membrane 3. The nonannealing Al_2O_3 paste was used for insulating. After the membrane warming up to 800^{0}C the vacuum-tight composition is obtained. The box (I), the heating system and all diagnostics are floating-potential. We can change the box potential in the range ±400V.
The membrane temperature is measured by three thermocouples mounted in the centre, middle and border of the membrane outlet surface and by the distance thermometer through the window (O). The membrane *area* temperature is changed in the range ±50^{0}C and the mean temperature is shown on all figures. The membrane is heated by external current and by plasma current. The external current can be changed in time linear by different speed.
For plasma flow production the precise DECR (Distributed ECR) source is used. The main parameters of this source are shown in TABLE 1. The source plasma parameters

(electron temperature, T_e, electron density, n_e, plasma potential, φ) are measured by movable Langmuir probe (L). The ion flow density on the membrane is measured by plane probe 5, which is placed close to membrane on box (I).

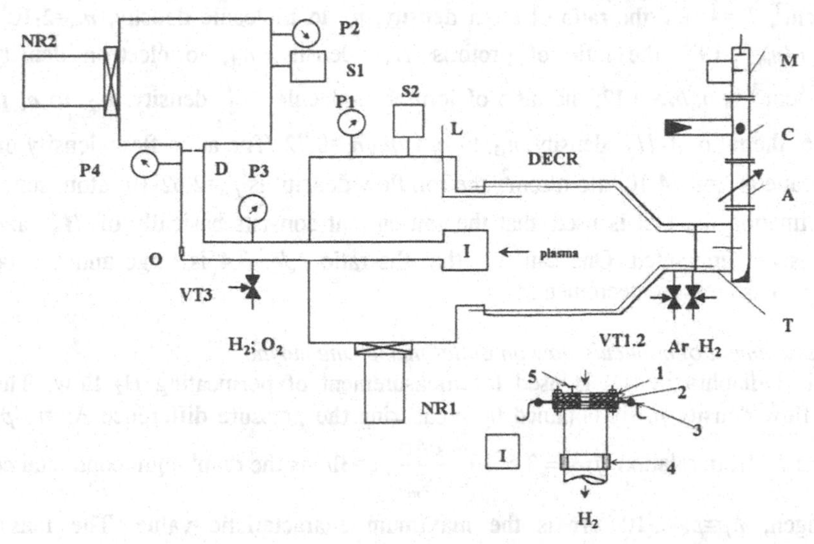

Figure 1. Schematic of the plasma-membrane set-up. DECR distributed electron cyclotron resonance source: (M) CW magnetron, (C) circulator, (A) attenuator, (T) transformer & divider, (L) Langmuir probe; Vacuum System: (NR) turbo-molecular pump, (P) vacuum-meter, (S) mass-spectrometer, (D) metrical diaphragm, (VT) gas controlling valve, (O) window; (I) permeation investigative box: 1 stainless steel flange, 2 thermostable insulator, 3 Nb membrane, 4 insulator, 5 probe, arrow shows plasma flow.

TABLE 1. DECR source parameters

Parameter	Value
Maximum flow density, ma/cm^2	10
Flow energy, eV	>10
Flow size (diameter),cm	15
Flow inhomogeneity, %	2
Flow time stability, %/hour	5
Flow fluctuations,%	5
Magnetron frequency, GHz	2.45
Magnetron power, kW	1
Number of waveguide electrodes	8
Type of permanent magnets	NdFeB
Gas pressure, torr	10^{-5}-10^{-3}
Ultimate vacuum, torr	10^{-7}
Vacuum chamber	without warming

All measurements were carried out with the continuous gas puffing and pumping. Bottled Ar and hydrogen (H_2) gases are used without special cleaning.

2.1.1. *Determination of atom & ions flows on entry membrane surface*

The previously developed and proved particles balance model for DECR plasma source was applied to determine the atoms and ions flows to the membrane. The experimentally measured values of gas pressure, p, probe current density, $j(V)$, and T_e are employed by code OGRAS to determine the plasma composition and particles streams. For an illustration we shall result a characteristic example. In hydrogen DECR discharge with parameters p=5.5·10^{-4}torr, ion current density at floating potential j_i=1.7ma/cm^2, T_e=4.1eV the ratio of atom density, n_a, to molecule density, n_m=2·10^{13} cm^{-3}, is n_a/n_m=3·10^{-3}; the ratio of protons H_1^+ density, n_{i1}, to electron density n_e=1.3·10^{10}cm^{-3} is n_{i1}/n_e=0.12; the ratio of ionised molecule H_2^+ density, n_{i2}, to n_e is n_{i2}/n_e=0.16; the ratio of H_3^+ density, n_{i3}, to n_e is n_{i3}/n_e=0.72. The atom flow density on the membrane is j_a=0.74·10^{16}atom/scm^2, the ion flow density is j_{ia}=2.52·10^{16}atom/scm^2. In last estimation the fact is used, that the ion current consists basically of H_3^+ and three atoms are implanted. One can see, that the ratio j_{ia}/j_a=3.4 is large and can be increased by gas pressure decrement.

2.1.2. *Measurement of molecule flow on outlet membrane surface*
The metrical diaphragm (D) is used for measurement of permeating H_2 flow. The molecule flow density j_{H2} is obtained by measuring the pressure difference $\Delta p=p_4-p_2$ (see *Figure 1.*) from relation $j_{H2} = 3.5 \cdot 10^{19} \dfrac{c\Delta p}{S_m}$, c=3l/s is the diaphragm conductance for hydrogen, $\Delta p \cong p_4$~2·10^{-3}torr is the maximum characteristic value. The mass-spectrometers (S) are used to control, that the pressure increment is only due to hydrogen.

2.1.3. *Efficiency of permeation*
The permeation efficiency,η, is determined by relation (2). Here 2 are due to two atoms in molecule.

$$\eta = \frac{2 j_{H2}}{j_a + j_{ia}}. \qquad (2)$$

2.1.4. *Nb membrane conditioning*
Two steps conditioning was used. (1). At the first stage of experiments the original Nb membrane (99,9% in purity and 25μm in thickness) was cleaned in acetone and was annealed in vacuum at 800^{0}C for about 10 hours. But the largest value of permeation efficiency was small, η~4%, and can be realised only at large temperature, T>700^{0}C. The paladization of the outlet membrane surface (1μm) has resulted only in some reduction of threshold temperature.
(2). The superpermeation, η~25%, at low temperature, T≥300^{0}C, was obtained only after intensive argon plasma cleaning of membrane both sides. The membrane was placed in DECR plasma wholly and on it was sent negative potential ~-300V. The Ar ion current density ~5ma/cm^2 was used for about 5 h. The spattering has resulted in reduction of membrane thickness up to 22μm. After such conditioning the outlet surface can be covered with 1μm layer of sputtered Pd (such membrane is marked as Nb/Pd).

2.2. NB MEMBRANE PROPERTIES EXPERIMENT

The another investigative box (I) was applied (see *Figure 2.*). The Nb membrane sample 2 ($8\times70mm^2$) was used for measuring resistance, *R*, and acoustic resonance at different temperature and plasma flow. The longitudinal acoustic resonance in membrane at the frequency $f\sim19kHz$ is investigated by means of the piezoceramics exciter and receiver 4. The temperature dependence of ratio $\Delta f/f$ (resonance width, Δf, and resonance frequency, f) is obtained. The membrane temperature is measured by thermocouple.

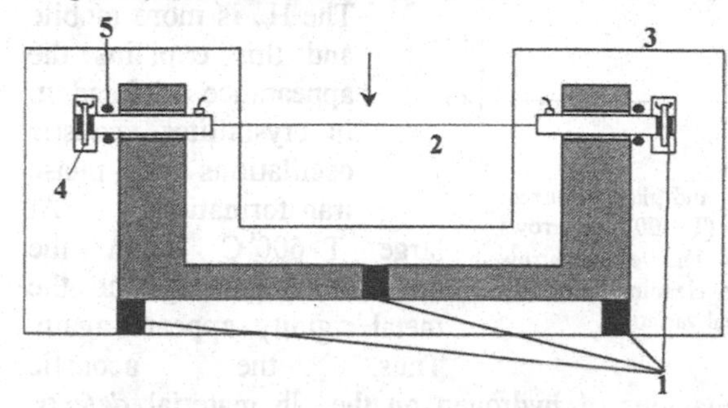

Figure 2. Schematic of the membrane properties investigative box (I): 1 insulators, 2 Nb membrane sample, 3 water cooled shield, 4 piezoceramics, 5 spring.

3. Experimental results

The steady state permeation at the arbitrary low membrane temperature ~400C^0 with ~25% efficiency after the Ar plasma conditioning (see 2.1.4) is obtained. The T dependencies of permeation efficiency, η, for two types of Nb membranes are shown on *Figure 3.* The strong temperature dependence of the permeation at low temperature is detected. The threshold temperature T_t~400°C can be slowly changed. The measured η is independent on the ion current density and on the hydrogen pressure in the outlet surface volume. The potential dependence is insignificant. *Figure 4.* shows, that at potential decrement the permeation flux increases, at first, approximately proportional to the ion flow increment. But at following potential increment the permeation flux falls up to the smaller values. The transit Ar cleaning of *entry* surface repairs the previous values.

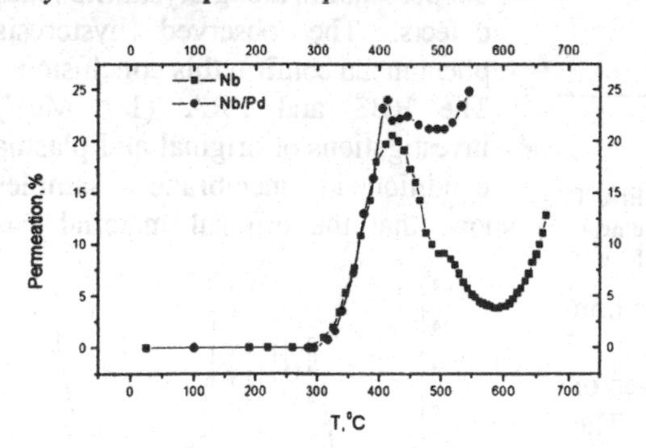

Figure 3. Permeation efficiency on temperature (floating potential).

At change of a direction of the temperature variation the *hysteresis* is observed (see *Figure 5).* The measurements of Nb sample properties show, that the resistance fluently increases on T increment and slowly depends on a plasma radiation. Another behaviour is developed for the $\Delta f/f(T)$ dependence (see *Figure 6.*) The step-like function is observed at critical temperature T_c=180÷200°C.

4. Discussion

The obtained behaviour of $\Delta f/f(T)$ function is connected with the hydrogen phase transformation in Nb material [11]. The measured T_c characterises the known β→α

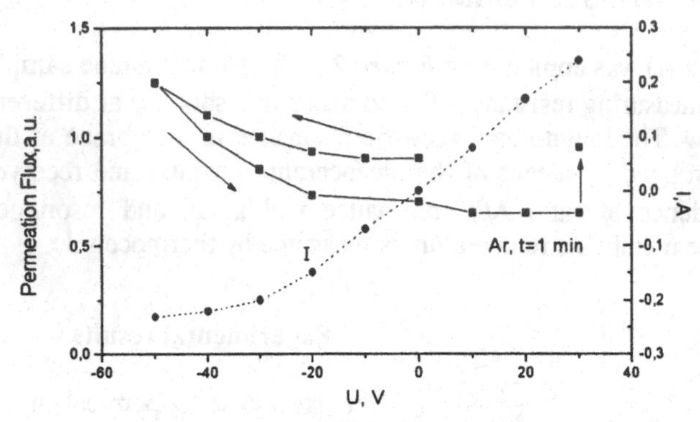

Figure 4. Permeation flux (solid line) and plasma current (dashed line) on membrane potential (T=400°C). Arrows show the order of potential variation. The vertical arrow shows the effect of 1 minute Ar cleaning of entry membrane surface after 1 hour potential variation.

phase transformation of diluted hydrogen in Nb. In [12] experimentally was shown, that H_β is concentrated on the *planes of crystallites*, that form a polycrystal membrane.

The H_α is more mobile and this explains the appearance of freedom in crystallites acoustic oscillations at phase transformation. At large T>600°C begin the intensive H gassing and the metal rigidity appears again. Thus, the acoustic measurements are explained by behaviour of hydrogen on the Nb material *defects*, namely, on the crystallite planes.

The defects play a determining role in an explanation of observed H permeability at arbitrary low temperature. The crystallites in fine membrane are textured. Hence, after the $\beta\rightarrow\alpha$ phase transformation the another *percolation* phase transformation can appear. The threshold temperature T_t~400°C on *Figure3.* characterises the percolation along crystallite sides defects. The observed hysteresis phenomena confirm this conclusion.

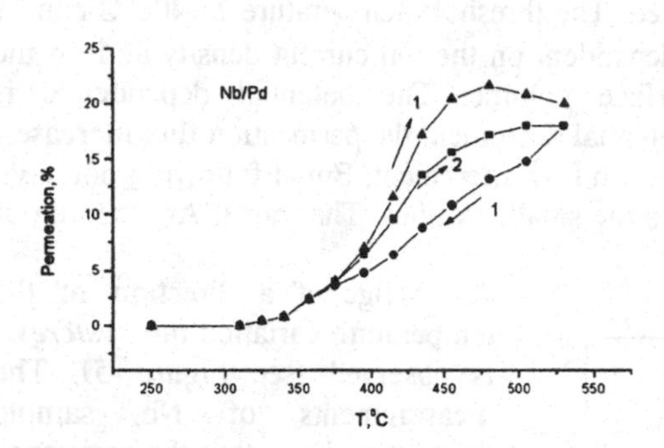

Figure 5. Hysteresis of permeation: 1 first T increment / decrement, 2 second T increment. Arrows show direction of changing in an hour.

The RBS and ERA (1.7 MeV) investigations of original and plasma conditioned membrane samples show, that the original material has 1μm surface layer with the composition 58% Nb, 9% O, 33% C after first step of conditioning. The second, Ar plasma step of conditioning makes away this layer. The quantity of hydrogen in samples, obtained from membrane used in permeation experiment, is very small (~2%).

The electron microscope pictures show the clean membrane surfaces. Only on the clean surfaces the superpermeation

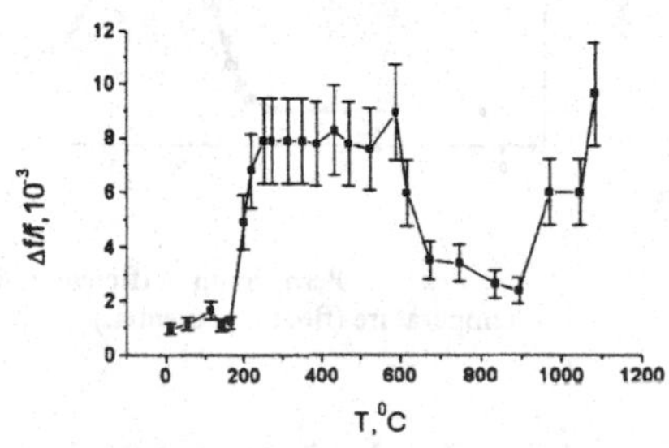

Figure 6. Acoustic resonance width on T.

was observed. Moreover, the transit Ar cleaning of entry surface rebuilds permeation (see *Figure 4.*). All these facts are in contradiction with traditional explanation of superpermeation, but agree with percolation predictions. In a similar way it is possible to explain experiments [9].

5. Conclusion

The traditional (without defects) picture of superpermeation is justified only to large temperature range, T>700^{0}C. At low temperature the volume defects in material play important role and the superpermeation can be realised for CW plasma streams. Percolation theory is more productive for observed phenomena explanation.
Acknowledgements. We thank V. Kurnaev, A. Pisarev and V. Somenkov for helpful discussions and P. Kosarev for help in measurements.

References

1. Kulygin, V.M., Skovoroda, A.A., Zhil'tsov, V.A.: Plasma Neutraliser for ITER NBI, *Plasma Devices and Operations* **6** (1998), 135-147.
2. Livshits, A.I.: Interaction of partitions with nonequilibrum gases, Sov. *J. Tech. Phys.* **20** (1975), 1207-1212.
3. Livshits, A.I., Pustovoit, Yu.M., Svishchov, V.S.: Membrane superpermeation and fusion problems, *Voprosy Atomnoi Nauki i Tekhniki, Ser.: Termoyad. Sintez (Topics in Atomic Science and Technology, Ser.: Thermonuclear Fusion, in Russian)* **2(8)** (1981), 45-49.
4. Livshits, A.I., Notkin, M.E., Pustovoit, Yu.M., Yakovlev, S.V.: Niobium membrane superpermeation by hydrogen atoms and ions in energy range 2-4000eV, *Voprosy Atomnoi Nauki i Tekhniki, Ser.: Termoyad. Sintez (Topics in Atomic Science and Technology, Ser.: Thermonuclear Fusion, in Russian)* **2(10)** (1982), 77-79.
5. See *Proceedings of Japan-CIS Workshops on Interactions of Fuel Particles with Fusion Materials* (every year since 1992).
6. Livshits, A.I., Obyabu, N., Nakamura, Y.: Absorption and permeation of energetic hydrogen particles in the group Va metals: surface effects and applications in fusion devices, *Proc. of 2nd Japan – Russia Symp. on Interaction of Fast Charged Particles with Solids*, Nagoya (1998), 341-355.
7. Pisarev, A.A., Smirnov, V.M.: Permeation of ions and atoms through membrane at different temperatures, *Atomnaya Energiya (Atomic Energy, in Russian)* **61** (1986), 178-182.
8. Bandourko, V., Yamawaki, M., Yamaguchi, K., Kurnaev, V., Levchuk, D., Pisarev, A.: Deuterium permeation through Nb during low energy ion irradiation at controlled surface conditions, *J. of Nuclear Materials* **233-237** (1996), 1184-1188.
9. Fujii, Y., Ishizuka, H., Nakano, H., Okamoto, M.: Hydrogen permeation enhanced by electrons, *Proceedings of Japan-CIS Workshops on Interactions of Fuel Particles with Fusion Materials*, Tokyo (1992), 58-67.
10. Bacal, M., Livshits, A., Notkin, M., Solovyov, M.: Plasma driven superpermeation trough V-group metals: first observations and characteristic properties, *Proceedings of Japan-CIS Workshops on Interactions of Fuel Particles with Fusion Materials*, Moscow (1997), 24-40.
11. Zacharova, G.V., Popov, I.A. et al.: *Niobiy i ego splavi (Niobium and its alloys, in Russian)*, GNTIL chernoi i cvetnoi metallurgii, Moscow, 1961.
12. Somenkov, V., Schilschtein, S.: Hydrogen phase transformations in metals, *Review IAE Kurchatov Institute, in Russian*, Moscow (1978), 1-81

was observed. Moreover, [illegible] surface reliefs [illegible] (see Figure 4). All these facts are in contradiction with traditional explanation of superpermeation, but agree with percolation predictions. In a similar way it is possible to explain experiments [9].

5. Conclusion

The traditional (without defects) picture of superpermeation is justified only for large temperatures (e.g., T>700 °C). At low temperatures [illegible] defects in material play important role and the superpermeation can be realized for CW [illegible] and strongly [illegible] more productive for [illegible]

[illegible] We thank V. Kurnaev, V. [illegible] and [illegible] for helpful discussions and P. Kassam [illegible] for help [illegible]

References

1. Kolyan, V.A., [illegible], A.A., Zhilisov, V.A. [illegible] Plasma [illegible] 6 [illegible] 135-147.
2. Livshits, A.I. Interaction of [illegible] with [illegible] J. Nucl. Mater. [illegible] 1301-1325.
3. Livshits, A.I., Notkin, [illegible] V.S. Membrane superpermeation and [illegible] (1986?) 43-47.
4. Livshits, A.I., [illegible], V.H., Pisarev, [illegible] Niobium membrane superpermeation by [illegible] Fusion in Russia [illegible]
5. See Proceedings of [illegible] Conference [illegible] of [illegible] Particles with Fusion Materials (every year since [illegible])
6. Livshits, A.I., Ohno, [illegible] Absorption and [illegible] of [illegible] particles in [illegible] surface effects and applications in fusion devices [illegible] (1993) 341-3.
7. Pisarev, A.A., [illegible] Permeation of [illegible] through [illegible] at different temperatures [illegible] (1986) 175-182.
8. [illegible], V., [illegible], M., [illegible], K., [illegible], V., [illegible] Deuterium permeation [illegible] during low [illegible] implantation at controlled surface conditions, J. of Nuclear Materials [illegible]
9. [illegible] enhanced by electron, [illegible] Interaction of [illegible] Tokyo (1992) 586-[illegible]
10. [illegible], Solovyev, [illegible] superpermeation [illegible] first observations [illegible] properties [illegible] Interactions of Fast Particles with Fusion Materials, Moscow (1993) [illegible]
11. Zakharova, G.V., [illegible] (1961).
12. [illegible], V., Schlichting?, [illegible] in metals [illegible] Moscow (1978) [illegible]

PHENOMENOLOGY MODEL OF HYDROGEN EVACUATION BY METAL MEMBRANES

A.V. PERESLAVTSEV, YU.M. PUSTOVOIT, V.S. SVISCHOV
RRC "Kurchatov Institute", Moscow, Russia

The physical model of the hydrogen diffusion through metal membranes is offered. The solutions of the diffusion equation with square law boundary conditions are resulted. The solutions of the diffusion equation are considered in stationary case for one layer metal membrane, for two layer metal membrane for metal membrane at plasma presence. The possibilities to create the hydrogen pumps on the base of metal membranes are discussed. It is shown the offered physical model gives the quantitative description of diffusion processes in the system " hydrogen - metal membrane "

The problem of hydrogen evacuation by the metal membranes is being discussed last decades [1-4]. There are no satisfactory descriptions of the membrane pumping speed dependence behaviour on pressure in a wide pressure range. One can use the phenomenology [5] of absorption-desorption processes in the system hydrogen -- nonevaporated getter for the description of absorption-desorption processes in the system hydrogen - metal membrane.

We shall not consider the adsorption and desorption kinetics at an atomic or molecular level, we shall be restricted to the consideration of the diffusion equation with some initial and boundary conditions. There are the difficulties of diffusion equation solving because of the nonlinearity of the boundary conditions. It is connected with the nonlinearity of a gas flux desorbed by a surface. Approximately, it is possible to describe the diatomic gas flux by the square law $q_{out} = \beta c^2$, where c is the surface concentration of gas atoms in metal, and β is the desorption coefficient. Such a model of the dehydridisation of metals was offered by Sieverts for a wide range of hydrogen concentration. According to the Sieverts law, the equilibrium gas pressure near the metal surface is proportional to the square of atomic gas concentration in metal $p = K \cdot c^2$, where K is the equilibrium constant, $K = K_0 \cdot exp(-E_s/RT)$, E_s is the activation energy of solubility. One can determine $\beta = \alpha K/\sqrt{2\pi mkT}$.

The gas flux adsorbed by a surface can be calculated by the formula $q_{in} = \alpha p/\sqrt{2\pi kmT} = \alpha n\bar{v}/4$, where p and T are the pressure and the temperature of the gas, m is the gas molecule mass, n is the gas density, $\bar{v}$ is the average molecular speed, $\alpha < 1$ is the sticking coefficient taking into account the existence of a surface energy barrier limiting the adsorption of molecules by a surface:

C.H. Wu (ed.), Hydrogen Recycling at Plasma Facing Materials, 185–194.

$\alpha = \alpha_0 \cdot exp(-E_a/RT)$, E_a is the activation energy for the adsorption of molecules. In a common case, the gas flux q_{in} adsorbed by a surface, consists of the various parts taking into account in particular the interaction of atoms and ions with metal. Atoms and ions have own "sticking coefficients", and these coefficients differ from the molecular ones.

1. ONE-LAYER MEMBRANE.

The hydrogen concentration $c(x,t)$ in a membrane (Fig. 1) at time t in a position x is described by the diffusion equation:

$$\frac{\partial c(x,t)}{\partial t} = D\frac{\partial^2 c(x,t)}{\partial x^2} \tag{1}$$

with boundary conditions:
on a surface 1 ($x=0$):

$$-D\cdot\frac{\partial c(0,t)}{\partial x} = q_{1in} - q_{1out}, \tag{2}$$

on a surface 2 ($x=l$):

$$-D\cdot\frac{\partial c(l,t)}{\partial x} = q_{2out} - q_{2in}. \tag{3}$$

where, $D = D_0 \cdot exp(-E_d/RT)$ is the diffusion coefficient E_s is the activation energy of diffusion. One can reduce the solution of the diffusion equation (1) with boundary conditions (2) and (3) to the solution of the system (4) of nonlinear integral equations [6] ($q_{2in} = 0$):

$$\begin{cases} c(0,t) = \int_0^t d\tau K_1(t-\tau)\left[q_{1in} - \beta_1 c^2(0,\tau)\right] - \int_0^t d\tau K_2(t-\tau)\beta_l c^2(l,\tau) \\ c(0,t) = \int_0^t d\tau K_2(t-\tau)\left[q_{1in} - \beta_1 c^2(0,\tau)\right] - \int_0^t d\tau K_1(t-\tau)\beta_l c^2(l,\tau) \end{cases}, \tag{4}$$

where

$$K_1(t) = \frac{1}{l}\left[1 + 2\sum_{k=1}^{\infty} exp(-k^2\pi^2 Dt/l^2)\right] \text{ and } K_2(t) = \frac{1}{l}\left[1 + 2\sum_{k=1}^{\infty}(-1)^k exp(-k^2\pi^2 Dt/l^2)\right].$$

The asymptotics for small times ($t \to 0$ [6]) is

$$c(0,t) \approx 2q_{1in}\sqrt{t/\pi D} = c_\infty\sqrt{t/\tau_0}. \tag{5}$$

Where the time $\tau_0 = \pi D/4\beta_1 q_{1in}$ establishes the limit of applicability of asymptotic expression (5) (while $t < \tau_0$ it is possible to neglect the gas desorption of a getter). c_∞ is the hydrogen concentration at large times ($t \to \infty$): $c_\infty = \sqrt{q_{1in}/\beta_1} = \sqrt{p_1/K}$. Hydrogen flux pumped by the membrane at small times is $q(0,t) \approx q_{1in}(1 - t/\tau_0)$. The gas flux passed through the membrane at small times [6] is expressed by

$$q(l,t) = \beta_l c^2(l,t) \approx \frac{1}{\pi} \cdot q_{1in} \cdot \varepsilon^{-2} \cdot \left(4Dt/l^2\right)^3 \cdot exp(-l^2/2Dt), \tag{6}$$

$\varepsilon = D/l\sqrt{\beta_0 q_{1in}}$ is the dimensionless parameter that characterises the hydrogen diffusion through the membrane.

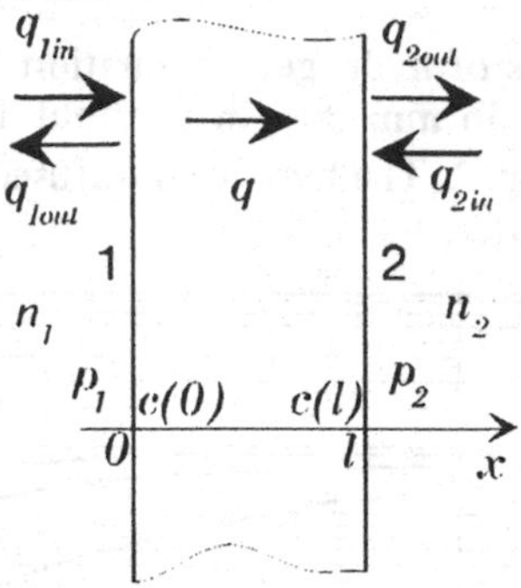

Fig. 1. One-layer membrane

In the case of small diffusion ($\varepsilon << 1$) the gas flux passed through the membrane at large times ($t \to \infty$) [6] is expressed by

$$q(l,t) \approx \varepsilon\, q_{1in}\left[1 - exp\left(-\pi^2 Dt/l^2\right)\right]. \tag{7}$$

In the case of high diffusion ($\varepsilon.>> 1$) the gas flux passed through the membrane at large times ($t \to \infty$) [6] is expressed by

$$q(l,t) \approx \left(q_{1in}/2\right)\left[1 - exp\left(-\sqrt{2}\, t/\tau_\infty\right)\right], \tag{8}$$

where $\tau_\infty = l/2\sqrt{\beta_1 q_{1in}}$ is the characteristic time that enables us to evaluate the applicability of expression (8) ($t/\tau_\infty \geq 1$).

The diffusion equation solution of the hydrogen evacuation by metal membrane for the stationary case ($t \to \infty$ and $q = Const$) is considered in works [6,7]. The hydrogen flux pumped out by a membrane is described by the equation:

$$\sqrt{p_1 - \frac{q}{\alpha_1}\sqrt{2\pi\, mkT_{g1}}} - \sqrt{\frac{q}{\alpha_2}\sqrt{2\pi\, mkT_{g2}} + p_2} - \frac{ql\sqrt{K}}{D} = 0. \tag{9}$$

The evacuation probability of hydrogen molecule by a membrane $\chi = q\sqrt{2\pi\, mkT_g}/p_1$ at $T_{g1} = T_{g2}$ is described by the equation:

$$\sqrt{1 - \frac{\chi}{\alpha_1}} - \sqrt{\frac{\chi}{\alpha_2} + \frac{p_2}{p_1}} - \chi\frac{l}{D}\frac{\sqrt{K}\sqrt{p_1}}{\sqrt{2\pi\, mkT_g}} = 0 \tag{10}$$

The equation (10) has two asymptotic solutions.

1) The hydrogen flux pumped out by a membrane is limited by the diffusion process a in a membrane, $p_1 >> p_2$ and $\chi << \alpha_1$:

$$\chi = \frac{\sqrt{2\pi\, mkT_g}}{\sqrt{p_1}} \cdot \frac{D}{l\sqrt{K}} \tag{11}$$

2) The hydrogen flux pumped out by a membrane is limited by an adsorption at low pressure $\chi\sqrt{p_1} \to 0$, ($p_1 >> p_2$):

$$\chi = \left(1/\alpha_1 + 1/\alpha_2\right)^{-1}, \tag{12}$$

for $\alpha_1 = \alpha_2 = \alpha$, $\chi = \alpha/2$.

The experimental results of hydrogen evacuation by the cylindrical palladium membrane (the diameter is 56 mm, length is about 100 mm and thickness is 20 microns) [7] are shown in Fig. 2. The hydrogen diffused through a membrane were oxidized by oxygen up to water.

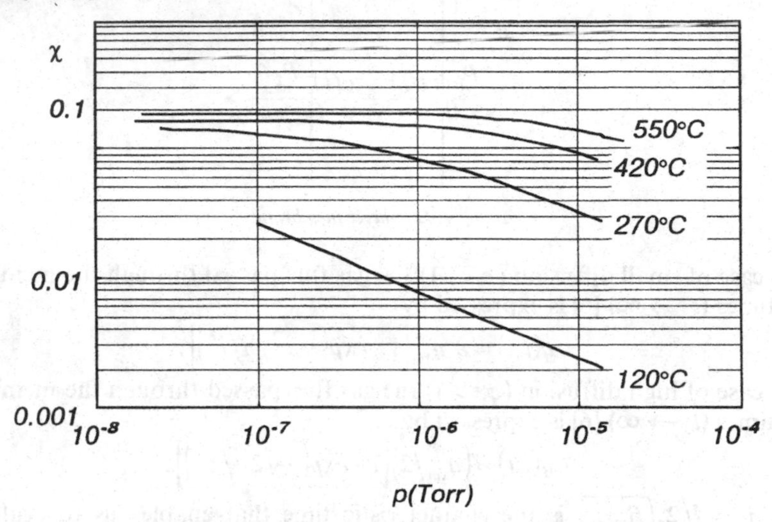

Fig. 2. Experimental dependence of the probability of hydrogen evacuation by a palladium membrane on pressure.

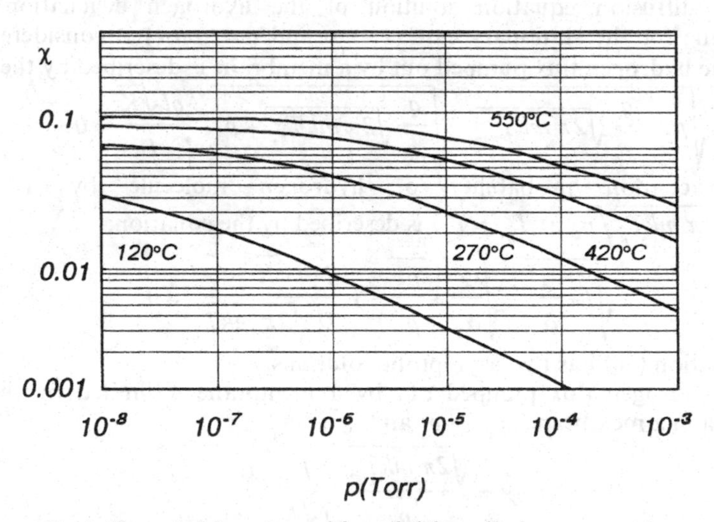

Fig. 3. Theoretical dependence of the probability of hydrogen evacuation by a palladium membrane on pressure.

One can determine the hydrogen adsorption probability $\alpha = 0.85 \cdot exp(-670/T)$ at $\alpha_1 = \alpha_2 = \alpha$, the hydrogen diffusion coefficient $D = 7 \cdot 10^{-6} \cdot exp(-4700/T)$ m²/s, (at $K = 31 \cdot exp(-2000/T)$ Pa/(m^{3}Pa/kg)^2 [8]) by means of these data [6] and to calculate the dependence of evacuation probability of hydrogen molecule by the mentioned above palladium membrane on pressure (Fig. 3).

2. TWO-LAYER MEMBRANE.

The hydrogen concentration $c(x,t)$ in a two-layer membrane (Fig. 4) at time t in a position x is also described by the diffusion equation (1) with boundary conditions (2) on surface 1 ($x=0$):

$$-D_I \cdot \frac{\partial c_I(0,t)}{\partial x} = \frac{\alpha_1}{\sqrt{2\pi mkT_g}} \left[p_1 - K_I c_I^2(0,t) \right], \tag{13}$$

and (3) on surface 2 ($x=l_I+l_2$):

$$-D_{II} \cdot \frac{\partial c_{II}(l_I + l_{II}, t)}{\partial x} = \frac{\alpha_2}{\sqrt{2\pi mkT_g}} \left[K_I c_{II}^2(l_I + l_{II}, t) - p_2 \right]. \tag{14}$$

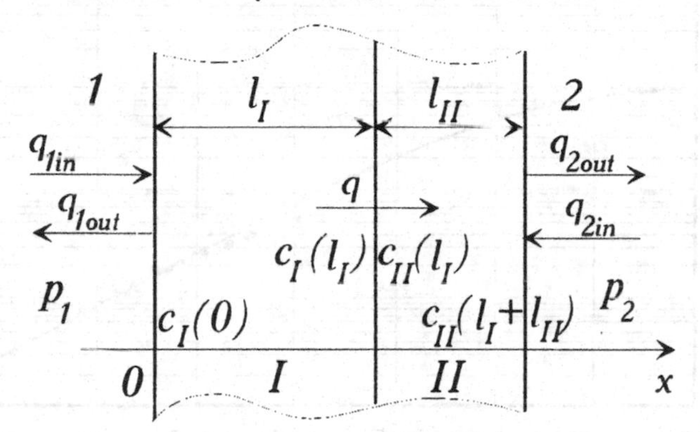

Fig. 4. Two-layer membrane

The boundary condition between the layer I and the layer II is

$$D_I \cdot \frac{\partial c_I(l_I, t)}{\partial x} = D_{II} \cdot \frac{\partial c_{II}(l_I, t)}{\partial x} \tag{15}$$

Where $c_I(x,t)$ and $c_{II}(x,t)$ are the hydrogen concentrations in layer, l_I and l_{II} are the thicknesses of layers, D_I and D_{II} are the diffusion coefficients in layer material, K_I and K_{II} are the equilibrium constants of hydrogen and layer material for the layer I and layer II accordingly.

There is no diffusion flux in a membrane in $p_1 = p_2 = p$ case and the equilibrium takes place $p = K_I \cdot c_I^2 = K_{II} \cdot c_{II}^2$, $c_I = c_{II} \cdot \sqrt{K_{II}/K_I}$ and $c_{II} = c_I \cdot \sqrt{K_I/K_{II}}$.

One can write the equation (10) for a two-layer membrane in a stationary case ($t \to \infty$, $q(x) = Const$):

$$\sqrt{p_I - \frac{q}{\alpha_I}\sqrt{2\pi mkT_{gI}}} - \sqrt{\frac{q}{\alpha_2}\sqrt{2\pi mkT_{g2}} + p_2} - q\left(\frac{l_I\sqrt{K_I}}{D_I} + \frac{l_{II}\sqrt{K_{II}}}{D_{II}}\right) = 0 \qquad (16)$$

and the equation (11) for evacuation probability of hydrogen molecule by a two - layer membrane at $T_{gI} = T_{g2}$:

$$\sqrt{1 - \frac{\chi}{\alpha_I}} - \sqrt{\frac{\chi}{\alpha_2} + \frac{p_2}{p_I}} - \chi\frac{\sqrt{p_I}}{\sqrt{2\pi mkT_g}}\left(\frac{l_I\sqrt{K_I}}{D_I} + \frac{l_{II}\sqrt{K_{II}}}{D_{II}}\right) = 0 \qquad (17)$$

The equation (17) has two asymptotic solutions by the analogy equation (11):

1) The hydrogen flux pumped out by a membrane is limited by the diffusion process in a membrane, $p_1 >> p_2$ and $\chi << \alpha_1$:

$$\chi = \frac{\sqrt{2\pi mkT_g}}{\sqrt{p_I}} \cdot \left(\frac{l_I\sqrt{K_I}}{D_I} + \frac{l_{II}\sqrt{K_{II}}}{D_{II}}\right)^{-1} \qquad (18)$$

2) The hydrogen flux pumped out by a membrane is limited by an adsorption, the probability of hydrogen evacuation is determined by the expression (12).

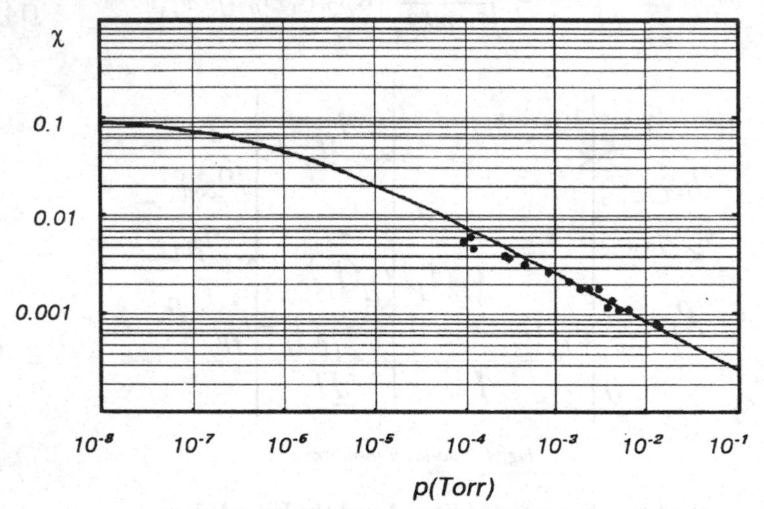

Fig. 5. The dependence of probability of hydrogen evacuation by the two-layer membrane (titanium-palladium) on pressure at the temperature 750°C : —— theory, • - experiment.

The experimental model of two-layer membrane supposed by us was used for the experimental researches of hydrogen evacuation. The extension surface of palladium tube (the diameter of tube is 6 mm, the length is 95 mm, the wall thickness is 0.2 mm) was covered by the layer of titanium nonevaporated getter with the thickness of 1 mm. The hydrogen evacuated was adsorbed by titanium getter, diffused through the tube's wall and was oxidized on internal palladium surface by oxygen. The tube is heated by electrical current. The experimental results of measurement of evacuation probability of hydrogen upon pressure at temperature 750°C are shown on Fig. 5. The dependence of hydrogen evacuation probability on pressure calculated by the equation (17) is plotted by the continuous line on Fig. 5. An equilibrium constant of hydrogen - titanium obtained by the data of solubility in [8] is $K = 4{,}839 \cdot exp(-9525/T)$ Pa/(m^{3}Pa/kg)^2

The coefficient of hydrogen diffusion in titanium a getter covered the palladium tube in the experiment is $D = 0{,}8 \cdot 10^{-6} \cdot exp(-5200/T)$ m^2/s.

The value of diffusion coefficient is less than usual diffusion coefficient for titanium nonevaporated getter ($D = 1{,}07 \cdot 10^{-5} \cdot exp(-5200/T)$ m^2/s).

One can explain the low diffusion coefficient by the technology used for the covering of the tube by getter. The getter was rendered on a palladium tube as the mixture of a powder titanium with organic binder. The binder was "burnt out" in vacuum furnace at the temperature 800°C. The accepted probability of hydrogen adsorption on a surface titanium getter was $\alpha_1 = 0.125$.

3. EVACUATION OF HYDROGEN FROM PLASMA BY METAL MEMBRANES

The hydrogen flux adsorbed by the surface 1 of a membrane (Fig. 1) q_{in} consists of the hydrogen molecule flux $q_{H_2} = \alpha_l\, p_l / \sqrt{2\pi\, mkT_g}$ (per unit surface area), the hydrogen's ion flux H^+ $q_{H^+} = j_{H^+}/2e^-$, the hydrogen's molecular ion flux H_2^+ $q_{H_2^+} = j_{H_2^+}/e^-$ and hydrogen atoms' flux $q_{H^0} = p_{lH^0} / 2\sqrt{\pi\, mkT_g}$ at the presence of plasma. j_{H^+} and $j_{H_2^+}$ are the currents of ions per surface area unit, e^- is the charge of electron. We assume the absence of surface absorption barrier for atoms and ions i.e. there are no power expenses on the dissociation of molecules. One can write the flux $q_{l\,in}$ in a boundary condition (2) as:

$$q_{l\,in} = \left(\alpha_l\, p_l + p_{lH^0}/\sqrt{2}\right) \Big/ \sqrt{2\pi\, mkT_g} + j_{H^+}/2e^- + j_{H_2^+}/e^- \tag{19}$$

The boundary condition (2) is:

$$-D \cdot \frac{\partial c(0,t)}{\partial x} = \frac{\alpha_l}{\sqrt{2\pi\, mkT_{gl}}} \left[p_l + p_{lH^0}/\sqrt{2}\alpha_l - K\, c(0,t)^2\right] + \left(j_{H^+} + 2 j_{H_2^+}\right) / 2e^- \tag{20}$$

We assume there is no plasma on the side 2 of membrane.

One can write the equation (10) at the plasma presence as:

$$\sqrt{p_l + \frac{p_{lH^0}}{\sqrt{2}\alpha_l} + \frac{\sqrt{2\pi\, mkT_{gl}}}{2\alpha_l}\left(\frac{j_{H^+} + 2 j_{H_2^+}}{e^-} - 2q\right)} - \sqrt{\frac{q\sqrt{2\pi\, mkT_{g2}}}{\alpha_2} + p_2} - \frac{ql\sqrt{K}}{D} = 0 \tag{21}$$

We shall consider some practical case: Let the hydrogen pressure at the side 2 of a membrane is the function of a pumping speed $S_0 = Const$ (per unit area of membrane surface). The pressure p_2 is determined by the expression $p_2 = p_{02} + q/S_0$, where p_{02} is the initial pressure at membrane surface 2. We can write the equation (21) in this case as:

$$\sqrt{p_1+\frac{p_{1H^0}}{\sqrt{2}\alpha_1}+\frac{\sqrt{2\pi\, mkT_{g1}}}{2\alpha_1}\left(\frac{j_{H^+}+2j_{H_2^+}}{e^-}-2q\right)}-\sqrt{\frac{q}{\alpha_2}\sqrt{2\pi\, mkT_{g2}}+\frac{q}{S_0}+p_{02}}-\frac{ql\sqrt{K}}{D}=0 \tag{22}$$

The results of the calculation of hydrogen flux pumped by niobium membrane with thickness 20 microns at T=600°C in the dependence on hydrogen ion current are shown on Fig. 6. The hydrogen pressure at the side 1 of the membrane is 10^{-5} Torr, the initial pressure at the side 2 of the membrane is 10^{-4} Torr, the pumping speed of hydrogen at the side 2 of the membrane is 1 m^3/m^2, the current of hydrogen molecular ions on the membrane is a half of hydrogen ion current, the pressure of atomic hydrogen is 5% of pressure of molecular hydrogen. We accepted the diffusion coefficient of hydrogen in niobium is $D=5\cdot 10^{-8}\, exp(-1200/T)$ m^2/s (in the accordance with [9]), the equilibrium constant of hydrogen - niobium is $K=12800\, exp(-10370/T)$ $Pa/(Pa.m^3/kg)^2$ (obtained by the data of hydrogen solubility of in niobium in [9]), the adsorption coefficients are $\alpha_1=0.5$ and $\alpha_2=0.2$. We accept the adsorption coefficients in the assumption of the existing of niobium oxide on the surface of niobium and absence of niobium oxide on the niobium surface at the hydrogen plasma interaction.

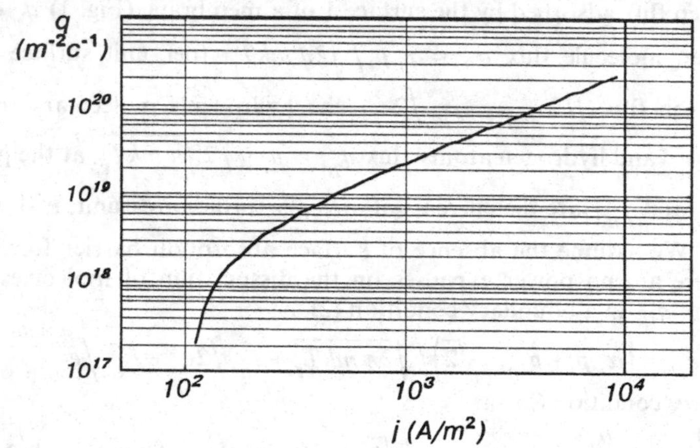

Fig. 6. Dependence of hydrogen flux through niobium membrane on current of hydrogen ions.

The threshold on the Fig. 6 is

$$j_{H^+}+2j_{H_2^+}=\frac{2\alpha_1 e^-}{\sqrt{2\pi\, mkT_{g1}}}\left[p_{02}-\left(p_1+\frac{p_{1H^0}}{\sqrt{2}\alpha_1}\right)\right]$$

The threshold takes place when $p_{02}>p_1+p_{1H^0}/\sqrt{2}\alpha_1$

4. POSSIBLE PERSPECTIVES

The vacuum pumps transfer gas in continuous operation mode from small pressure area to high pressure area. The vacuum pumps on the base of permeable membranes are not the exception from this point of view. It is necessary to support the gradient of hydrogen concentration in the membrane in the evacuation direction to pump the hydrogen by the permeable membrane.
There are two ways to support the necessary gradient.

- The first one is based on the decreasing of the hydrogen pressure on the output side of the membrane. It could be achieved by the oxidation of hydrogen on the heated surface of palladium. The formation of water vapour will accompany this mechanism.
- The second mechanism is based on the increasing of hydrogen concentration in the surface stratum of membrane on its input side by the organization of atoms and ions' fluxes directed to the membrane. It is possible to achieve at the presence of plasma or directed ion fluxes. There is no stage of the dissociation of hydrogen molecules in this case.

We shall consider the possible areas of applications of vacuum pumps on the base of both mechanisms.
The researches executed show the creation of vacuum pumps on the base of palladium membranes is rather problematic because of that the necessary parameters of working pressure range may be achieved at the small thickness of palladium membrane [7]
In this case the main problem is the durability of a membrane. The using of two-layer membrane enables us to solve this problem. One can use the non-evaporated getter with a higher value of hydrogen solubility and diffusion coefficient as the material of the first layer. It allows to increase the membrane thickness, that is to increase the membrane durability. As the second layer material one can use a palladium. It ensures the necessary conditions of the catalytic oxidation of hydrogen. The application area of such vacuum pumps is similar to the application area of nonevaporated getters. The vacuum pumps on the base of two-layer' membranes can compete with nonevaporated getters at a long evacuation of rather high hydrogen fluxes because of the necessity to regenerate nonevaporated getters in such conditions.
The creation of vacuum pumps on the base of plasma systems is problematic because of the rather high power expenses on the formation of plasma. At the same time one can use this mechanism in the plasma installations.
To make the reliable evaluations the perspectives and applicability of vacuum pumps on the base of metal membranes it is necessary to execute a series of experimental and theoretical researches of materials for membranes. A main problem is the reliable determination of constants of materials researched: the constant of solubility, the diffusion coefficient and the probability of hydrogen adsorption by the metal.

REFERENCES

1. Young, J.R. (1963) Palladium - Diaphragm Hydrogen Pump., Rev.of Sci. Instr., 34, (№ 4), 374-377.
2. Balovnev, Yu.A. (1969) Hydrogen diffusion in palladium at low pressures, Journ. of Phys. Chem., 43, (№10), 2461-2464 [in Russian].
3. Livshitz A.I. Notkin M.E., Pustovoit Yu.M. and Samartsev A.A. (1979) Superpermeability of solid membranes and gas evacuation. Part I and Part II., Vacuum,, 29, (№3), 103-124.
4. Bacal M., Balgiti-Sube F.El., Livshits A.I. et.al. (1998) Plasma driven supermeation and its possible application to ion sources and neutral beam injectors Rev. Sci. Instr. 69, (№ 2), 935-937,
5. Knize R.J. and Cecchi J.L. (1983) Theory of bulk gettering., J. Appl. Phys., .54,(6), 3183-3189.7.
6. Lukianov A.A., Pereslavtsev A.V., Fedotov V.Yu. (1998) Hydrogen sorption at low pressures and diffusion in getters and membranes. Problems of Nuclear Science and Technique, Series: Thermonuclear Synthesis, (№1-2) , 45-51. [in Russian].
7. Karaulov V.N., Pereslavtsev A.V., Pustovoit Yu.M., Svischov V.S. (1998) About the possibility of palladium membrane using for the hydrogen pumping in controlled thermonuclear synthesis installations. Problems of Nuclear Science and Technique, Series: Thermonuclear Synthesis, (№1-2), 52-57, [in Russian].
8. Dushman S., (1962) Scientific Foundation of Vacuum Technique, John Willey&Sons inc., N.Y.
9. Telkovskiy V.G., Pisarev A.A., Tsyplakov V.N., Ygritskiy A.N. (1985) The selection of materials and temperature regime of reception plate of fast ion injector, Atomic Energy, (№ 3), 209-214, [in Russian].

HYDROGEN RECYCLING AND WALL EQUILIBRATION IN FUSION DEVICES

PETER MIODUSZEWSKI
Oak Ridge National Laboratory
Oak Ridge, Tennessee 37716, U.S.A.

1. INTRODUCTION

The design of fusion devices for steady state operation requires the knowledge of relaxation processes and time constants affecting the stationary conditions. Among the more prominent candidates are the confinement times, current relaxation times, and the wall equilibration times. Typical confinement times of today's midsize devices are in the order of 1 second, current relaxation times are, depending on the plasma parameters, in the range of 5 seconds, and wall equilibration times are commonly assumed to be in the range of hundreds of seconds. This wall equilibration time obviously determines the pulse length necessary to achieve stationary density conditions. On the other hand, it is not very well defined and depends to a large degree on the wall state and plasma conditions. In order to shed some light on the involved processes, this paper investigates wall equilibration processes which are controlled by hydrogen recycling.

2. PARTICLE BALANCE BETWEEN PLASMA, WALL, EXTERNAL EXHAUST

Of the total amount of gas puffed into a tokamak plasma discharge, typically only 10-20% ends up in the plasma itself, whereas the rest is implanted into the walls. If the device is equipped with external particle exhaust via divertors or pump limiters, a certain fraction of the total gas can also be exhausted from the system. To obtain a better understanding of the distributions of the gas input, we construct a particle balance including the plasma inventory, wall inventory, and the exhausted gas. In the following equations, N_{pl}, N_w, and N_{pmp} are the inventories of the plasma, the wall, and the exhaust respectively:

$$\frac{dN_{pl}}{dt} = \eta\ \Phi(t) - \frac{N_{pl}}{\tau_p} + R[(1-\eta)\ \Phi(t) + \frac{N_{pl}}{\tau_p}] \quad (1)$$

$$\frac{dN_w}{dt} = (1-R)\ [(1-\eta)\ \Phi(t) + \frac{N_{pl}}{\tau_p}] \quad (2)$$

$$\frac{dN_{pmp}}{dt} = \varepsilon \cdot R[(1-\eta)\ \Phi(t) + \frac{N_{pl}}{\tau_p}] \quad (3)$$

Here η is the fueling efficiency, Φ is the gas input (TorrL/s), τp is the particle confinement time, R is the recycling coefficient, and ε is the fraction of the gas flow exhausted by the divertor or pump limiter. If Φ is taken as a step function, puffing gas

C.H. Wu (ed.), Hydrogen Recycling at Plasma Facing Materials, 195–201.

between t=0 and t=5s, the particle balance as a function of time is obtained by integrating equations (1) to (3). The result is depicted in fig. 1 for a fixed recycling coefficient of R=0.9. It can be seen that at a pulse length of 5 s the density is not stationary. The particle confinement time is taken as $\tau p=0.250$ s, so that the global confinement time with the given R=0.9 is $\tau p^*=\tau p/(1-R)=2.5$ s. If on the other hand, the pulse length is 25 s, as indicated in fig.2, then the density is stationary, i.e. if the pulse length is large compared to τp^*, the balance between plasma density and wall pumping is stationary. This can also be achieved with a lower recycling coefficient, since it reduces τp^* and the critical parameter is the ratio of the pulse length to τp^*.

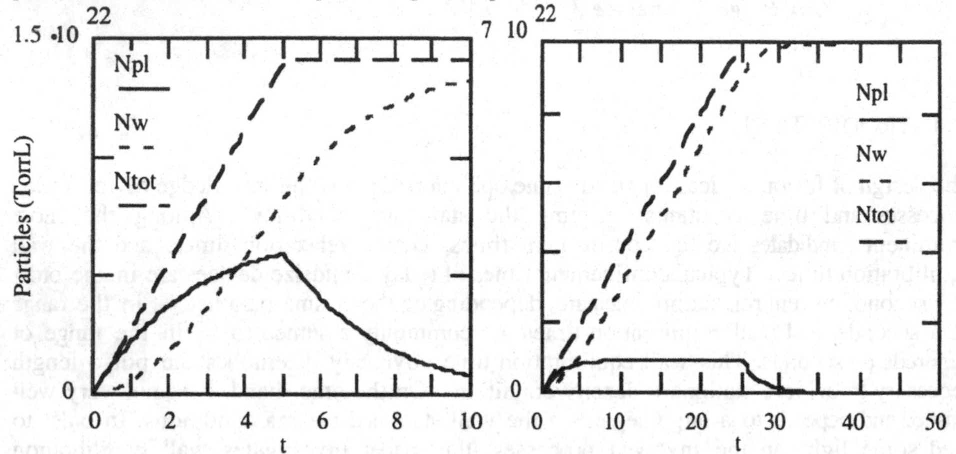

Fig.1 Particle balance with pulse length of twice τp^*

Fig.2 Particle balance with pulse length of ten times τp^*.

However, in reality the recycling coefficient is not constant and will change as a function of time during the discharge. In the next paragraph we will develop a simple analytical formula that can be used to describe the time-dependence of the recycling coeffient in our balance equations.

3. TIME-DEPENDENT RECYCLING COEFFICIENT

During a plasma discharge the recycling coefficient starts with a certain value, depending on the history of the wall, and then evolves as a function of the trapped fluence which is usually a function of time. So, for a given incident flux, the recycling coefficient evolves as a function of time. We have constructed an analytical formula for the recycling coefficient which reproduces TRIM calculations of trapped deuterium versus incident fluence for Maxwellian energy distributions, published by W. Eckstein [1]. The recycling coefficient as a function of time is:

$$R(t)=\frac{2}{\pi}\cdot a\tan(c\cdot t+R_0), \qquad c=\frac{a\cdot\phi}{2\cdot 10^{17}} \tag{4}$$

The parameter c includes the flux ϕ and an adjustable parameter a which can be used to adjust for various particle energies. R_0 can be used to fix the initial recycling coefficient. Assuming an incident constant flux of 10^{15} cm^{-2} s^{-1}, and a particle energy of 100 eV, the recycling coefficient is shown in fig. 3. The number of trapped particles resulting from this recycling coefficient as a function of incident fluence is depicted in fig. 4.

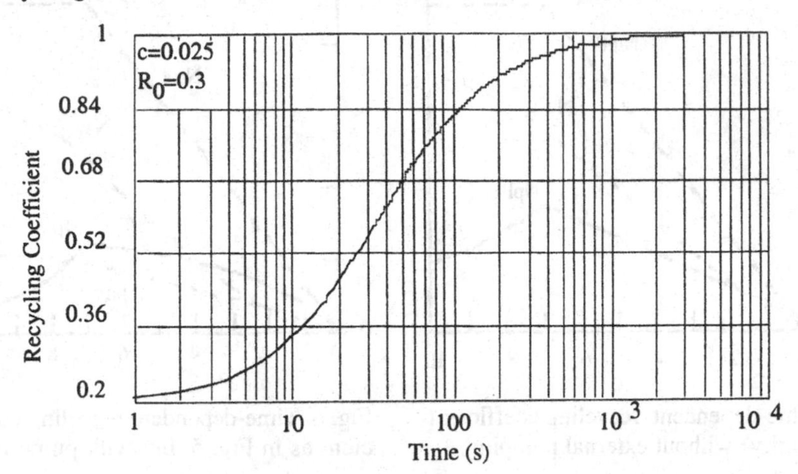

Fig. 3 The recycling coefficient as a function of time for a constant incident flux of 10^{15} cm^{-2} s^{-1}, with an initial value of 0.2.

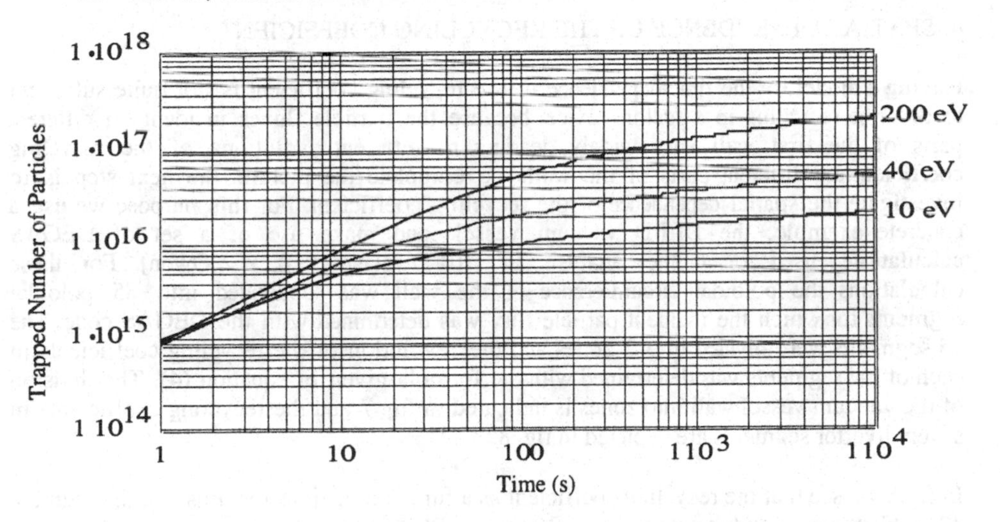

Fig.4 Number of trapped particles as a function of time (incident particle fluence) and particle energy.

It is obvious that different parts of the vacuum vessel will behave differently with respect to recycling. If we introduce the formula in equation (4) into the balance equations (1) through (3), we get the density evolution with a variable recycling coefficient. Assuming high particle fluxes and a relatively high recycling coefficient which is getting close to unity during the pulse, the resulting particle balance without external pumping is shown in fig. 5, whereas including external pumping using $\varepsilon = 0.1$ will result in the density evolution shown in fig. 6.

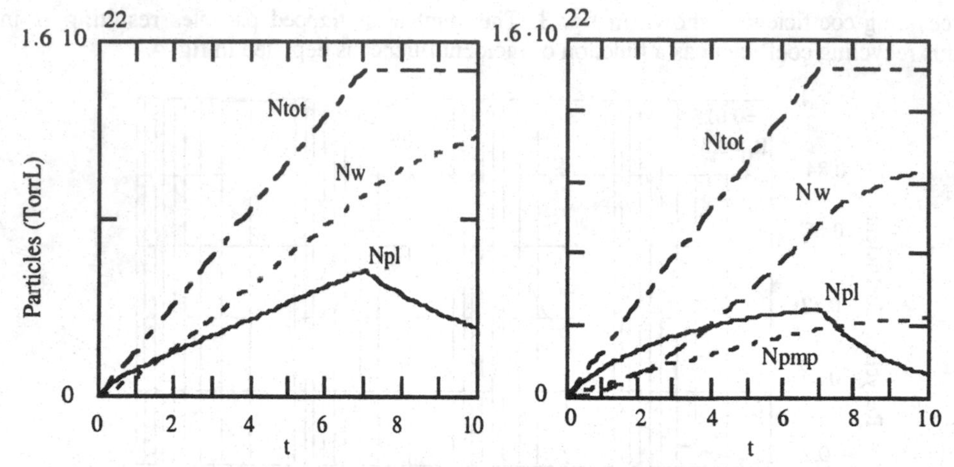

Fig. 5 Time-dependent recycling coefficient (close to unity) without external pumping.

Fig. 6 Time-dependent recycling coefficient as in Fig. 5, but with pumping.

It can be seen that the same trend prevails as with fixed recycling coefficients, external pumping forces the equilibration of the particle balance.

4. SPATIAL DEPENDENCE OF THE RECYCLING COEFFICIENT

Having a model for the time-dependence of the recycling coefficient is not quite sufficient to describe recycling in a fusion device, because the particle fluxes incident on different parts of the first wall vary widely, leading to different evolutions of the recycling coefficient on different parts of the wall. To complete the picture, the next step is to investigate the spatial dependence of the recycling coefficient. For this purpose we use a concrete example, the DIII-D vacuum vessel, and make use of a set of DEGAS calculations, used for another analysis of DIII-D plasmas [L.W. Owen]. For these calculations the poloidal circumference of the wall was subdivided into 85 poloidal segments for which the incident particle flux was determined with the DEGAS code. The 85 segments were divided into 5 zones and the evolution of the recycling coefficients in each of the segments was determined with the formula given in equation (4). The division of the vacuum vessel wall into zones is indicated in fig. 7 and the recycling coefficients in seven divertor segments are depicted in fig. 8.

In fig.8 we see that the recycling coefficient as a function of time remains low in segment 42, which corresponds to the private flux zone (PFZ) where the ion flux is zero and the neutral flux has a minimum too. The three highest recycling coefficients correspond to the segments of the peak fluxes in the two strike-points and the divertor baffle. When we combine all segments into the 5 zones and add up all the pumped particle fluxes as well as all the totally pumped fluences in each zone, we find that different parts of the vacuum vessel play different roles during different parts of the discharge. The pumped particle fluxes for a given time and for each segment are depicted in fig. 9.

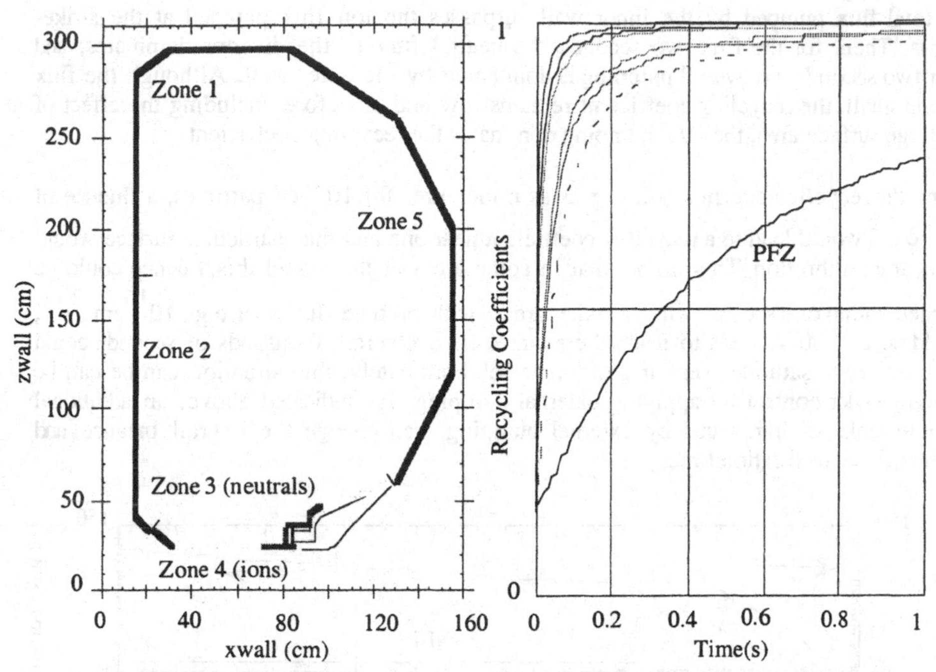

Fig. 7 Five zones of the DIII-D vacuun vessel

Fig. 8 Recycling coefficients of 7 seg- ments in the divertor area as function of time.

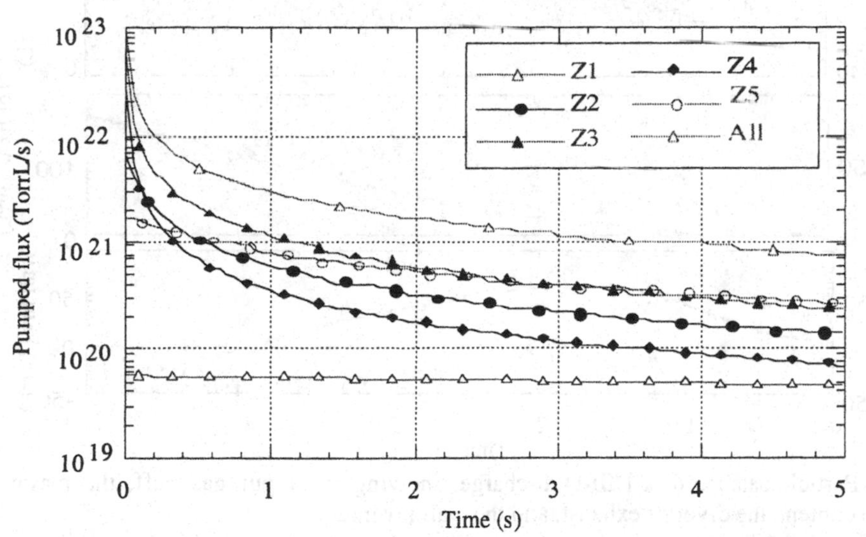

Fig. 9 Pumped particle fluxes in zone 1 through 5 (Z1=top wall, Z2=inner wall, Z3=divertor neutrals, Z4= divertor ions, Z5=outer wall)

The total flow of pumped particles is dominated by the neutrals in the divertor, due to the large surface area and high flux. The strike-point ion flux is the highest and, hence, the recycling coefficient saturates first and the pumped flux drops fastest. After about 0.1 s the total flux pumped by the inner wall surpasses the ion flux pumped at the strike-points. Then, for the first two seconds, the neutral flux in the divertor dominates, and after two seconds the overall pumping is dominated by the outer wall. Although the flux here is small, the recycling coefficient remains low and, therefore, including the effect of the large surface area, the total pumping dominates the recycling coefficient.

Using the recycling coefficient in fig. 3 as a measure for 100 eV particles, a fluence of 10^{19} cm^{-2} would lead to a recycling coefficient near one and that particular surface would be near equilibration. This means that in some areas of the vessel this fluence could be reached within one second, whereas other areas, with particle fluxes of e.g. $10^{16} cm^{-2} s^{-1}$, would need 1000 seconds to reach the required flux. Overall, thousands of seconds could be necessary to saturate every area of the vessel. Fortunately, this situation can be can be brought under control by applying external pumping. As indicated above, an additional particle sink, as introduced by external pumping, can change the overall balance and shorten the equilibration time.

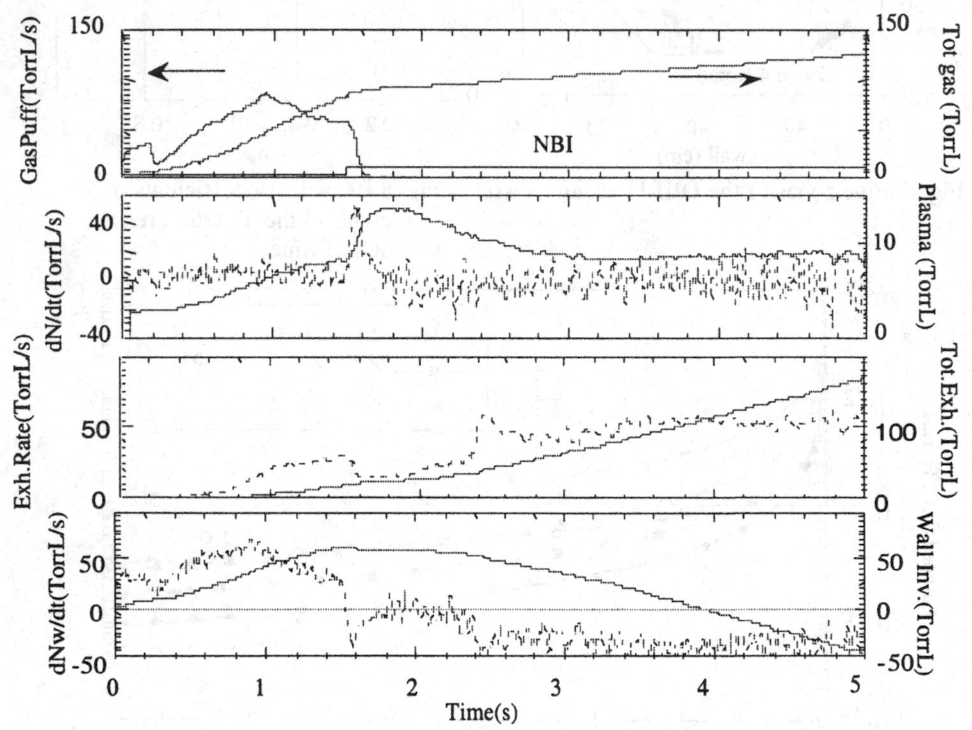

Fig.10 Particle balance in a DIII-D discharge, showing the input gas puff, the plasma particle content, the divertor exhaust, and the wall inventory.

5. EXTERNAL CONTROL OF DYNAMIC WALL INVENTORY

In addition to adding external sinks to the overall particle balance, it has been shown that the change of the wall inventory during a plasma discharge can be controlled in either direction, i.e. it can be either increased or decreased. Without external pumping, the inventory will eventually have to increase until the walls are saturated. However, when external pumping is applied, the wall inventory can be depleted and the particles removed from the walls are eventually removed from the system. This leads to a totally different picture of wall equilibration. In a plasma discharge with external pumping, the walls respond to externally imposed sources and sinks. The wall inventory becomes dynamic. As a consequence, the recycling coefficient and, therefore, the wall equilibration time can be controlled externally.

This depletion of the wall inventory has been demonstrated in several tokamaks such as Tore Supra, DIII-D, and TEXTOR. As an example, we show in fig. 10 the evolution of a DIII-D discharge with gas puff, plasma inventory, particle exhaust, and wall inventory [2]. The wall inventory is usually obtained from the particle balance. The left hand side of the plots depicts the particle rates in TorrL/s and the right hand side shows the particle inventories in TorrL. The figure shows clearly, that, with sufficient external pumping, the external gas source can be balanced and the exhausted amount of gas can actually surpass the gas input. For constant plasma density, these extra particles can only originate from the walls. Similar control of the wall inventory by external exhaust has been shown in other machines.

6. SUMMARY AND CONCLUSION

Wall equilibration due to recycling can take hundreds or even thousands of seconds if no external particle exhaust is provided. However, providing an external particle sink to the particle balance, can force the equilibration to shorter times. In addition, it has been demonstrated that the wall inventory can be dynamically changed with external exhaust such as provided by pumped limiters or divertors, so that the recycling coefficient becomes one the controllable parameters in a magnetic fusion device.

[1] Eckstein. W., *Garching Report* IPP 9/33, October 1980.
[2] Mioduszewski , P.K., Hogan, J.T., Owen, L. W., Maingi, R., et al., *J. Nucl. Mater.* 220-222 (1995) 91-103

Mechanisms of tritium retention in, and the removal of tritium from plasma facing materials of the Fusion Devices

C. H. Wu

NET Team, Max-Planck Institut für Plasmaphysik, D-85748 Garching, Germany

In the next step D/T fusion device, the accurate predictions of tritium retention in Plasma Facing Component is important both from the point of view of the safety and the physics performance.
However, for reliable prediction of T-retention, a detailed knowledge and understanding on the mechanisms, which are responsible for T-retention are essential. The mechanisms are quite complex, for example, in fusion devices; the PFC is in contact with plasma and PFC surface exposed to large flux of ion and neutral particles. The large flux of ions may be retained in the PFC materials by implantation in the depth of ion range, the particles will diffuse to the bulk of materials and eventually, the particles are trapped. In addition, the ion and neutral particles cause the sputtering of surface. The sputtered particles are ionized in plasma and redeposited somewhere on the surface within the fusion devices. The sputtering and deposition lead to T-retention via co-deposition.

After a decade experimental investigation and theoretical study on D/T plasma wall interaction, D/T transport in materials as well as neutron effects, the progress on understanding of the complex of T-retention is significant, so that the prediction of T-retention can be performed more accurate .
In this paper, a detailed analysis on the mechanisms of tritium retention has been performed, based on:

1. Tritium retention via implantation,
2. Bulk retention via bulk diffusion,
3. Tritium retention via neutron transmutation,
4. Tritium retention via co- deposition,
5. Tritium retention via neutron damage induced trapping sites and
6. Tritium retention in dust and flakes,

To assure an optimum plasma performance, to meet physics and engineering goals, it seems necessary to remove effectively the tritium and co-deposited layers as well. To develop effective in situ removal technique, an extensive R & D in framework of EU Fusion Program have been initiated.

In this paper, the recent results of EU R & D on Mechanisms and removal of tritium as well as co-deposited layers are presented. The results, implications and consequences for next step fusion devices are discussed.

C.H. Wu (ed.), Hydrogen Recycling at Plasma Facing Materials, 203–212.

1. Introduction

The Characteristics of tritium retention and release of plasma facing materials are extremely important from an environmental, safety, economical and plasma density control point of view.

As next step fusion devices, i.e. the International Thermonuclear Experimental Reactor (ITER), unlike present tokamaks devices, will be long pulse, high heat flux, higher PFC temperature and they will produce intensive neutron fluxes. Substantial R & D is still needed to elucidate the influences of long pulse, high heat flux, high target temperature and intense neutron on PFC tritium retention and eventually on the total tritium in the devices.

The present fusion devices with short pulse (< 20 second), and low target temperature, the tritium retention will be dominated by implantation, whilst the bulk diffusion will be the major mechanism of tritium inventory for next step devices with long pulse ~1000 seconds and higher target temperature >350 °C. The bulk retention will contribute significant to total tritium inventory. Erosion and redeposition of PFC during normal and off-normal operation caused tritium retention via co-deposition of tritium in plasma and redeposited particles, which based on present knowledge, will play also dominant role.

Intense neutron will induce damage on the microstructure and critical properties of PFC materials e.g. thermal conductivity, swelling and consequently the behavior of tritium trapping will be changed. Depending on neutron damage, they could limit the utilization of PFC materials in the next step fusion devices. The another issue is tritium production by neutron reaction (transmutation) with PFC materials, which, depending on PFC materials, may be significant.

The other important mechanism is tritium retention in dust or flakes. During the normal operation, the PFC will be sputtered by plasma and redeposition occurred and dust is formed at PFC surface. Flakes are formed during the off-normal events; e.g. disruption, ELMs or run-away electrons by peel-off of layers over saturated with hydrogen isotope.

In this paper, a detailed analysis on the possible mechanisms of tritium retention has been performed and relevance of results is discussed

2. Results of analysis

2.1 IMPLANTATION

The tritium retention via implantation is special relevant to present tokamak device, because it is short pulse machine and low PFC temperature. The most of particles are stopped at the depth of ion range, which is dependent on impinging particle energy. After the implantation there is almost no further diffusion to the bulk. At the relevant energy for next step PFC (≤ 200 eV), the ion range D into beryllium or carbon is around 5-10 nm. The saturation level of retention is a function of energy and target temperature. Database is very well established. Fig. 1 and Fig.2 show the ion range of D into Be and carbon as a function of ion energy (1). It is seen that the two low Z materials behave very similar. The total amount of retention via implantation is proportional to the fluency and impinging particle energy.

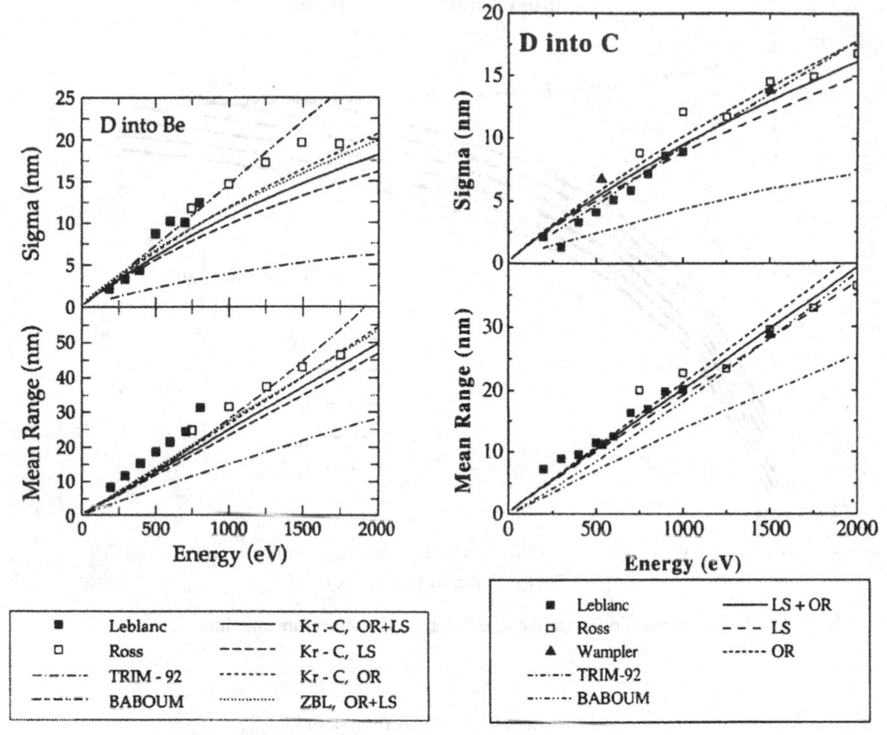

Fig. 1 Mean ion ranges of deuterium in beryllium

Fig. 2 Mean ion ranges of deuterium in carbon

2.2 RETENTION VIA BULK DIFFUSION

In the next step fusion devices, the operation conditions will be significantly different from the present tokamaks: long pulse ~500-1000s, or steady state operation, higher target temperature (because of higher heat loading $\geq$ 10 MW/m^2). Therefore, the role of bulk diffusion will make an important contribution to total tritium retention. The diffusvity D (cm^2/s) is expressed as : $D = D_0 \exp(-E_d/RT)$, where D_0 is pre-exponential factors and Ed is the activation energy. The total amount of retention via bulk diffusion is correlated to the solubility of materials which is expressed as $S = K_s\, p^{1/2}$ where K_s is Sieverts constant and $K_s = K_{s0} \exp(-E_s/RT)$. K_{s0} (at.fr.Pa$^{-1/2}$) is a pre-exponential factor, Es is activation energy solubility. In general, the lower the temperature, the higher the retention via solution. However, at low temperature, the expected diffusivity is low, the time needed to reach the saturated values is longer. Fig. 3 and 4 show the retained values as a function of exposure time and target temperature. It is seen in Fig. 3, at target temperature lower than 1600 K, the bulk saturation has not been reached even the exposure time as long as 10^8 seconds, because the low diffusivity of H-isotope in carbon based materials. Fig. 4 shows that the saturation value of retention in beryllium has been reached even after only 10^3 seconds.

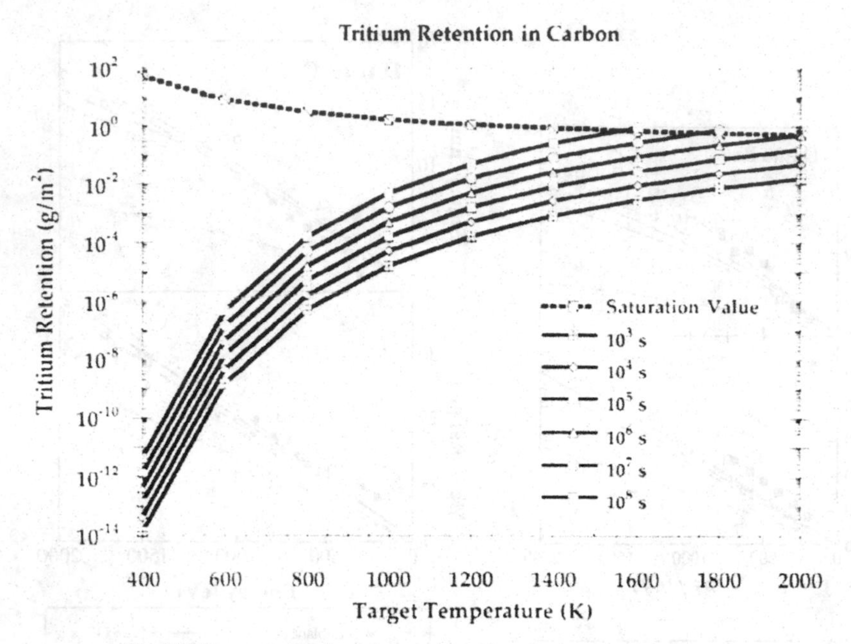

Fig. 3 Tritium retention in carbon as a function of temperature and time

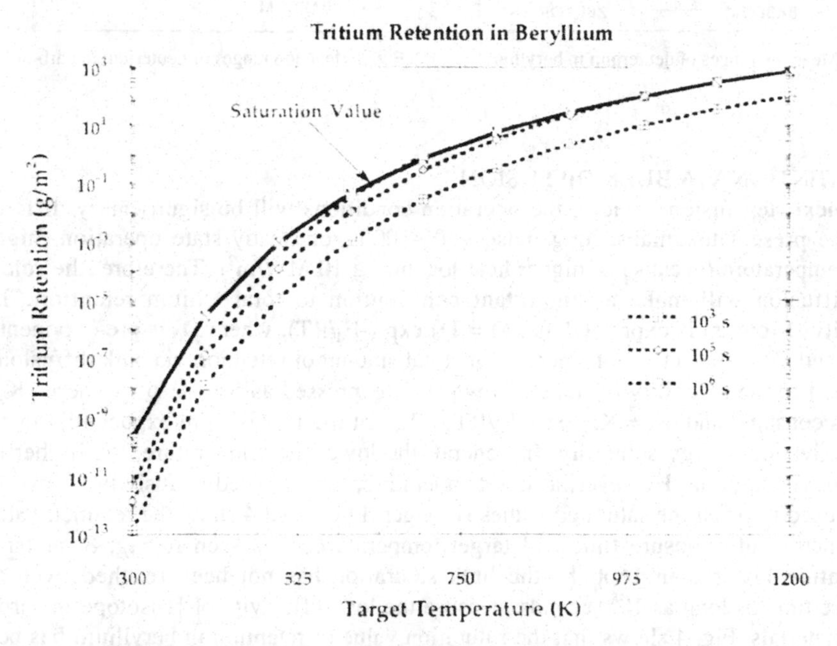

Fig. 4 Tritium retention in beryllium as a function of temperature and time

2.3 TRITIUM RETENTION VIA NEUTRON TRANSMUTATION

In the next generation fusion devices, i.e. the International Thermonuclear Experimental Reactor (ITER) will produce intense neutron fluxes via D-T reaction, which react with wall materials and may produce tritium, depending on concentration and target temperature, tritium will retain in the bulk of wall materials. For example, at present design of ITER, Be will be the first wall material. Several nuclear reactions contribute the tritium production, and depending on reaction time, the role of each reaction is significant different. In the first 24 hours, the $^9Be(n,t)^7Li$ reaction contributes more than 99.85% of total tritium production, whilst the reaction $^9Be(n\alpha)^6He(\beta)^6Li(n,\alpha)t$ is dominant after 3 years. The contributions of all reactions as a function of reaction time are given in Table 1. The total tritium production via transmutation in ITER Be first wall as a function of reaction time is given in table 2. It is seen, that the tritium production in Be first wall via transmutation can be as much as 1.5 kg.

Reaction	Reaction Time				
	24 h	1a	2a	3a	5a
$^9Be(n,t)^7Li$	99,85	66,87	52,57	44,60	35,92
$^9Be(n\alpha)\ ^6He(\beta)^6Li(n,\alpha)t$	-	31,95	44,88	50,30	54,77
$^9Be(n,X)t(\beta)^3He(n,p)t$	-	0,88	1,96	2,8	4,38

Table 1 Tritium Production via Nuclear Reactions (%)

	Reaction Time			
	24h	115.74 d (1.10^7Sec)	1 a	3 a
Tritium (g)	0.675	90.624	360.838	1544.440

Table 2 Tritium Production in the ITER Be First Wall

2.4 TRITIUM RETENTION VIA CO-DEPOSITION

During the normal operation in fusion devices, the surface of wall materials will be sputtered constantly by bombarding with plasma (mainly D, T ions). During the deposition of the sputtered particles of C, Be will be co-implanted with plasma, which leads to tritium retention. Depending on target temperature or oxygen concentration in plasma (particular for beryllium). The retention can be as high as D/C= 0.4 and D/Be = 0.37 at 300 K. Fig 5 show the D/T retention via co-deposition as a function of target temperature and oxygen concentration (2, 3)

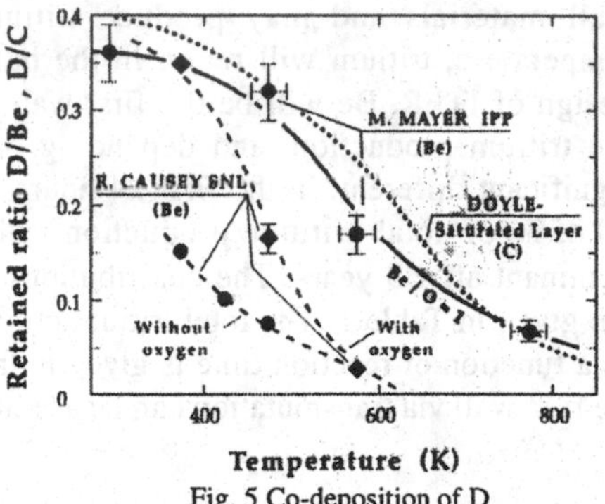

Fig. 5 Co-deposition of D

2.5 TRITIUM RETENTION VIA NEUTRON DAMAGE INDUCED TRAPPING SITES

In several studies on the effects of neutron induced damage on the tritium retention have shown, that the neutron damage on wall materials (C, Be) will increase the tritium trapping sites (4, 5). The tritium retention increases with increasing of neutron damages. Fig. 6

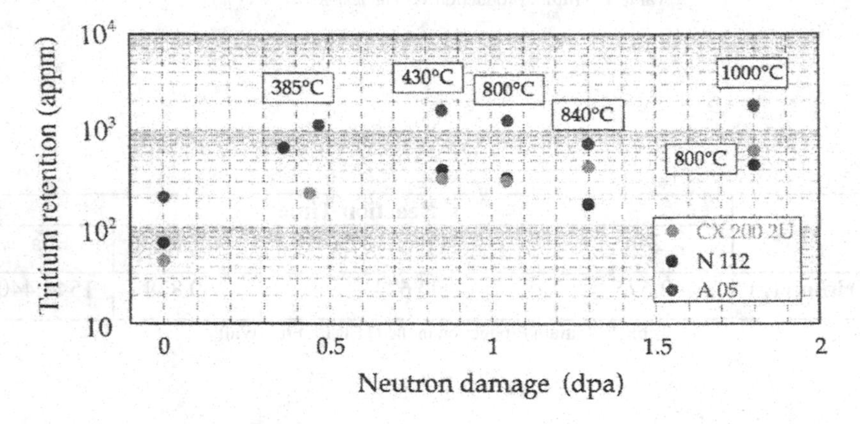

Fig. 6 Tritium retention as a function of neutron damage and irradiation temperature

shows the tritium retention as a function of neutron damage and irradiation temperature. It is seen, almost independent on irradiation temperature, at a damage of 0.5 dpa, a saturation value of 1000 appm has been reached for carbon, whilst a saturation value of beryllium is 10000 appm. The comparison on the neutron damage induced tritium trapping sites in C and Be is given in table 3. It is clear, under same neutron irradiation conditions, the neutron damage induced tritium trapping sites in Be is almost one order of magnitude higher than that of in carbon.

Material	Loading atmosphere		Loaded Tritium	
			Not irradiated	Irradiated
Be	H_2 + 5 ppm T_2, 2 bar, 850°C, 6-7h	(Bq/g) (T/Be)	~ 1 x 10^6 ~ 0.8 x 10^{-8}	~ 3 x 10^7 ~ 2.5 x 10^{-7}
	T_2, 2bar, 850°C, 6-7 h	(T/Be)	~ 1.6 x 10^{-3}	~ 5 x 10^{-2} (~ 10000 appm)
Graphite	T_2, 0.8bar, 850°C, 10h	(T/C)	~ 1 x 10^{-4} (100 appm)	~ 1 x 10^{-3} (~ 1000 appm)

Table 3 Neutron induced tritium trap sites in Carbon and Beryllium

2.6 TRITIUM RETENTION IN DUST AND FLAKES

Dust and flakes are formed either by sputtering at normal operation or at off-normal operation, i. g. disruption, run-away electron events. The results of a systematic study show, that the tritium retention in dust decreases with increasing particle size, and also decreases with increasing temperature. The tritium retention in dust as a function of particle size and temperature is given in table 4. The results of detailed study on dust formation in JET showed, that average particle size of dust is about 27 μm (6, 7).

Tritium presure ≈ 3.5 x 10^{-3} Pa
Exposure time ≈1.0 x 10^5 s

Size (μm)	BET (m^2g^{-1})	25°C	600°C	800°C
4.5	36.7	1.54	6.11	3.22
20	15.9	0.45	1.7	0.83
80	11.1	0.35	1.6	1.1

Tritium retention in carbon dust mg/kgC

Table 4 Experimental investigation of T-retention in dust as a function of particle size (BET) and temperature

3. Removal of tritium from plasma facing materials

To develop an effective in situ removal technique, extensive R & D in the framework of the EU Fusion Program into removal of tritium and co-deposited layers via exposure to air, oxygen, water vapor and by ECR oxygen and deuterium discharge as a function of temperature and partial pressure (flux density) has been performed (3).
While exposure to nitrogen and heating in vacuum at or below 570 K was found to have no effect on the release of deuterium from D-implanted layers, exposure of D-implanted layers, as well as TFTR codeposits, to water vapor did results in D removal, but with no evidence of C erosion, suggesting that D is removed via isotope exchange between the impacting H_2O and the trapped D in the film. However, after about 1 h, no further D removal was observed with water, leaves a considerable amount of residual D in the specimen. Oxygen was seen to be considerably more effective in removing the trapped D than water vapor by factors of 4-8, under similar conditions (16 torr and 520-620 K).
Table 5 shows measured C erosion and derived D removal rates during oxygen exposure for several codeposits obtained from major tokamak devices: TFTR, JET and DIIID. Table 5 contains D removal rates for TFTR specimens during water and air exposure, and an ASDEX-Upgrade specimen during atmospheric exposure. C erosion rates in oxygen or air (16 torr) are strongly temperature dependent, ranging from about 0.1 – 1 µm/h at 523 K to tens of µm/h at 623 K.

a: Temp. (K)	TFTR tile edge (~50µm) 101 kPa air	TFTR N3-15 (~50µm) 2.1 kPa O_2		TFTR N3-24 (~0.15µm) 2.1 kPa O_2	
	derived erosion rate (µm/h)	derived erosion rate (µm/h)	D removal rate (10^{21} D/m²h)	derived erosion rate (µm/h)	D removal rate (10^{21} D/m²h)
523		0.8	20	0.07	2
573		3.5	100		
623	> 50	> 10	> 250	> 0.5	15

b: Temp. (K)	ASDEX-U (0.75µm) 101 kPa air	JET divertor 6A (~2µm) 2.1 kPa O_2		DII-D 6410#B (~ 2µm) 2.1 kPa O_2	
	measured erosion rate (µm/h)	derived erosion rate (µm/h)	D removal rate (10^{21} D/m²h)	derived erosion rate (µm/h)	D removal rate (10^{21} D/m²h)
523		0.5	10	0.23	2.7
623	0.3 (650K)	3.3	70	3.5	41

Table 5 Erosion and D removal rates from co-deposited layers on (a) TFTR bumper limiter tiles and (b) tokamak divertor tiles

The results show that exposure to oxygen is most effective to remove tritium and co-deposited layers. The target temperature and partial pressure of oxygen is the major controlling parameters for optimum removal rate. For example, at a temperature of 623 K, and oxygen pressure of about 2000 Pa, a removal rate as high as ≥ 50 µm has been obtained for co-deposited layers of TFTR CFC tiles. The ECR discharge investigation shows very promising results (8). Fig. 7 show the Removal rates of co-deposited layers via O_2, D_2, H_2 ECR discharge, at T = 300 K and Usb = 0 V.

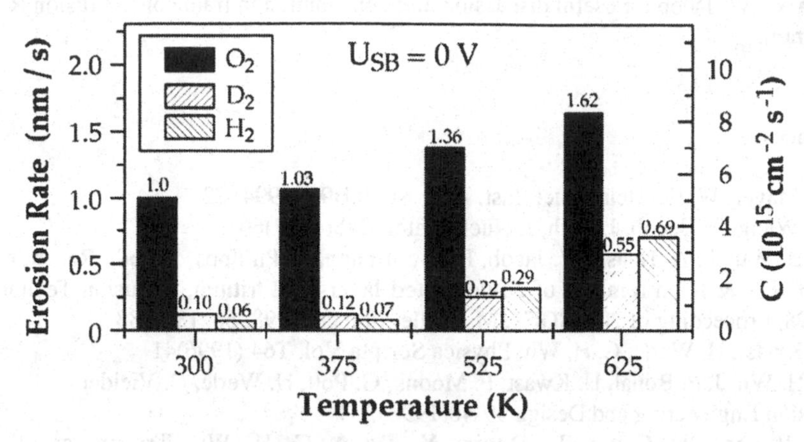

Fig. 7 Erosion rates by O_2, D_2 and H_2 ECR discharge as a function of temperature

4. Conclusions

In this paper, a detailed analysis on the mechanisms of tritium retention in plasma facing materials has been performed, and the EU experimental results on the removal of tritium from plasma facing components via exposure to air, water vapor, oxygen as well as O_2, D_2, H_2 ECR discharge have been discussed and the following conclusions can be made:

A. MECHANISMS OF TRITIUM RETENTION IN PLASMA FACING MATERIALS;
The tritium retention via implantation is a dominant mechanism for present tokamak device, because the low wall temperature and short pulse operation. In addition, the D/T retention in flakes may be important, depending on plasma condition. However, as next generation devices, i.e., the International Thermonuclear Experimental Reactor (ITER) will produce intense neutron fluxes, the temperature of plasma facing components is high, long pulse or steady state operation, the mechanisms of retention fusion devices are a complex. The bulk diffusion, neutron damage induced trapping sites, co-depositions will play dominant role and contribute major fraction to total tritium inventory.

B. REMOVAL OF TRITIUM
It has been demonstrated that exposure to oxygen is most effective to remove tritium and co-deposited layers. The target temperature and partial pressure of oxygen are the major controlling parameters for optimum removal rate. For example, at a temperature of 623 K, and oxygen pressure of about 2000 Pa, a removal rate as high as ≥50 mm can be obtained. Another interesting aspect is, exposure to O_2 ECR discharge, tritium can be removed effectively, even the target temperature as low as ≤500 K.

Acknowledgements

The author would like to thank M. Mayer, W. Eckstein, J. Roth, H. Werle, A.A. Haasz, W. Jacob for useful discussion and contribution in frame of EU fusion R & D Program.

References

1. M. Mayer, W. Eckstein, Nucl. Inst. And Meth.B94 (1994) 22
2. W. Wang, W. Jacob, J. Roth, J. Nucl. Mater. 245(1997)66
3. C. H. Wu, A. A. Haasz, W. Jacob, R. Moormann, V. Philipps, J. Roth, Recent results of EU R & D on removal of co-deposited layers and tritium, in:Fusion Technology 1998, Proceeding of 20th SOFT, Marseille, France, 1998, pp. 185-188
4. H. Kwast, H. Werle, C. H. Wu, Physica Scripta Vol. T64 (1996)41
5. C. H. Wu, J. P. Bonal, H. Kwast, F. Moons, G. Pott, H. Werle, G. Vieider Fusion Engineering and Design 39-40(1998)263
6. D. Boyer, Ph. Cetier, L. Dupin, Y. Belot, C. H. Wu, Tritium sorption in carbon dust, in: Fusion Technology 1996, Proceeding of 19th SOFT, Lisbon, Portugal 1996, pp. 1751-1754
7. C. H. Wu, Ph. Cetier, A. Peacock to be published
8. W. Jacob, B. Landkammer, C. H. Wu, J. Nucl. Mater. 266-269 (1999) 552

STUDIES ON TRITIUM INTERACTIONS WITH PLASMA FACING MATERIAL AT THE TRITIUM PROCESS LABORATORY OF JAERI

T. Hayashi, S. O'hira, H. Nakamura, T. Tadokoro, W.M. Shu, T. Sakai, K. Isobe and M. Nishi,
Tritium Engineering Laboratory, Japan Atomic Energy Research Institute
Tokai-mura, Naka-gun, Ibaraki-ken, 319-1195 JAPAN

1. Introduction

From a view point of safety evaluation of fusion reactors, estimation of permeation amounts of tritium to coolant through plasma facing components (PFC) is one of the critical issues. Moreover, estimation of tritium inventory in plasma facing materials is important subject not only for safety, but also for tritium accountability in fusion reactors [1].

In TPL/JAERI, to simulate tritium permeation behavior in PFC materials, such as Be and W, implantation driven permeation (IDP) behavior have been investigated for deuterium, as well as the other pure metals [2-3]. Furthermore, IDP study using pure tritium ion beam was just started to check the isotope effect.

On the other hand, to obtain tritium inventory information in PFC materials, such as Be, W and CFC, thermal desorption behavior have also been investigated after various hydrogen beam irradiation source. Moreover, tritium micro-autoradiography investigation was just started to investigate depth profile of tritium in the bulk of PFC materials.

In this paper, we summarize recent topics on tritium permeation and retention study in TPL/JAERI.

2. Studies on tritium permeation evaluation for plasma facing materials

2.1 Tritium ion driven permeation through tungsten (W)

Implantation driven permeation (IDP) behavior of tritium and deuterium

C.H. Wu (ed.), Hydrogen Recycling at Plasma Facing Materials, 213–221.

implanted into pure Tungsten was investigated.

The ion source was a modified quartz capillary duoPIGatron in order to produce low energy ions with high intensity designed by Isoya [4]. More than 90% of the ion species in the extracted beams was founded to be atomic ion. The other species were di-atomic and tri-atomic ions (D+ or T+). The monochromatic atomic ion energy could be varied from 100 eV to 2 keV. The effective implantation area of the target is the central region of 25 mm in diameter. More detail information of IDP apparatus was reported in the previous paper [5]

Pure tungsten (>99.5%) metal samples from Nilaco corporation were cut into discs 34 mm$^{\phi}$ from a foil of 25 μm thickness, for use as the target for IDP experiments. one sample was used for a target specimen without annealing and polishing, the other sample was annealed at 1273 K for 3hours in a furnace under vacuum below $1x10^{-5}$ Pa before the experiment. Each specimen was heated up to about 670 K for more than several hours under vacuum for degassing after clamping on the target flange. After degassing of the sample, D^+ or T^+ ions with energy of 1 - 2 keV were implanted onto the target at about 660 K for about several hours. After the above sample treatment, permeation flux through the target reaches steady state value and the reproducibility of the IDP data become to be good enough. The series of IDP experiments were performed as a function of target temperature (540-660 K), incident ions flux (0.2-1.2 x 10^{19} ions/m^2 s) and incident ion energy (400-2000 eV).

The results of the incident ion flux dependence of the IDP flux of deuterium and tritium at the steady state (640K, 1000 eV) shows that the permeation flux significantly depended upon the incident ion flux according to the following relationship : $\phi_p=\alpha(\phi_i)^n$. where ϕ_p is a permeation flux, ϕ_i is an incident flux, and α is a constant. The n value was found to be unity (about 0.8–1.2) by least square fitting. From the steady state model proposed by Doyle and Brice [6,7] and Kerst et al.[8], this result means that the controlled process of this IDP through W was DD or RR regime.

The results of the incident ion energy dependence of IDP (640 K, 9.0x10^{18} ions/m^{2}s) showed that there was clear energy dependence of IDP flux for tritium and deuterium both case, which the IDP flux increased linearly with the ion energy. Generally, when the incident ion energy becomes to be larger, the implantation depth (r) from the target surface would be larger. From the above steady state, the permeation flux only at the DD regime has the linear dependence of r, if the relationship between the diffusion coefficient of the incident side and that of the bulk side would not be changed under the above ion energy conditions.[9]

Figure 1 shows the results of the temperature dependence of the permeation flux for annealed and un-annealed specimen ($9.0x10^{18}$ D^+/m^2s, 1000 eV). The permeation flux of deuterium was found to be almost constant, the activation energies of permeation was found to be about – 0.8 and 0.6 kJ/mol for annealed and un-annealed specimen respectively. Namely, there was little clear temperature dependence. The permeation flux of tritium was also found to be corresponding to the result of deuterium experiment. The IDP flux through each specimen at steady state was found to be $3x10^{-5}$ - 10^{-4} of incident ion flux for each experimental conditions. The result of target temperature dependence of IDP supports that the IDP of deuterium and tritium through tungsten could be controlled by DD process, assuming the diffusion coefficient at the implanted and permeated side are almost equal.

It was concluded that the IDP at steady state was controlled by DD regime from the results of IDP behavior at steady state. Also, there was almost no isotope effect for IDP through W. [9]

2,2 Ion driven permeation of deuterium through beryllium (Be)

The IDP behavior of deuterium implanted into beryllium membrane specimen, which are no annealed and annealed (1173 K, 1 hr) was investigated using same apparatus inTPL. The used specimen was pure beryllium (>99.6%, BeO 0.9%) metal from Nihon Gaishi corporation, the size is 34 mm in diameter x 50 μm thickness. Each specimen was heated up to about 695 K for more than several hours under vacuum for degassing. After degassing of the sample, D^+ ions with energy of 1 - 2 keV were implanted onto the specimen at about 660 K for about several hours as same as W experiment. The series of the IDP experiments were carried out as a function of target temperature (620 – 695 K) and incident ions flux ($0.8\text{-}1.1x10^{19}$ $ions/m^2$ s) at incident ion energy of 1000 eV. At the surface even after polishing, oxide layer about 50 –100 nm was found to be existed by XPS analysis.

Observable IDP was not observed for as received specimen even with after over 40 hours D^+ implantation under conditions of incident ion energy 1 - 2 keV and incident ion flux of about $1x10^{19}$ D^+/m^2s at a temperature over 675 K. On the other hand, permeation was observed for the annealed specimen. However, the IDP flux was less than 10^{-5} of the incident ion flux. It was ascertained that the permeation would be enhanced by annealing as well as tungsten.

Figure 2 shows the temperature dependence of the IDP of deuterium through the annealed specimen at incident ion energy of 1000 eV and incident ion flux of $9x10^{18}$ D^+/m^2s. The activation energy was calculated about 68 kJ/mol from fig. 2.
The incident ion flux dependence was measured at incident ion energy: 1000 eV and

specimen temperature: 675 K. It was found that the IDP was proportional to the incident ion flux at the steady state. From the steady state model proposed by Doyle and Brice [6,7] and Kerst et al.[8], this result means that the controlled process of this IDP through Be was DD or RR regime. On the other hand, when the IDP was controlled by diffusion process or surface recombination process, transient behavior of the IDP would be represented by an equation either diffusion theory or surface recombination theory.

Observed IDP flux was well fitted with a curve expressed by folowing eq.

$$\frac{\phi_p(t)}{\phi_p(0)} = \left(\frac{d}{(kr_1 + kr_2)C_0 t + d}\right)^2$$

It implies that the IDP was controlled by surface recombination process rather than controlled by diffusion process.[10]

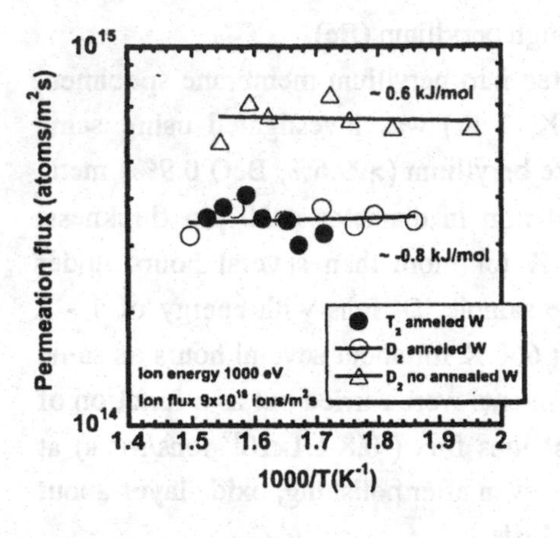

Figure 1 Temperature dependence of IDP flux of D,T implanted into W (0.025mm)

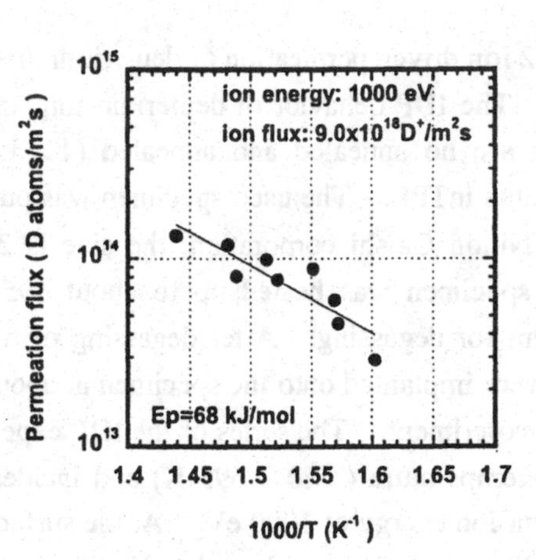

Figure 2 Temperature dependence of IDP flux of D implanted into Be (0.05mm)

3. Studies on tritium retention evaluation for plasma facing materials

3.1 Tritium retention in tungsten

Tritium retention behavior of wrought tungsten samples was investigated by exposure to various hydrogen isotope irradiation sources under controlled sample characterization and temperature. It is shown that the thermal desorption behavior of deuterium implanted in the bulk was strongly correlated to the change in the microstructure of tungsten. Especially, dislocations produced during fabrication play a significant role in increasing hydrogen retention. From the thermal desorption experiments using Tritium Plasma Experiment (TPE @TSTA/LANL) and 3 keV deuterium ion beam, annealing wrought tungsten at 1773 K reduced tritium retention to about one fifth of that in unannealed tungsten. It is shown that thermal desorption of deuterium implanted in the bulk was strongly correlated to change in microstructure of tungsten. Especially, dislocations produced during sample fabrication process play a significant role to increase hydrogen retention. However, even without annealing at 1773 K, the ratio of (retained atoms)/(incident ions) of hydrogen isotope for wrought tungsten was 5×10^{-6} at an incident ion fluence of 7×10^{25} ions/m^2 and a temperature of 523 K, was lower than the value reported in the past.

Figure 3 shows the thermal desorption spectra of tritium released from the unannealed and annealed wrought tungsten samples irradiated with the D/T ion beam. It is obvious that annealing at 1773 K significantly reduced hydrogen isotope retention in wrought tungsten. In the TDS of the unannealed sample, two major maximum peaks of hydrogen isotope release rate were observed, one at around 780 K and the other at around 1300 K. In the annealed sample, a small peak at around 900 K was observed, and the tritium concentration still increased at the higher temperature region above 1000 K. However, the temperature of another desorption peak was out of the outgassing system range (<1473 K). The total amount of retained tritium and hydrogen isotopes derived were 184.4 μCi and 3.6×10^{20} atom/m^2 for the unannealed sample, and 34.3 μCi and 6.7×10^{19} atom/m^2 for the annealed one against the total amount of irradiated tritium and hydrogen isotope, 36 Ci and 7×10^{25} ions/m^2. To identify the states of hydrogen isotopes which contributed to those peaks of deuterium release rate from unannealed and annealed wrought tungsten, exposure of a 3 keV deuterium ion beam to another wrought tungsten samples of higher purity was carried out, and TDS of deuterium were observed.

The effects of annealing at 1773 K on TDS of wrought tungsten samples are shown in Figures 4 with TDS of the CVD and single crystal tungsten samples. In both TDS taken with annealed wrought samples irradiated at 300 K, there was no peak observed in the high temperature region over 1000 K, while a small peak and was observed in the case of irradiation at 300 K. In the TDS of the CVD and single crystal tungsten samples, there was no peak observed in the high temperature region over 1000 K either. In the TDS of the CVD sample, there was a peak at around 820 K in both cases of irradiation at 300 K. In the TDS of the single crystal tungsten sample, there was almost no peak observed for irradiation at 300 K.

According to the results obtained above, it is clear that trapping sites, which existed specially in a "as received" wrought tungsten sample, capture the implanted hydrogen isotopes and could be eliminated by annealing at a temperature of 1773 K. The ratios of the retained hydrogen isotope amounts in the unannealed wrought sample to that in the annealed one were 5.4 for TPE and 4.8 for 3 keV ion irradiation at 473 K. These values are very close to the one of the trap densities for unannealed and annealed wrought tungsten estimated by Anderl et. al. [11]. It was reported that vacancies in tungsten start moving in the temperature range above 473 K, dislocations in wrought tungsten start moving above 673 K [12,13]. Therefore, the peak of deuterium desorption rate maximum at 600 K in the TDS of the unannealed tungsten sample can be attributed to vacancy motion, which transfer hydrogen isotope to the surface, and the other peak at 1400 K can be attributed to dislocation motion which might be associated with channelling of deuterium release. Production and motion of vacancies in an unannealed wrought tungsten may be strongly correlated with dislocations, because the TDS was quite reproducible even when a sample which had been re-used after a TDS measurement associated with heating up to a temperature 1473 K, which was enough high to eliminate vacancies. This implies that vacancies may be reproduced as far as the dislocations exist in wrought tungsten.

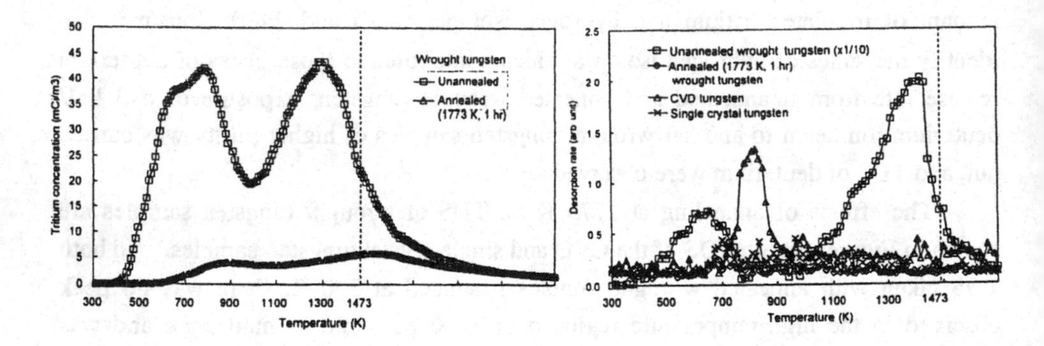

Figure 3 TDS of tritium from W irradiated 100eV-D/T ion beam of TPE

Figure 4 TDS of D from vaious W irradiated 3 keV-D ion beam at 300 K

3.2 Depth profile of tritium in plasma exposed CX-2002U

Carbon Fiber Composite (CFC) is one of the plasma facing material candidates because of its low atomic number and good thermal properties. Autoradiography is a technique, which can give us a visual proof of real tritium behavior inside the bulk of materials when the position of tritium is well preserved through the process of charging tritium and observation [14]. Using this technique, visual examinations of tritium behavior in 2-dimensional CFC (CX-2002U) samples exposed to a high flux D/T particles are being carried out at TPL. Depth profiles of the Ag grains density in the autoradiographs, which represents that of tritium concentration in the CX-2002U samples exposed to a high flux D/T particles under various conditions, were examined and the apparent diffusion coefficients estimated from the profiles, were obtained. Before plasma or gas exposure, sample disks of CX-2002U (Toyotanso.Co, 10 mmϕ, 4 mm^t) were degassed at 1150 K for one hour under vacuum below 10^{-4} Pa. D/T RF (13.56 MHz, 50 W) discharge plasma was generated by introduction of $D_2/T_2(D_2:T_2=1000:1)$ mixture gas supplied at a flow rate of 1 cm^3/min into the quartz plasma tube [15]. Two runs of D/T RF-plasma exposure were performed for 36 hours at temperatures of 293 and 573 K individually. D/T gas exposure also for 2160 was carried out with the same gas flow rate and tritium concentration at 293 K.

Measurement procedures for autoradiography taken in this experiment are shown in figure 5. The average diameter of Ag grains converted from AgBr by a β-particle was 0.17 μm and the range of 18 keV electron in the sample is about 5 μm, so it was possible to observe the spatial distribution of tritium with an accuracy within 5 μm. As an example of autoradiograph, that of an edge part of the cross-section of the samples exposed to D/T plasma at 293 K is shown also in figure 5. White spots correspond to Ag grains formed by β-particle from tritium.

Some comparisons were made by visual examinations of the samples exposed gas, D/T plasma and oxygen plasma to confirm differences in tritium behavior under the different treatments.

Figure 6 shows the depth profiles of the Ag grains density in the CX-2002U samples exposed to plasma at 293 and 573 K as a function of depth from the plasma exposed surface. Based on the data of the Ag grains density depth profile, which represents the tritium concentration profile, an approach to obtain apparent diffusion coefficients at 293 K and 573 K which may include both bulk and pore diffusion processes was taken with an one dimensional diffusion model with an assumption of a constant tritium concentration at the sample surface. The equation of calculation for the tritium concentration profile was given by,

$$C(x) = C" \cdot (1 - \mathrm{erf}(x / 2\sqrt{(D_t \cdot t}).$$

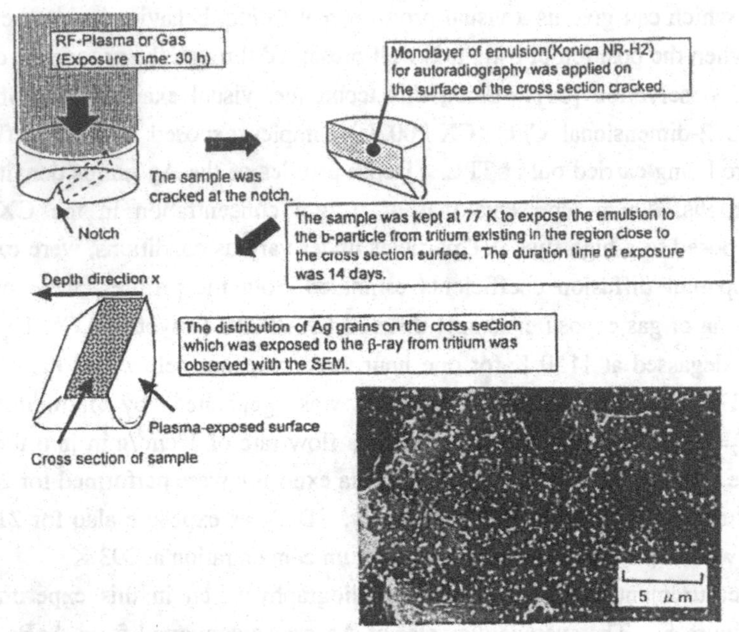

Figure 5 Measurement procedures for autoradiography & typical cross section results of radiography

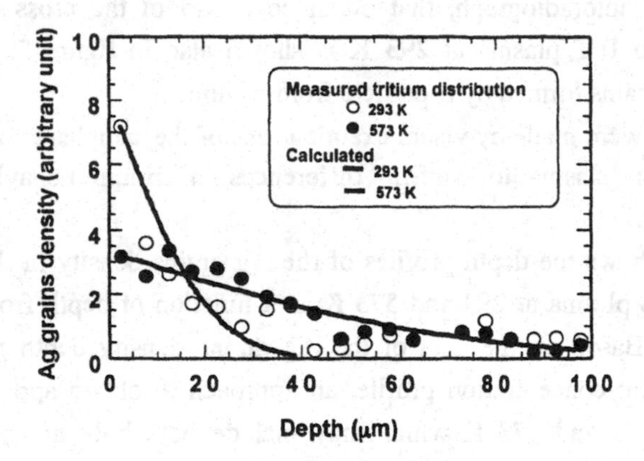

Figure 6 Ag grains density depth profile in CX-2002U exposed to D/T RF-plasma at 293 and 573 K

Where C(x) is tritium concentration at a depth x from the exposed surface (atoms/m^3), C" is constant concentration (atoms/m^3), D_t is bulk and pore diffusion coefficient (m^2/sec), t is exposed time (sec) and erf() is the error function. This equation was applied between the surface and the point where the amount of concentration was 1/10 of the value at the region near the surface. In this case we fitted C(x) to measured

concentrations in the region of 2.5– 40 μm at 293 K and 2.5– 100 μm at 573 K, and derived D_t = 1.7×10^{-16} m^2/sec at 293 K and 2.3×10^{-15} m^2/sec at 573 K from the fitting curves shown in figure 6. The values obtained here are much larger those reported as bulk diffusion coefficients in previous researches [16-21]. The processes, to which this fact can be attributed, we suppose, are diffusion of atom or molecule radical of hydrogen through the pore.

References

[1] ITER, ITER EDA Series No.7, IAEA, Viena, (1996)
[2] T. Hayashi, K. Okuno, K. Yamanaka, and Y. Naruse, J. Alloys and compounds 189 (1992) 195.
[3] W. Shu, K. Okuno, Y. Hayashi and Y. Naruse, J. Nucl. Mater., 203 (1993) 50.
[4] A. Isoya, Helva. Phys. Acta. 59 (1986) 632.
[5] K. Okuno, S.O'hira, H.Yoshida, Y. Nause, et al., Fusion Technol. 14 (1988) 713.
[6] B. L. Doyle, J. Nucl. Mater. 111/112 (1982) 628.
[7] D. K. Brice and B. L. Doyle, J. Vac. Sci. Technol. A5 (1987) 2311.
[8] R. A. Kerst, and W. A. Swansiger, J. Nucl. Mater., 122/123 (1984) 1499.
[9] H. Nakamura, T. Hayashi et al, private communication, to be published.
[10] H. Nakamura, T. Hayashi et al, to be published to JAERI Research (1999).
[11]R.A. Anderl, D.F. Holland, G.R. Longhurst, et al., Fusion Technol., 21 (1992) 745.
[12]L.A. Neimark, R.A. Swalin, Trans. AIME., 218 (1960) 82.
[13]R.H. Atkinson, G.H. Keith and R.C. Koo, "Refractory Metal and Alloy" (1961) 319.
[14] K.Isobe, Y.Hatano and M.Sugisaki, J. Nucl. Mater., 248, (1997) 315.
[15] T.Tadokoro, S. Ohira, M. NIshI and K. Isobe, J. Nuci. Mater., 258-263, (1998) 1092.
[16] T.S. Elleman, unpublished work at General Atomic, San Diego, CA, USA.
[17] H. Atumi, S. Tokura and M. Miyake, J. Nucl. Mater. 155-157 (1988) 241.
[18] M. Saeki, J. Nucl. Mater. 131 (1985) 32.
[19] R.A. Causey, T.S. Elleman and K. Verghere, Carbon 17 (1979) 323.
[20] K. Morita, K. Ohtsuka and Y. Hasebe, J. Nucl. Mater. 162-164 (1989) 990.
[21] T. Tanabe and Y. Watanabe, J. Nucl. Mater. 179-181 (1991) 218.

TRITIUM RECYCLING AND INVENTORY IN ERODED DEBRIS OF PLASMA-FACING MATERIALS

AHMED HASSANEIN
Argonne National Laboratory
Argonne, IL 60439, USA

1. Abstract

Damage to plasma-facing components (PFCs) and structural materials due to loss of plasma confinement in magnetic fusion reactors remains one of the most serious concerns for safe, successful, and reliable tokamak operation. High erosion losses due to surface vaporization, spallation, and melt-layer splashing are expected during such an event. The eroded debris and dust of the PFCs, including trapped tritium, will be contained on the walls or within the reactor chamber; therefore, they can significantly influence plasma behavior and tritium inventory during subsequent operations.

Tritium containment and behavior in PFCs and in the dust and debris is an important factor in evaluating and choosing the ideal plasma-facing materials (PFMs). Tritium buildup and release in the debris of candidate materials is influenced by the effect of material porosity on diffusion and retention processes. These processes have strong nonlinear behavior due to temperature, solubility, and existing trap sites. A realistic model must therefore account for the nonlinear and multidimensional effects of tritium diffusion in the porous-redeposited and neutron-irradiated materials. A tritium-transport computer model, **TRAPS** (TRitium Accumulation in Porous Structure), was developed and used to evaluate and predict the kinetics of tritium transport in porous media. This

C.H. Wu (ed.), Hydrogen Recycling at Plasma Facing Materials, 223–237.

model is coupled with the **TRICS** (Tritium In Compound Systems) code that was developed to study the effect of surface erosion during normal and abnormal operations on tritium behavior in PFCs.

2. Introduction

Damage to plasma-facing and nearby components as a result of various plasma instabilities that cause loss of plasma confinement remains a serious obstacle to successful tokamak operation. Plasma instabilities take various forms such as hard disruptions, which include both thermal and current quench (sometimes producing runaway electrons), edge-localized modes (ELMs), and vertical displacement events (VDEs). Plasma instabilities can cause surface and bulk damage to plasma-facing and structural materials. Surface damage includes high erosion losses from surface vaporization, spallation, and melt-layer erosion. In addition to these effects, the transport and redeposition of the eroded surface materials to various locations on plasma-facing and nearby components are of major concern for plasma contamination, for safety (dust inventory hazard), and for successful and prolonged plasma operation following instability events [1]. Figure 1 schematically illustrates possible erosion products and effects of plasma instabilities on target surface materials.

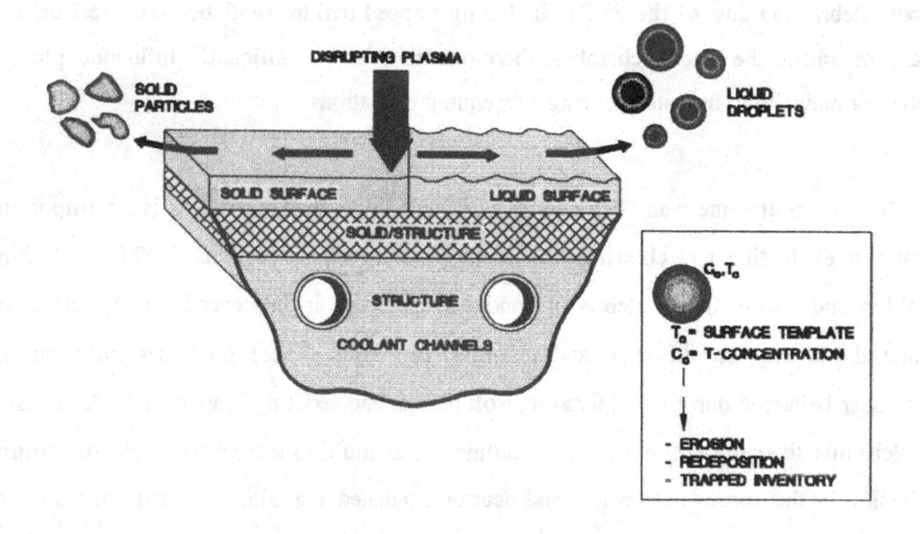

Fig. 1. Schematic illustration of erosion products as a result of plasma instabilities.

Four key factors are known to significantly influence the overall erosion lifetime of plasma-facing materials (PFMs) as a result of the intense deposited energy during loss of plasma confinements [2]: (a) characteristics of incident plasma flow (i.e., particle type, kinetic energy, energy content, deposition time, and location) from the scrape-off-layer to the divertor plate; (b) characteristics of the vapor cloud that develops from the initial phase of energy deposition on target materials and its turbulent hydrodynamics; (c) generated-photon radiation and transport in the vapor cloud and to nearby regions; and (d) characteristics of plasma/solid/melt-layer interactions.

The comprehensive **HEIGHTS** computer simulation package was developed to evaluate in detail various effects of sudden high-energy deposition of different sources on target materials and to calculate erosion products and materials lifetime [3]. The package consists of several integrated models and codes that follow the beginning of the disrupting plasma from the scrape-off layer up to the transport of the eroded debris and splashed target materials to nearby components. Factors that influence the lifetime of target materials and nearby components, such as loss of vapor-cloud confinement and vapor removal duc to MHD effects, damage to nearby surfaces due to intense vapor radiation, melt splashing, and brittle destruction of target materials, are also studied with the HEIGHTS package. The first part of the present work focuses mainly on modeling the erosion behavior of a metallic surface with a liquid layer subject to various internal and external forces during the energy-deposition phase, as well as the explosive erosion and characteristics of brittle-destruction erosion of carbon-based materials (CBMs). Lifetime predictions due to disruption erosion in a tokamak device are also presented.

Hydrogen isotope trapping and release by plasma facing and eroded/redeposited materials will influence fuel retention and recycling and plasma operation in future fusion reactors. Understanding particle and energy flow between PFMs and plasma is also necessary to optimize plasma performance, as well as to ensure safe and reliable reactor operation. Extensive studies of hydrogen diffusion and retention in fusion reactor materials have been carried out in recent years [4-6]. One important characteristic of the eroded materials is the porous nature of the redeposited layers. Erosion from plasma-facing and nearby components as a result of plasma instabilities is enough to form a redeposited layer on these components. Figure 2 schematically illustrates a typical layer structure of a porous material that may form as a result of dust deposition. The dust and debris of eroded materials may also be redeposited in different chemical forms (for example, CT_4). During reactor operation, therefore, new materials can be produced. The structure, chemical composition, transport, and retention properties of this material are quite different from the initial material. It is also very important to realize that the redeposited or codeposited layers can be porous and inhomogeneous, possibly resulting in more retained amount of tritium than in the initial PFM. Porosity is an important parameter that influences tritium buildup and release in candidate materials because of its nonlinear effect on the diffusion and retention of hydrogen isotopes. Diffusion in porous materials consists of three different and distinct diffusion processes: along grain boundaries, along microcrystallite boundaries, and diffusion in pure structure crystallites [7].

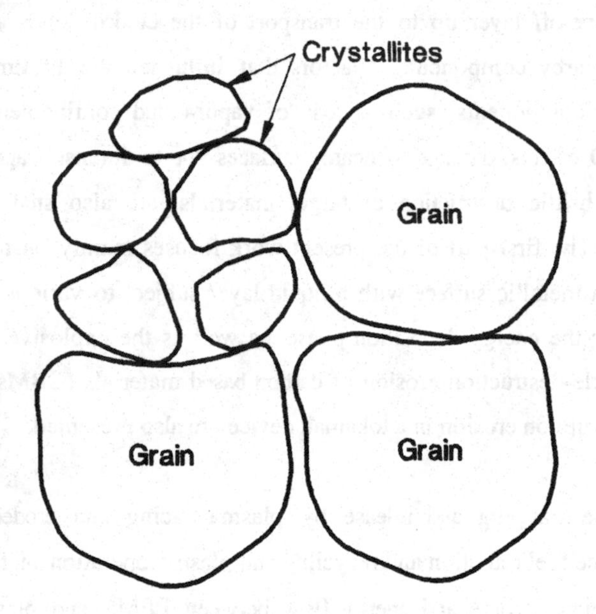

Fig. 2. Schematic view of porous structure of redeposited layers.

Comprehensive and reliable models of these phenomena do not fully exist and only few experimental data are available. Therefore, theoretical results of hydrogen isotopes retention and transport in PFM are not fully understood and do not always agree with available data. This discrepancy is especially due to the fact that diffusion in real candidate materials (graphites, Be, W) is much different (several orders of magnitude) than that of ideal homogeneous crystals, which is better described by theory. Existing models do not take into account, in a self-consistent manner, the strong inhomogeneous structure of real PFMs. The second part of the present work focuses on models of diffusion into nonhomogeneous porous materials implemented in TRAPS computer code. The effect of surface erosion during normal and abnormal operations on tritium diffusion and inventory is implemented in the TRICS code [7].

3. Surface Erosion Mechanisms

During the thermal-quench phase of a tokamak plasma disruption, part of the core plasma energy (>50% of total thermal energy) is delivered from the tokamak core to the SOL and then carried to the divertor plate by energetic plasma ion and electron fluxes. Therefore, the power load to the surface is very high, reaching hundreds of GW/m^2 and is capable of causing significant damage [1]. However, because of the developed vapor cloud of surface material in the early stages of a disruption above the divertor plate, the original surface is shielded from the incoming energy flux and the heat load onto the divertor plate surfaceis reduced significantly. Calculations with HEIGHTS predict that radiation power W_s onto the divertor plate surface is <10% of the original incident power because of the shielding effect [3].

Models for surface vaporization, material cracking and spallation, and liquid-metal ejection of melt layers have been developed for various erosion-causing mechanisms and implemented in the comprehensive HEIGHTS computer package [3]. Below are brief descriptions of some of these models and mechanisms used to study surface erosion and to predict component lifetimes.

3.1 Erosion From Surface Vaporization

Detailed physics of various interaction stages of plasma particles with target materials must be correctly modeled to evaluate the initial response of PFCs. Initially, the incident plasma particles of the disrupting plasma will deposit their energy at the surface of the target material. As a result, a vapor cloud of surface material quickly forms above the bombarded surface and in front of the incoming plasma particles. Depending on a range of parameters such as incident plasma power, magnetic field structure, geometrical considerations, vapor diffusion and motion, etc., the developed vapor cloud can significantly shield the original exposed areas from the incoming plasma particles and therefore further reduce surface damage.

To calculate the effectiveness of vapor-cloud shielding in protecting PFMs, detailed physics of plasma/vapor interactions have been modeled. The models include plasma particle slowdown and energy deposition in the expanding vapor, vapor heating, excitation, and ionization, and vapor-generated photon radiation. The detailed vapor motion above the exposed surface is calculated by solving the vapor MHD equations for conservation of mass, momentum, and energy under the influence of a strong magnetic field [8]. A significant part of the incident plasma kinetic energy is quickly transformed into vapor-generated photon radiation. Finally, multidimensional models for photon transport throughout the expanding vapor cloud have been developed to calculate the net heat flux that reaches the original disruption surface of plasma-facing components (PFCs), as well as the radiation heat load reaching various nearby components. It is the net heat flux reaching the surface that will further determine most of the response and the net erosion from surface vaporization, as well as from liquid splashing and brittle destruction of PFCs during these instabilities.

Figure 3 shows a typical time evolution of a tungsten-surface temperature, melt-layer thickness, and vaporization losses during a disruption for an incident plasma energy of 10 MJ/m^2 deposited in a disruption time of 1 ms, as predicted by HEIGHTS package [3]. An initial magnetic field strength of 5 T with an incident angle of 2° is assumed in this analysis. The sharp initial rise in surface temperature results from the direct energy deposition of incident plasma particles at the material surface. The subsequent decrease in surface temperature was caused by the shielding effect of the eroded material accumulated above the target surface. The subsequent behavior of target is determined mainly by the energy flux from the emitted photon radiation in the vapor cloud, as discussed above, and by vapor-electron heat conduction. As more vapor accumulates above the surface, the vapor becomes more opaque to photon radiation and therefore less energy is transmitted to the target surface, in turn resulting in less splashing and vaporization.

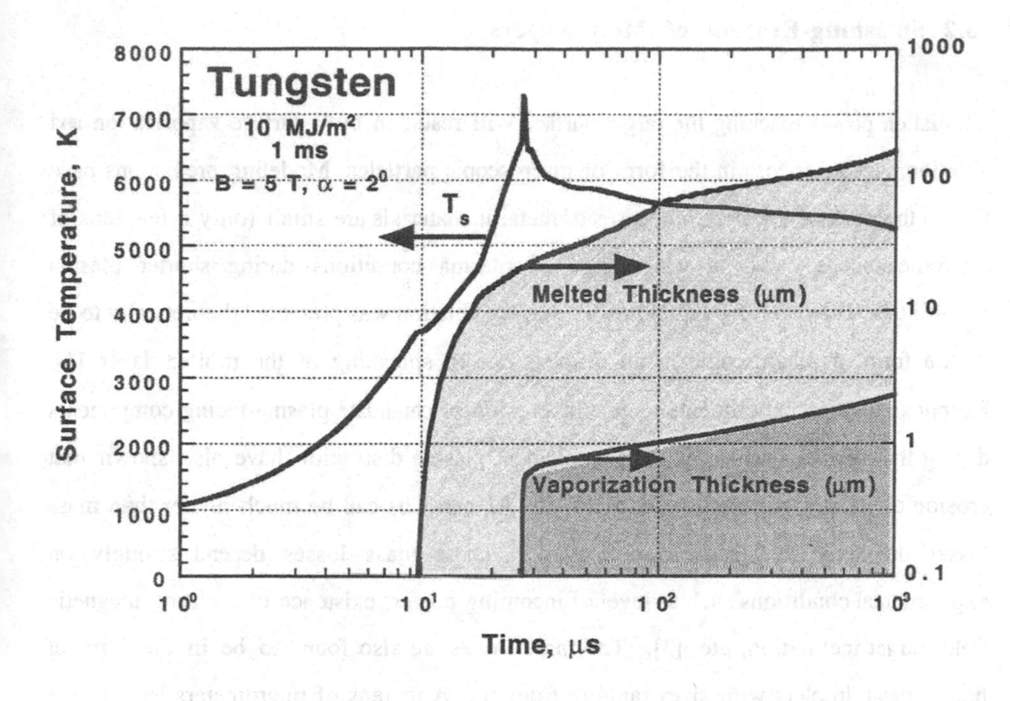

Fig. 3. Time evolution of tungsten temperature, melt layer, and vaporized thickness.

Calculation of photon radiation transport in this non-local thermodynamic equilibrium (non-LTE) vapor plasma is complex, tedious, and requires much computer time for reasonable accuracy. This is quite important because the results of this calculation will determine the net radiation power flux reaching the divertor surface and other nearby components. This net power flux will eventually determine PFC erosion lifetime. It follows from detailed calculations that the radiation power reaching the plate surface, W_{in}, is ≤10% of the incident total plasma power. There is a tendency for increasing radiation power at the plate surface with decreasing atomic number for lighter elements. Because the vapor cloud is more transparent in such cases, vapor shielding is less effective and therefore more radiation power reaches the target surface, causing more splashing [9]. The shape of the radiation power to the surface will also determine the spatial distribution of the damage profile [3].

3.2 Splashing-Erosion of Melt Layers

Radiation power reaching the target surface will result in both surface vaporization and ablation, i.e., mass loss in the form of macroscopic particles. Modeling predictions have shown that surface vaporization losses of metallic materials are small (only a few tens of micrometers deep) over a wide range of plasma conditions during shorter plasma instabilities. However, for liquid metals, surface ablation was predicted theoretically to be in the form of macroscopic metal droplets due to splashing of the molten layer [1]. Recent simulation experiments to predict erosion of candidate plasma-facing components during the thermal quench phase of a tokamak plasma disruption have also shown that erosion of metallic materials (such as W, Be, Al, and Cu) can be much higher than mass losses due only to surface vaporization. These mass losses depend strongly on experimental conditions such as level of incoming power, existence of a strong magnetic field, target inclination, etc. [3]. The mass losses are also found to be in the form of liquid-metal droplets with sizes ranging from 100 Å to tens of micrometers leaving the target surface with velocities $V \approx 10$ m/s. Such ablation occurs as a result of splashing of the liquid layer due to several mechanisms [1]. Splashing erosion can occur due to boiling and explosion of gas bubbles in the liquid, absorption of plasma momentum, hydrodynamic instabilities developed in the liquid layer from various forces, runoff of melt layers over the structure, and mechanical vibration of the machine during the disruption. One main splashing mechanism results from the hydrodynamic instabilities developed in the liquid surface (such as Kelvin-Helmholtz and Rayleigh-Taylor instabilities). It was shown that Kelvin-Helmholtz instability can occur if the vapor plasma is not well-confined by the magnetic field and vapor flow occurs along the target surface. Another splashing mechanism is from volume bubble boiling, which usually occurs from overheating of the liquid metal above the vaporization temperature T_V, i.e., the temperature at which saturation pressure is equal to the outer pressure of the vapor plasma above the divertor plate surface. Therefore, the erosion energy is roughly equal to the sum of the thermal energy (required to heat the liquid above a certain temperature, i.e., melting temperature for hydrodynamic instabilities and vaporization temperature for bubble boiling), melting energy (i.e., heat of fusion), and kinetic energy of the droplets.

The kinetic energy of the splashed droplets is determined from the surface tension of the liquid metal.

To correctly predict melt-layer erosion, a four-moving-boundaries problem is solved in the HEIGHTS package [8]. The front of the vapor cloud, generated from the initial plasma power deposition, is one moving boundary determined by solving vapor hydrodynamic equations. The second moving boundary due to surface vaporization of the target is calculated from target thermodynamics. Immediately following the surface vaporization front is a third moving boundary due to the melt-splashing front. Finally, the fourth moving boundary is at the liquid/solid interface and further determines the new thickness of the melt layer. These moving boundaries are interdependent, and a self-consistent solution must link them dynamically and simultaneously. It is the third moving boundary (the liquid splashing front), however, that determines the extent of metallic PFC erosion and lifetime due to plasma instabilities. The SPLASH code (part of the HEIGHTS package) calculates mass losses by using a splashing-wave concept as a result of each erosion-causing mechanism [3]. Thus, total erosion can be calculated from the sum of all possible erosion mechanisms.

3.3 Macroscopic Erosion of Carbon-Based-Materials

Nonmelting materials such as graphite and CBMs have also shown large erosion losses significantly exceeding that from surface vaporization. Models were developed to evaluate erosion behavior and lifetime of CBMs of plasma-facing and nearby components due to brittle destruction during plasma instabilities [3]. The macroscopic erosion of CBMs depends on three main parameters: net power flux to the surface, exposure time, and threshold energy required for brittle destruction. The required energy for brittle destruction is critical in determining the net erosion rate of CBMs and is estimated to be ≈10 kJ/g, or ≈20 kJ/cm^3. Therefore, for a net power flux to the material surface during the disruption of ≈300 kW/cm^2, the deposited energy for a time of 1 ms is ≈0.3 kJ/cm^2, resulting in net erosion of ≈150 μm per disruption. This is much higher than that predicted from pure surface vaporization of ≈10 μm per disruption for CBMs [3]. A sacrificial coating/tile thickness ≈1 cm thick would last fewer than 70 disruptions. Again, this is far fewer than the current expectation of several hundred disruptions during the reactor lifetime. Longer disruption times can also significantly reduce disruption lifetime. Therefore, more relevant experimental data and additional detailed modeling are needed to evaluate the erosion of CBMs, which strongly depends on the type of carbon material.

4. Droplets and Macroscopic Shielding Concept

Complete and accurate calculation of mass losses during plasma instabilities requires a full MHD description of the vapor media near the target surface; the media consist of a mixture of vapor and droplets moving away from surface. Photon radiation power from the upper regions of the vapor cloud will then be absorbed by both the target surface and the droplet cloud, resulting in the surface vaporization of both target and droplet surfaces. Therefore, in such a mixture of erosion products, further screening of the original target surface occurs because of the splashed droplets or macroscopic CBM debris. This has the effect of reducing photon radiation power to the target surface. Such screening can be called "droplet shielding" in an analogy to the vapor shielding effect [9]. Features of this droplet shielding and its influence on total mass loss are calculated for the cases of volume bubble boiling with homogeneous velocities of droplets in momentum space, and in the case of Rayleigh-Taylor instability with droplets that move normal to the surface preferentially [9].

To summarize the simulation results, due to overheating of the divertor plate surface, macroscopic particles and droplets are ejected/splashed upstream and away from the surface. These particles then absorb some part of the incoming vapor radiation. The net fraction of radiation power reaching the divertor plate surface is determined only by the ratio of vaporization to splashing energies. The distance at which macroscopic particles are completely vaporized is calculated to be about 100 times the initial radius of droplets. Because the initial droplet radius is small (≤10 μm), the droplets completely vanish at distance L ≤ 1 cm. Therefore, the mixture of vapor and macroscopic particles exists only very near the divertor plate surface. Because the vaporization energy is much higher than the energy required for splashing/destruction, this means that most radiation power from the upper vapor cloud is expended in vaporization. Therefore, despite the large initial splashing erosion, total erosion of the divertor plate is defined only by vaporization losses, including both divertor plate vaporization and macroscopic-particle vaporization. Again, this is true only if both the vapor cloud and the splashed droplets are well-confined in front of the incoming disrupting plasma.

5. Tritium Transport Mechanisms in the 2-D TRAPS Model

Two- and three-dimensional effects of a PFM porous-structure and the redeposited dust and debris of these materials on tritium diffusion and inventory were previousely discussed in more detail [7]. An important feature of this model is enhanced in the 2-D TRAPS code, i.e., modeling of diffusion in grain-structured materials. It is known that tokamak candidate materials (C, Be, W) will consist of grains with size, L_g, in the order of a few micrometers and are separated from each other by an intergrain substrate with width, d_g, where $d_g \approx 0.1\ L_g$. The grains in turn consist of crystallites with size L_c on the order of 50-100 Å, that are also separated from each other by intercrystallite substrates with width $d_c \approx 0.1\ L_c$, as schematically illustrated in Fig. 4 [9].

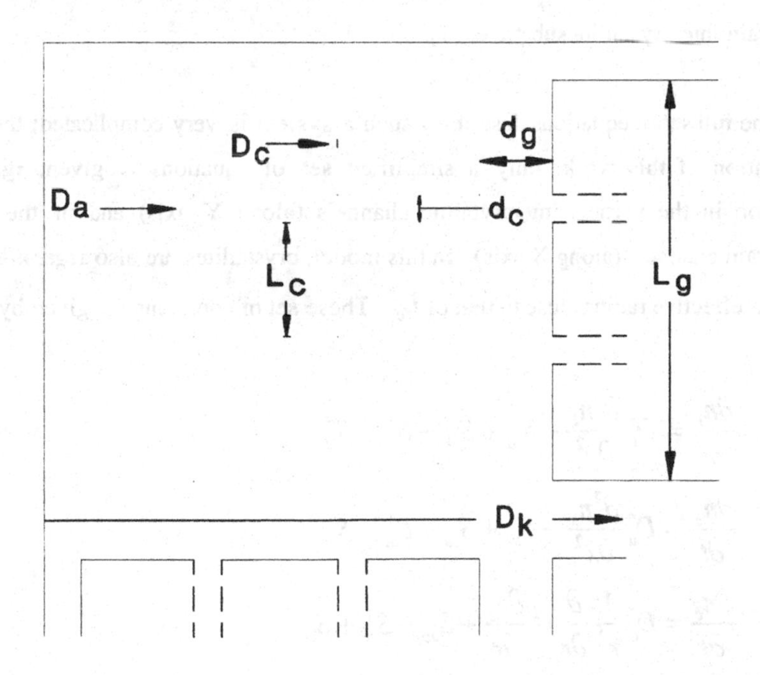

Fig. 4. Numerical simulation model of a porous structure used in TRAPS.

The diffusion coefficients of both the intergrains substrate, D_k, and intercrystallite substrate, D_a, are much greater than the diffusion coefficient, D_c, of the crystallites, which is similar to that of an ideal crystal. This means that implanted plasma particles (T, D) will diffuse mainly along these substrates. After saturation of the substrates, the implanted gas diffuses into the crystallites. Because of the small size, L_c, of crystallites, the diffusion time into the crystallites, τ_c, becomes comparable to or smaller than the diffusion time along the substrates:

$$\tau_c = \frac{L_c^2}{D_c} \approx \frac{L_g^2}{D_{k/a}}, \quad L_c \ll L_g \quad , D_k \approx D_a \tag{1}$$

To describe diffusion in such a medium, direct numerical modeling requires much computer time. Therefore, we use a set of equations that regards this inhomogeneous discrete system of grains and crystallites as a continuous medium but nevertheless takes into account the inherent discrete structure, i.e., existence of grains, crystallites, and intergrain/intercrystallite substrates [7].

The full set of equations describing such a system is very complicated; therefore, for illustration of this work, only a simplified set of equations is given, ignoring the diffusion in the vertical intercrytallite channels (along Y axis) and in the horizontal intergrain channels (along X axis). In this model, crystallites are also regarded as spheres with an effective radius close to that of L_c. These set of equations are given by [7]:

$$\begin{aligned}
\frac{\partial n_k}{\partial t} &= D_k^* \frac{\partial^2 n_k}{\partial y^2} - S_{ka} + S_{ak} - S_{kt}^+ + S_{kt}^- \\
\frac{\partial n_a}{\partial t} &= D_a^* \frac{\partial^2 n_k}{\partial x^2} - S_{ac} + S_{ca} - S_{at}^+ + S_{at}^- \\
\frac{\partial n_c}{\partial t} &= D_c^* \frac{1}{r^2} \frac{\partial}{\partial r} r^2 \frac{\partial n_c}{\partial r} + S_{ac} - S_{ct}^+ + S_{ct}^- \ ,
\end{aligned} \tag{2}$$

where subscripts k, a, c are those relating to intergrains substrate, intercrystallites substrate, and the crystallites, respectively, and D_α, (α = k, a, c) is the corresponding diffusion coefficient defined from l_α and τ_α, i.e., the corresponding path length and collision time of medium α respectively, and $n_{\alpha s}$ is maximum dissolved density. The linear number of crystallites in one grain is given as $N_c = L_g / L_c$. Fluxes in, $S^+_{\alpha x}$, and out, $S^-_{\alpha x}$, (x = k, a, c) of intergrains, intercrystallites, crystallites, and traps can be given by certain distribution functions [7,10]. Results of such modeling indicated that three distinctive regimes exist with three effective diffusion coefficients (D_1, D_2, D_3) corresponding to diffusion along grain boundaries, diffusion along crystallite boundaries, and into crystallites, taking into account the 2-D structure of the channels. The analysis indicates strong dependence on the porous structure parameters and dimensions [7].

To model tritium behavior in the original PFM as a result of erosion, the self-consistent computer model TRAPS is used [7]. The code can include up to four different tritium trapping-sites with different spatial and trapping energy distribution in the coating as well as in the substrate materials. Inhomogeneous trap distribution is necessary to account for traps created by the implanted flux, which is near the surface region, and those traps created by neutron irradiation distributed throughout the coating and the substrate structure.

Surface erosion in the TRICS code is implemented in two ways [7]: continuous erosion rate as a result of normal operation, and pulsed erosion rate due to abnormal events. Continuous erosion rate is due mainly to physical and chemical sputtering of the surface coating material. As a result of the continuous erosion, the surface location is a moving coordinate with time. The surface temperature will continue to decrease because the thickness of the material is decreasing and the incident heat flux is presumed to be time-independent. Figure 5 illustrates these processes [7]. At each time-step, the coating thickness is decreased by a certain amount, depending on the erosion rate. The decrease in the surface temperature as a result will have two main effects, i.e., reduction in recombination flux at the surface, and a resulting increase in permeation flux at the coolant side. The reduction in the coating thickness will also result in a reduction in tritium inventory and in the diffusing flux under typical tokamak conditions [7].

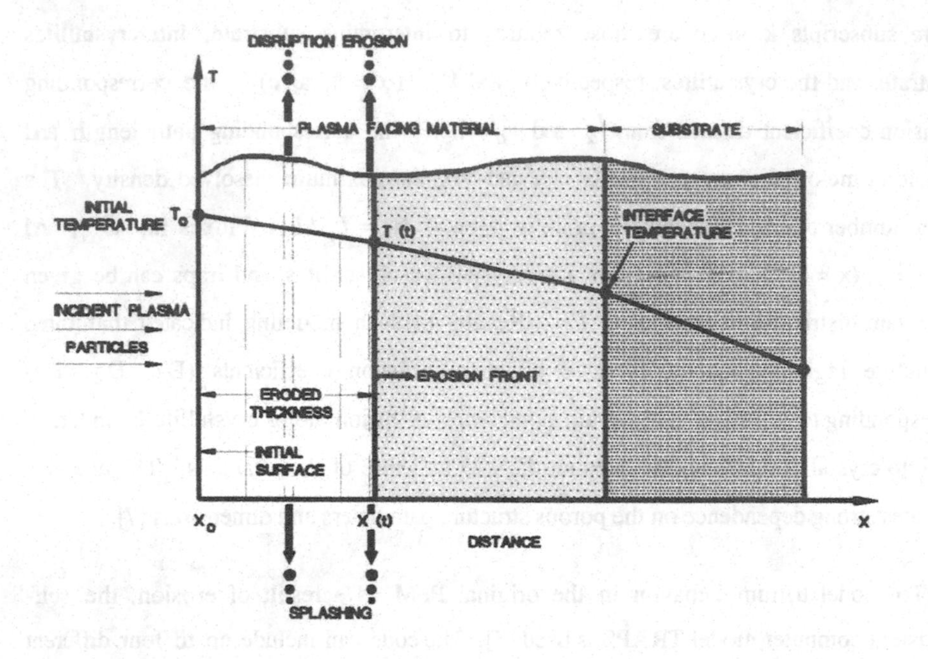

Fig. 5. Illustration of erosion processes and moving surface coordinates.

In addition, the TRICS code can account for the effect of sudden erosion due to abnormal events such as plasma disruptions and vertical displacement events. In such cases a surface layer of a certain thickness is removed with all its tritium inventory and trap concentrations. The tritium contained in the eroded materials will decrease as time passes because more tritium is diffusing to the coolant side as a result of reducing the total coating thickness. The code can also model tritium behavior in eroded or splashed material, as a result of plasma instabilities, and its redeposition on colder nearby surfaces. Future work will focus on complete and more accurate estimates of tritium diffusion and inventory for conditions resembling that of the tokamak Demo design.

6. Conclusions

The effect of porosity of redeposited dust and debris materials on tritium diffusion and inventory in such materials is quite important. The 2- and 3-dimensional effects of the porous structure of the redeposited dust and debris of PFMs (carbon, beryllium...) as well as that from plasma-spraying techniques or from original as-fabricated materials must be taken into account in the numerical models describing gas diffusion and gas release in porous two-level structural materials. Specifically designed laboratory experiments in which the dust and debris of candidate PFMs are produced and studied are required to correctly predict tritium behavior in these materials. Preliminary models of diffusion in porous materials, including diffusion along grain boundaries, intercrystallite substrates, and into crystallites, were developed. TRAPS calculations show that tritium filling of a porous material having a certain solubility and number of traps has the form of "filling-wave" that takes on a one-dimensional nature at times exceeding these of characteristic times of diffusion into separate grains and crystallites. TRAPS also takes into account other important processes such as recombination rates of atoms on material and pore surfaces and the existence of open and closed pores. Future work will evaluate and assess the total tritium inventory in porous materials as a function of reactor operating conditions. The TRICS code was developed to study the effect of surface erosion on tritium behavior in the original plasma-facing materials. Erosion generally will increase tritium concentrations in the coolant and in the eroded and splashed debris due to permeation and trapping.

7. Acknowledgment

Work is supported by the U.S. Department of Energy, Office of Fusion Energy Science, under Contract 20-31-109-Eng-38.

8. References

1. Hassanein, A. et al. (1998) Fusion Eng. Des. **39 & 40**, 201.
2. Hassanein, A. and Konkashbaev, I. (1996) J. Nucl. Mater., **233-237**, 713.
3. Hassanein, A. and Konkashbaev, I. (1999) J. Nucl. Mater. **273**, 326.
4. Andrew, P.L. and Pick. M.A. (1994) J. Nucl. Mater. **212-215**, 111.
5. Federici, G., et al. (1995) Fusion Eng. and Design **28**, 136.
6. Baskes, M.I. (1980) J. Nucl. Mater. **85 & 86**, 318.
7. Hassanein, A. et al. (1998) J. Nucl. Mater. **258-263**, 295.
8. Hassanein, A. (1996) Fusion Technol. **30**, 713.
9. Hassanein, A. et al. (2000) to be published in Fusion Eng. & Design.
10. Konkashbaev, I. et al. (1997) Fusion Technology, C. Varandas and F. Serra, eds. 1803.

EFFECTS OF Cu-IMPURITY ON RETENTION AND THERMAL RELEASE OF D IMPLANTED INTO Be

B. Tsuchiya, T. Horikawa* and K. Morita*
The Oarai Branch, Institute for Materials Research, Tohoku University, Oarai-machi, Ibaraki-ken 311-1313, Japan
**Department of Crystalline Materials Science, Graduate School of Engineering, Nagoya University, Furo-cho, Chikusa-ku, Nagoya 464-8603, Japan*

Abstract

Retention and thermal release of D implanted into 0.5 % and 2.3 % Cu-doped Be have been studied by means of an elastic recoil detection (ERD) technique combined with Rutherford backscattering (RBS) technique. It is found from the standard analysis of the ERD spectra that the saturation concentrations (D/Be) of D retained in the 0.5 % and 2.3 % Cu-doped Be, irradiated with 5 keV D_2^+ ions at room temperature, are 0.16 and 0.11, respectively, and rather lower than that in pure Be (D/Be=0.56). Re-emission experiments due to thermal anneal at temperatures from 323 K to 773 K for 10 min show that the re-emission of D retained in the specimens occurs in three stages : the lower temperature one at around 353 K and the higher temperature ones at 373~673 K and 673~773 K. The effects of Cu doped on the retention and thermal re-emission of hydrogen in Be are discussed in comparison with the experimental results with pure Be.

C.H. Wu (ed.), Hydrogen Recycling at Plasma Facing Materials, 239–245.

1. Introduction

Beryllium is one of the primary candidate plasma facing materials. However Be has a property of accumulating a large amount of hydrogen. The release of the retained hydrogen, due to the particle bombardment and the heat load from plasma, induces cool down of the core plasma. For achievement of controlled ignition and extended burn of deuterium-tritium plasmas at steady-state as a final goal, it is of essential importance to evaluate and predict the transient recycling fluxes of hydrogen from the plasma facing materials. So far, the processes for retention and the re-emission of hydrogen isotopes in Be have been studied by many authors [1-3]. Very recently, the present authors have studied the isochronal and isothermal annealing re-emissions of hydrogen isotopes retained in Be by means of the ion beam analysis technique, an elastic recoil detection (ERD) technique combined with Rutherford backscattering (RBS) technique, and have shown that the saturation concentration of D in Be (D/Be), implanted with 3 keV D_2^+ ions at room temperature, is 0.56 and 1.4 times as much as that in graphite and the re-emission of hydrogen isotopes occurs in three stages : the first one at around 353 K, the second one at 373~673 K and the last one at 673~773 K. We have also found from the experimental results that as the re-emission of D is described by the first order reaction kinetics, the activation energies for the rate constants of re-emission in the three stages are 0.14, 0.26 and 0.89 eV, respectively, and which might represent the dissociation energy from the Be-hydride clusters and the detrapping energies of hydrogen isotopes from the Be-oxide layers and the vacancy clusters such as void or bubble, respectively. It has been also shown that the ratio of saturation concentrations of H and D retained in Be by the simultaneous irradiation of H^+ and D^+ ions is well expressed using the mass balance equations where the elementary processes such as the trapping, detrapping and recombination are taken into account [4, 5].

On the other hand, in practical use Be is brazed with copper used as the heat sink. It is predicted that Be is contaminated with Cu by the interdiffusion under severe thermal heat load for long term discharge. Therefore, it is of essential importance to study the effects of Cu-impurity on the hydrogen retention properties of Be.

In this paper, we report the experimental results on retention and thermal release of D implanted into Cu-doped Be which were measured by means of the ion beam analysis techniques. The saturation concentration and the thermal re-emission curves of D in the Cu-doped Be are compared with those in pure Be.

2. Experiments

The specimens used for the retention and the thermal release measurements were 0.5 % and 2.3 % Cu-doped Be of 15 mm in diameter and 1.0 mm in thickness. The 2.3 % Cu-doped Be specimen was prepared in NGK Insulators Ltd. The 0.5 % Cu-doped Be specimen was prepared by flash heating of the 2.3 % Cu-doped Be specimen at 1073 K for 10 min under a pressure of 4.0x10^{-6} Pa. The specimens were irradiated up to saturation concentration of D with 5 keV D_2^+ ions at a flux of 4.0x10^{13} ions/cm^2 at room temperature. The incident angle was 10° to the surface normal. After the saturation implantation, the specimens were annealed at temperatures from 323 K to 773 K. The depth distributions of D retained in the specimens were measured by means of the ERD technique, where1.7 MeV He^+ ion beam was incident on the specement at 80° to the surface normal and the recoiled D^+ ions were detected at a forward recoil angle of 20° to the incident direction. Moreover, the RBS measurement was simultaneously performed in order to measure the amount of doped Cu and the thickness of oxide layers on the specimen surface, where the backscattering angle was 150° to the He^+ ion incident direction.

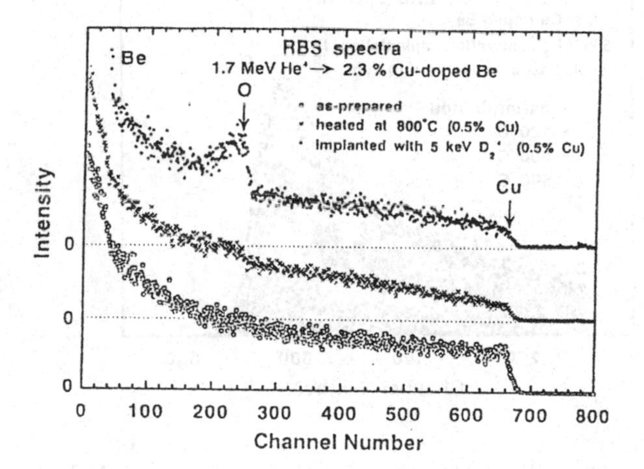

Fig. 1 RBS spectra of 1.7 MeV He^+ ions from a Cu-doped Be specimen as-prepared (2.3 % Cu) (○), heated flashily at 1073 K for 10 min (0.5 % Cu) (x) and implanted up to saturation concentration with 5 keV D_2^+ ions at room temperature (0.5 % Cu) (◆).

3. Experimental results and discussion

Typical RBS spectra of 1.7 MeV He^{+} ions from the 2.3 % Cu-doped Be are shown in Fig.1 where the spectra are shown for the specimen as-prepared (○), heated at 1073 K for 10 min (X) and irradiated with 5 keV D_2^{+} ions at room temperature (◆). It is seen from these RBS spectra that Cu, O and Be yields clearly appear at channel numbers of 660, 240 and 40. It can be seen from comparison of two spectra (○) and (X) in Fig. 1 that the Cu yield decreases after the flash heating. The fraction of Cu doped in Be was computed to be 0.5 % using the standard analysis of RBS measurement. Furthermore, it can be also seen from the spectrum (◆) of Fig. 1 that oxygen was simultaneously implanted into the specimen during irratiation with 5 keV D_2^{+} ions, because the ion gun without the mass analyser has been used to generate the D_2^{+} ion beam in present study. The oxide layers of BeO are estimated to be about 14 nm in thickness and are so stable as not to be changed due to the annealing at 1073 K.

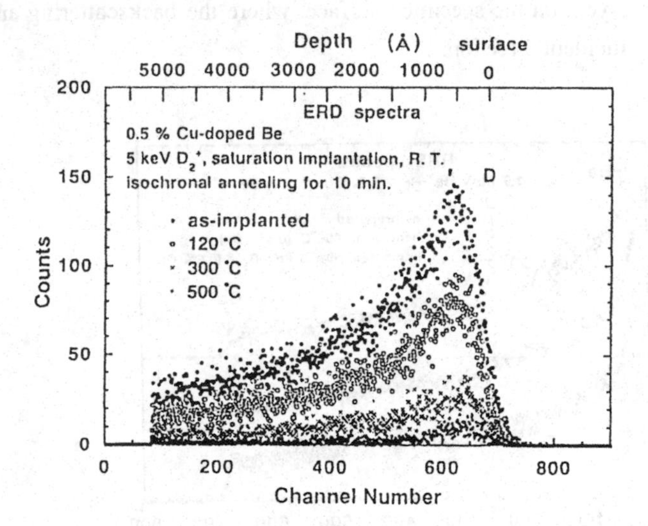

Fig. 2 ERD spectra of D recoiled from the 0.5 % Cu-doped Be specimen by 1.7 MeV He^{+} ions, which were implanted up to saturation concentration of D with 5 keV D_2^{+} ions at room temperature (●) and subsequently annealed at temperatures of 393 K (○), 573 K (X) and 773 K (◆) for 10 min.

Fig. 2 shows typical ERD spectra of D recoiled from the 0.5 % Cu-doped Be specimen, which were measured after the saturation implantation with 5 keV D_2^{+} ions at room temperature (●) and the annealings at 393 K (○), 573 K (X) and 773 K (◆)

for 10 min. The vertical axis corresponds to the concentration of D retained specimen. The horizontal axis corresponds to the energy of the recoiled D, from which the depth from the specimen surface is estimated, as shown at the upper part of Fig. 2. It is found that the saturation concentration (D/Be) of D in the 0.5 % Cu-doped Be specimen is 0.16, which was calculated from the standard analysis of the ERD spectrum the counts averaged over 40 channels around the peak in ERD spectrum (●) of Fig. 2 and the He^+ ion fluence, the recoil cross-sections of He^+ ion for D, the stopping cross-sections [6] for D^+ and He^+ ions in Cu-Be and the solid angle of detection using. For the 2.3 % Cu-doped Be specimen, the value of D/Be is determined to be 0.11 using the same analysis. These values are rather lower than that for the pure Be specimen (D/Be=0.56 [5]). The result indicates that the number of trap sites for D in Be decreases due to the existence of Cu. For the long term discharge, it may be predicted that the transient concentration of hydrogen retained in Be brazed with Cu is reduced by the contamination through the interdiffusion.

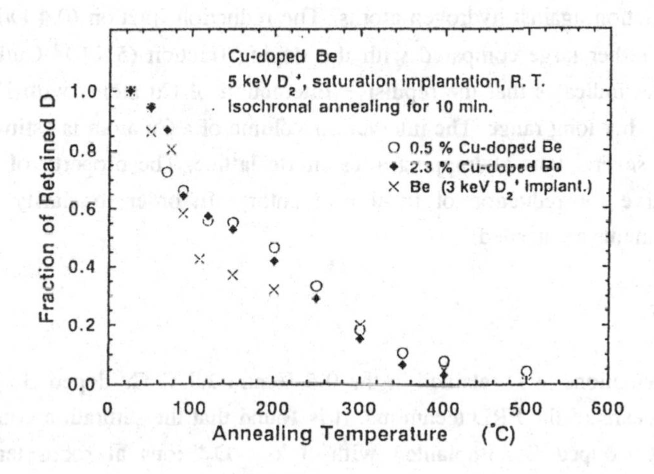

Fig. 3 Thermal re-emission curves of D in pure Be (X) and in 0.5 % (O) and 2.3 % (◆) Cu-doped Be, which were implanted up to saturation concentration and hereafter isochronally annealed from 323 to 773 K for 10 min.

Fig. 3 shows the thermal re-emission curves of D for 0.5 % (O) and 2.3 % (◆) Cu-doped Be and pure Be (X), implanted up to saturation concentration at room temperature and subsequently annealed at temperatures from 323 K to 773 K for 10 min. The vertical and the horizontal axes represent the fraction of retained D which is normalized by the saturation concentration and the annealing temperature, respectively.

One can see clearly from Fig. 3 that the thermal re-emission of D from the specimen occurs in three stages : the first one at around 353 K, the second one at 373~673 K and the third one at 673~773 K. This fact indicates that the Cu-impurity dose not influence the re-emission processes of D. It is also seen from Fig. 3 that the curves for Cu-doped Be correspond well with that for pure Be, except in the temperature range of 373~473 K. According to the previous interpretation proposed by the present authors [4, 5], the re-emissions at each stage are attributed to thermal dissociation of linear chain-like clusters of BeD_2 which might be a stable Be-D compound in the bulk [7], thermal desorption of D from the oxide layers in the surface region and thermal detrapping from the extended defects formed by the radiation damage, respectively. Therefore, increase in the fraction of retained D in the temperature range of 373~473 K in Fig. 3 is also attributed to increase in the thickness of oxide layers formed by implantation with D_2^+ ions at higher energy (5 keV), as shown in the RBS spectrum (◆) of Fig. 1. This fact indicates that the oxide layers play the role inportant in trapping of implanted D.

The reduction in the D retention due to existence of Cu-impurity is ascribed to its repulsive interaction against hydrogen atoms. The reduction fraction (0.4 D/Be) of the D retention is rather large compared with the doping fraction (5×10^{-3} Cu/Be) of Cu atoms. This fact indicates that the repulsive interaction of Cu atoms with D atoms is not short range, but long range. The interaction volume of a Cu atom is estimated to be extended over several tens of trapping sites in Be lattice. The property of Cu atoms may be effective for reduction of tritium inventory. In order to clarity it, further detailed experiments are needed.

4. Summary

The depth distributions of D retained in the 0.5 % and 2.3 % Cu-doped Be have been measured by means of the ERD technique. It is found that the saturation concentration of D in the Cu-doped Be, implanted with 5 keV D_2^+ ions at room temperature, decreases with increasing the amount of doped Cu. The thermal annealing data has been found to be essentially the same as those for pure Be which indicates that thermal release of D from Cu-doped Be takes place due to the dissociation from Be-hydride clusters at the lower temperature of 353 K and the detrapping from the Be-oxide layers on the surface and the extended defects such as void or bubble, formed by the radiation damage, at the higher temperature of 573±200 K.

5. Acknowledgement

The authors are grateful to Dr K. Nishida in NGK Insulators Ltd Co for supplying samples of Cu-doped Be used in this study.

6. References

Wampler, W. R. (1984) Retention and Thermal Release of Deuterium Implanted in Beryllium, *J. Nucl. Mater.* 122-123, 1598-1602.

Wampler, W. R. (1992) Trapping of deuterium in beryllium, *J. Nucl. Mater.* 196-198, 981-985.

Yoshida, N., Mizusawa, S., Sakamoto, R. and Muroga, T. (1996) Radiation damage and deuterium trapping in deuterium ion injected beryllium, *J. Nucl. Mater.* 233-237, 874-879.

Tsuchiya, B. and Morita, K. (1996) Retention and re-emission of hydrogen in beryllium studied by the ERD technique, *J. Nucl. Mater.* 233-237, 898-901.

Tsuchiya, B. and Morita, K. (1997) Retention of hydrogen isotopes in beryllium by simultaneous H^+ and D^+ irradiation, *J. Nucl. Mater.* 248, 42-45.

Andersen, H. H. and Ziegler, J. F. (1977) Hydrogen Stopping Powers and Ranges in All Elements, *Plenum Press, New York*,.

Bell, N. A., Coastes, G. E. and Emsley, F. W. (1966) Dimethylberyllium. PartIV.[1] Spectroscopic Properties of Some Methyl- beryllium Compounds, *J. Chem. Soc. London*, A, 49-52.

DEUTERIUM RETENTION IN BERYLLIUM AND BERYLLIUM OXIDE

V.Kh. Alimov* and A.P. Zakharov

Institute of Physical Chemistry, Russian Academy of Sciences
Leninsky prospect 31, 117915 Moscow, Russia

Abstract

Retention of deuterium in Be sample and BeO film implanted with D ions or exposed to D atoms at temperatures in the range 300 to 700 K have been studied using SIMS and RGA (residual gas analysis) measurements in the course of surface sputtering. The microstructure of implanted specimens was investigated by TEM.

(1) Be and BeO implanted with (3-9) keV D-ions accumulate deuterium in the form of both D atoms and D_2 molecules. In Be irradiated with D ions up to high fluences at which maximum D concentration is reached, the fraction of deuterium trapped in the form of D2 molecules is ~0.8 at irradiation temperature *Tirr* =300 K and ~0.6 at *Tirr* =700 K. As for BeO, the majority of implanted deuterium is retained as D atoms both at 300 and 700 K.

In Be sample and BeO film implanted with D ions at T_{irr} = 300 K, the D_2 molecules are present in tiny bubbles (Be and BeO) and in intercrystalline non-interconnected gaps (BeO). In Be irradiated at T_{irr} = 700 K, along with relatively small facetted bubbles (near the very surface), large oblate gas-filled cavities and channels forming extended labyrinths. The microstructure of BeO film subjected to D ion implantation at T_{irr} = 700 K is indistinguishable from that of BeO film irradiated at T_{irr} = 300 K.

The high concentration of D atoms in the ion stopping zone of Be matrix and BeO film after implantation to saturated fluences at T_{irr} = 300 and 700 K is attributed to deuterium (i) trapped in radiation vacancies, (ii) adsorbed on the walls of bubbles and channels and (iii) chemically bonded to O atoms adjacent to point defects in the BeO lattice.

(2) Due to the presence of oxygen containing molecules as trace impurities, the beryllium oxide layer is formed on the surface of Be sample exposed to D atoms at elevated temperatures. It has been found that the deuterium atom exposure leads to the deuterium retention in this layer. The majority of deuterium (> 90%) retains as D atoms, the other part is accumulated in the form of D_2 molecules. After termination of D atom exposure D concentration in BeO layer decreases with time at room temperature. It is supposed that the formation of beryllium hydroxide $Be(OD)_2$ under D atoms exposure of growing BeO layer takes place. The decrease of deuterium concentration in time is explained by dehydration of $Be(OD)_2$.

* Tel.: (095) 330 2192, fax: (095) 334 8531; e-mail: alimov@ipc.rssi.ru

C.H. Wu (ed.), Hydrogen Recycling at Plasma Facing Materials, 247–264.

1. Introduction

Present concepts for ITER consider beryllium as a candidate material for the plasma-facing component of the first-wall system [1]. Sputtering of Be tiles with hydrogen isotope ions and neutrals as well as by impurity atoms leads to codeposition of beryllium, carbon, oxygen and hydrogen isotope atoms at deposition dominated areas of the wall [2-4]. Due to a high affinity of Be to oxygen [5], beryllium oxide formation takes place. Therefore, reliable data on the hydrogen retention in beryllium and beryllium oxide, in particular, under irradiation are of great importance for the application of Be tiles in fusion devices.

Trapping characteristics of hydrogen implanted into beryllium have been reviewed by Wilson et al. [5]. The most systematic studies of hydrogen retention have been performed by Wampler [6,7], Möller et al. [8] and Kawamura et al. [9]. At irradiation temperature, T_{irr} = 300 K and low fluences, Φ, 100% of deuterium implanted into Be is trapped, while at high Φ the maximum gas concentration reaches 0.31±0.05 D/Be [6,9,10]. Wampler suggested two types of traps for D atoms with detrapping energies of 1 eV and 1.8 eV, correspondingly, and put forward an idea about the formation of deuterium bubbles [6]. At low Φ implanted D atoms were believed to be trapped in vacancies, while at high Φ deuterium was concluded to precipitate into gas bubbles [7]. Having in mind a very low solubility of hydrogen in Be [11,12], Pemsler and Rapperport [13] have shown using optical microscopy that, presumably, H_2 filled cavities were formed in this metal implanted with 7.5 MeV protons and annealed at 623 K and higher. Blister formation due to hydrogen ion implantation was reported in Ref. [14]. Thus, if even not direct, some facts in the literature pointed out that implanted hydrogen could reside in Be not only as separate atoms, but also in the form of molecules.

As for beryllium oxide, a wide range of values for the tritium diffusivity was observed by Fowler et al. [15] in studies of single crystal, sintered and powdered BeO with activation energies ranging from 0.7 to 2.3 eV. The value for the tritium diffusivity in sintered BeO, $D = 7\times10^{-6}\cdot exp(-2.1\ \mathrm{eV}/kT)\ \mathrm{m^2s^{-1}}$ [15], is generally used for BeO films thermally grown onto a Be substrate. The solubility of deuterium in BeO was determined to be $S \approx 10^{18}\cdot exp(-0.8\ \mathrm{eV}/kT)\ \mathrm{at.\cdot m^{-3}\cdot Pa^{-0.5}}$ [16]. The solubility seems to be related to the formation of hydroxide bonds, because of the similarity of the activation energy, -0.8 eV, to the O-H bond formation energy of -0.81 eV [17].

The trapping of ion implanted deuterium (5 keV D^+) in thermally grown BeO films at implant temperatures ranged from 140 to 470 K was investigated by Behrisch et al. [18] using the $D(^3He,H)^4He$ nuclear reaction. The ratio of implanted D to BeO molecules obtained at

saturation was found to be 0.34 and 0.24 for implant temperatures of 140 and 300-470 K, respectively.

The paper reviews results on the evolution of concentration profiles of D atoms and D_2 molecules in Be matrix and BeO films implanted with (3-9) keV D ions at 300 and 700 K and exposed to D atoms at 340 and 500 K.

2. Experimental

S-65B type beryllium (Brush Wellman), a powder metallurgy product that is vacuum hot pressed, containing about 1% of BeO, was used as a Be sample and as a substrate for the formation of BeO layer. According to the data of electron probe microanalysis (EPMA), the thickness of initial BeO layer formed on the surface of Be sample after electropolishing procedure was less than 6 nm.

The formation of thick beryllium oxide layer was done by annealing of Be electropolished plate at 1023 K for 60 min in air atmosphere at pressure of 0.5 kPa. The thickness of the BeO layer seems to be uniform and according to both an estimation by EPMA and an evaluation by the weight gain, is 120±20 nm.

D ion implantation and depth profile measurement were performed in a special two-chamber UHV system with a typical background pressure lower that 1×10^{-7} Pa. The samples were implanted in the first vacuum chamber with mass-separated D_2^+ ions at energy of either 18 keV (Be samples) or 6 keV (Be layers thermally grown onto Be substrates, below termed BeO/Be samples) at irradiation temperatures T_{irr} = 300 and 700 K.

A plasma source with heated cathode was used to produce deuterium atoms. During D atom exposure the positive potential of +80 V was applied to Be sample to screen the sample from positive D ions. The fraction of negative D ions in the incident deuterium stream was practically negligible. Therefore, the sample was mainly exposed to atomic and molecular deuterium and electron fluxes. The flux of incident D atoms was ~10^{20} $m^{-2}\cdot s^{-1}$ [19], the deuterium pressure in the chamber during the exposure was about 5×10^{-1} Pa. The D atom exposure was performed for 1 h at temperatures T_{exp} = 340 and 500 K. It should be noted that Be samples were also exposed simultaneously to oxygen and carbon containing impurities from residual gas (background pressure in the plasma source chamber was about 1×10^{-4} Pa).

After D ion implantation or D atom exposure the Be or BeO/Be samples were transferred into the second analytical chamber for SIMS measurements of H^- and D^- secondary ion yields and RGA measurements of the partial pressures of D_2 and HD molecules in the

course of sputtering of the surface with 4 keV Ar^+ [20,21]. All measurements were made at 300 K in 1-2 days after ion implantation. The use of two quadrupole mass spectrometers (QMS) made it possible to register SIMS and RGA signals simultaneously. Partial pressures of HD and D_2 molecules evolved were determined taking into account the background intensities of masses 3 and 4, which were measured with the Ar ion beam switched off. The sputtering rate of the sample was estimated with an accuracy of 30% as a ratio of the depth of a crater produced by sputtering on the surface to the sputtering time. It should be noted that the sputtering rate of BeO is less than that of Be by about 20%.

We attribute the appearance of the SIMS D^- signal to the existence of separate D atoms within the matrix. When recording D atom depth profiles, a special method was undertaken to minimize the influence of oxygen atoms on the D^- yield. It consisted in the simultaneous monitoring of the D^-/H^- signal ratio, as a correction factor to the experimental D^- yield. In the case Be sample, the underlying assumption is that during sputtering at constant parameters of the Ar ion beam, protium yield depends solely on vacuum conditions. Usually, the H^- yield is first relatively high until a layer of surface oxide is sputtered fully and then remains constant with an accuracy of 20%. As for BeO/Be sample, thermal oxidation of Be in the presence of H_2O vapors is accompanied by a substantial increase of the protium concentration in BeO film [22] and the protium yield from BeO is constant during Ar ion sputtering.

Due to recombination of sputtered D atoms with H atoms adsorbed on the inner surfaces of the vacuum chamber, the dependencies of the HD RGA signal on the depth are reminiscent of those of the D^- SIMS signal. A cause of the D_2 signal appearance seems to be the recombination of D atoms as well as the direct release of D_2 molecules from the sputtered layers. In order to distinguish the recombination and molecular fractions, the HD and D_2 RGA signals were measured simultaneously. We assume that in the absence of the molecular fraction, the intensity of the D_2 signal should be proportional to the square of the intensity of the HD signal. Such a proportionality is observed in fact after D ion irradiation of beryllium at 300 K up to fluences $\Phi < 1\times10^{20}$ D/m^2. The coefficient of the proportionality, K, depends on vacuum conditions and has much in common with that obtained in our earlier experiments with graphite [20]. Regarding K as a constant during sputtering in RGA measurements, one can determine the intensity of the signal I_{mol} caused solely by the release of molecular deuterium:

$$I_{mol} = I_{D_2} - K(I_{HD})^2, \quad (1)$$

where I_{D_2} and I_{HD} are the experimentally measured intensities of the D_2 and HD RGA signals, respectively.

The D-atom concentration was determined by comparison of the integral SIMS signal (all over the implantation depth) with the total amount of deuterium in the sample implanted at 300 K to $\Phi = 6.0\times10^{19}$ D/m^2. For the above fluence Be or BeO/Be sample retains 100% of implanted atoms [8,18] and the value of I_{mol} is zero. The amount of D_2 molecules released under sputtering was estimated from a I_{mol} signal value taking into account the sensitivity of RGA QMS and a known pumping speed. The total error of RGA QMS calibration was about 40%.

The amounts of oxygen and carbon within the surface films were determined by quantitative electron probe microanalysis (EPMA) method with making corrections, respectively, for oxygen and carbon present in the bulk of substrates.

The microstructures of Be and BeO/Be samples subjected to D ion implantation and D atom exposure were studied in analytical TEM EM-400T operated at 120 kV. The detailed information on TEM specimen preparation and methods of TEM data processing are given in Ref.[23]. Bright field and dark field images were analyzed. Small details were imaged using phase contrast and were recorded mainly under underfocusing conditions. Evaluation of the thickness of beryllium oxide layers on the surface of Be specimens was carried out by stereo-pair analysis or directly by viewing these specimens in cross-section when tilting them in the goniostage at a large angle. Selected area diffraction patterns (SADPs) were taken from areas of 0.5 or 1.0 μm in diameter.

3. Results and Discussion

3.1. Retention of ion implanted deuterium in beryllium

Depth profiles of deuterium implanted into Be as 9 keV D ions at T_{irr} =300 K and trapped in the implantation zone in the form of D atoms and D_2 molecules are shown in Fig. 1. With the fluence increase the D atom concentration, C_D, reaches the value of about 2×10^{27} atoms/m^3 (0.02 D/Be) at the depth of the mean projected range R_p and practically does not change any more. A significant feature is a much higher value of C_D near the surface compared to that in the bulk.

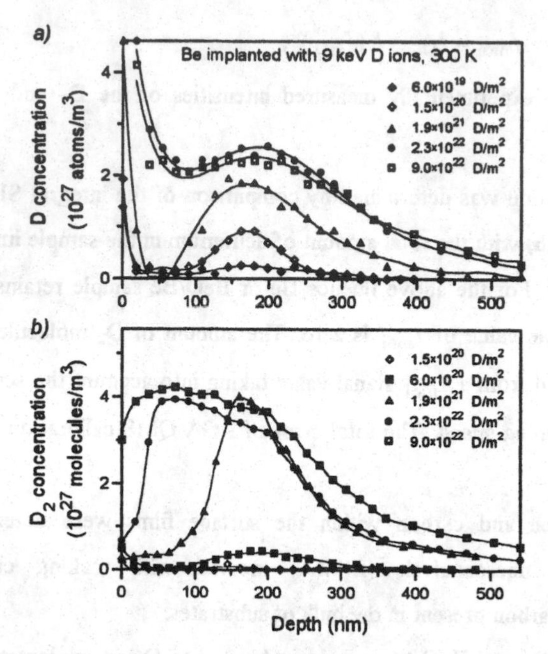

Fig. 1. Concentration profiles of deuterium trapped as D atoms (a) and in the form of D_2 molecules (b) in Be implanted with 9 keV D ions at 300 K.

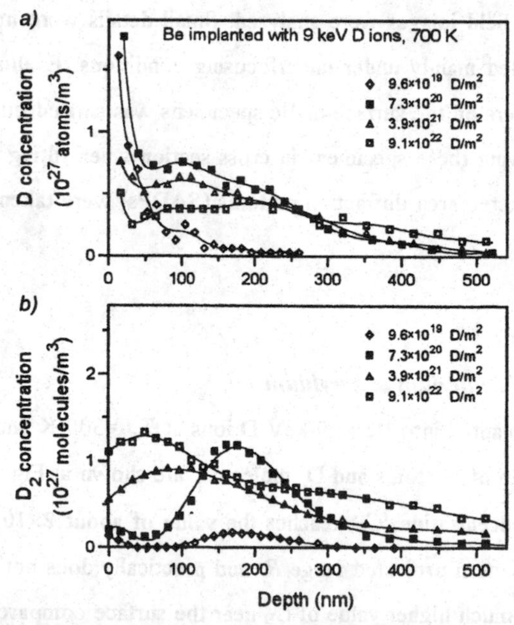

Fig. 2. Concentration profiles of deuterium trapped as D atoms (a) and in the form of D_2 molecules (b) in Be implanted with 9 keV D ions at 700 K.

Analysis of depth profiles of molecular deuterium shows that the formation of D_2 molecules in the ion stopping zone starts at a fluence $\Phi \geq 1.5\times10^{20}$ D/m^2 (Fig. 1b) that is at $C_D \geq 1$ at% (Fig. 1a). The concentration of molecules, C_{D_2}, increases nonlinearly with a fluence. While C_D (at the depth of R_p) only doubles when increasing Φ from 1.5×10^{20} to 1.9×10^{21} D/m^2 (Fig. 1a), the concentration of D_2 increases more than an order of magnitude reaching its maximum value of 4×10^{27} molecules/m^3 (0.04 D_2/Be). The further fluence increase leads to the broadening of C_{D_2} profile (Fig. 1b).

At maximal Φ and T_{irr} =300 K the concentration of deuterium present in both states (D atoms and D_2 molecules) at the depth of R_p is estimated to be about 0.1 D/Be. It should be noted that the concentration of 0.1 D/Be derived from our experiments is lower than the values determined in Refs.[6,9,10]. This discrepancy is attributed to the fact that the deuterium accumulated in Be has a tendency for a partial escape after the implantation completion. Within 14 h its concentration falls down from 0.36 to 0.24 D/Be, as it was observed in Ref.[9]. In the present work SIMS and RGA measurements were carried out in 1-2 days after ion implantation which can serve a plausible explanation for the magnitude of D/Be ratio value we obtained.

Quite different features of deuterium accumulation were found for T_{irr} = 700 K (Fig. 2). Depth profiles of C_D do not concur with the ion projected range distributions and are characterized by a gradual decrease with the depth. The maximal concentration C_D in the ion stopping zone (excluding surface layers) does not exceed 8×10^{26} atoms/m^3 (about 0.01 D/Be) (Fig. 2a).

In contrast to T_{irr} = 300 K, an active formation of D_2 molecules at 700 K starts already at Φ as low as 9.6×10^{19} D/m^2 (Fig. 2b). For fluences up to about 7×10^{20} D/m^2 the shape of C_{D_2} profiles agrees with that of the ion projected range distributions. At higher Φ the shape of D_2 profiles undergoes transformations, namely C_{D_2} decreases by about 30% at the depth of R_p. The C_{D_2} maximum of 1×10^{27} molecules/m^3 (0.01 D_2/Be) appears at the depth of 50-100 nm (Fig. 2b).

After T_{irr} = 700 K the maximal total concentration of deuterium (D atoms and D_2 molecules) in the ion stopping zone was estimated to be ≈ 0.03 D/Be.

The application of TEM made it possible to relate the appearance of D_2 molecules in the ion stopping zone to the formation of D_2 gas bubbles and cavities. Irradiation of Be with D

ions at 300 K leads to the formation of tiny bubbles of mean radius r_b = 1.0 nm and volume density $c_b = 3\times10^{24}$ m^{-3} with internal gas pressures near or higher than the equilibrium values of (5-9) GPa. When the total deuterium concentration in the ion stopping range reaches about 0.08 D/Be atomic ratio, bubble interconnection begins and extends towards the surface with increasing fluence.

In Be samples implanted with D ions at T_{irr} = 700 K the most of molecular gas is found in large oblate and flattened-out cavities forming at a depth of R_p complicated labyrinth system of channels which is interacted with the outer surface. A minor part of captured D_2 gas appears to be within smaller in size well facetted gas filled cavities found closer to the surface. The average lateral width of channels grows with the fluence increase from (150-250) nm at Φ = 2×10^{21} D/m^2 up to (300-500) nm at $\Phi = 4\times10^{21}$ D/m^2. The thickness of elements belonging to these structures in the direction of the surface was estimated as less than 100 nm and the values of gas swelling, calculating on the swelling layer, of more than 50%. Making use of a D_2 depth distribution for $\Phi = 4\times10^{21}$ D/m^2 (Fig. 2b) the mean inner D_2 pressure in the volume of closed labyrinth system was estimated as ≤ 50 MPa.

The D_2 depth profiles measured after T_{irr} = 300 and 700 K (Figs. 1b and 2b) reflect the processes of gas cavity evolution with fluence increase. For instance, the coalescence of gas bubbles under D ion implantation at 700 K and the development of interconnected channels explain the D_2 profile transformations at Φ above 7×10^{20} D/m^2 (Fig. 2b).

As to D atoms, we believe that at T_{irr} = 300 K they are trapped, first in radiation vacancies as it was proposed in Refs.[6,7] and checked experimentally in Ref.[23]. At T_{irr} = 700 K deuterium-vacancy complexes are unstable [6,7,23]. A relatively high concentration of D atoms in the surface oxide layer at T_{irr} = 300 and 700 K can be associated with the formation of beryllium hydroxide as a consequence of the reaction between D atoms and BeO [24]. A radiation enhanced growth of BeO surface layers (with a radiation modified microstructure) revealed in Ref.[23] may provide for the further increase of the amount of D atoms captured near the surface. Obviously, BeO inclusions in the bulk are strong traps for D atoms too. Another source of D atoms in D implanted Be at T_{irr} = 300 and 700 K is deuterium adsorbed on bubble and cavity walls [23,25]. Rough destruction in the near-surface layers and uncovering of interconnected wide channel systems occurring at T_{irr} = 700 K are possibly the reasons of the non-monotonous evolution of C_D and C_{D_2} profiles with the fluence increase at $\Phi \geq 3\times10^{21}$ D/m^2 (Fig. 2).

3.2. Retention of ion implanted deuterium in beryllium oxide

According to TEM studies, beryllium oxide layers thermally grown onto Be substrates consist of small cph-microcrystals of a size of 25-30 nm locally separated by not interconnected gaps (channels) of about 1 nm in width and about 10 nm in length. The estimated value of the free volume associated with these gaps is 7-10%.

Depth profiles of D atoms and D_2 molecules in BeO/Be sample irradiated with 3 keV D ions at T_{irr} =300 K are shown in Fig. 3. The maximum of the D atom distribution at $\Phi \leq 2\times10^{21}$ D/m^2 is at about 50 nm, while the thickness of BeO layer is about 120 nm. With the fluence increase the D atom concentration, C_D, reaches the value of about 1.3×10^{28} D atoms/m^3 (0.18 D/BeO) and practically does not change thereafter (Fig. 3a).

At low fluences, $\Phi < 2.2\times10^{20}$ D/m^2, depth profiles of D_2 concentration, C_{D_2}, are characterized by a gradual decrease with depth (Fig. 3b). At $\Phi = 6.7\times10^{20}$ D/m^2 C_{D_2} profile concurs with the ion projected range and the further fluence increase leads to broadening of C_{D_2} profile shape. The maximum value of D_2 concentration at T_{irr} = 300 K is about 8×10^{26} molecules/m^3 (0.01 D_2/BeO)

As is shown in Fig. 4, the maximum concentration of D atoms reaches 4×10^{27} atoms/m^3 (0.05 D/BeO) at T_{irr} = 700 K. Depth profiles of D_2 molecules do not concur with the ion projected range distributions and are characterized by a gradual decrease with the depth. The maximum concentration C_{D_2} in the ion stopping zone does not exceed 8×10^{26} molecules/m^3 (0.01 D_2/BeO).

Thus, the increase of irradiation temperature from 300 to 700 K leads to the decrease of D atom concentration by a factor of about 3 without marked change in the D_2 content.

At T_{irr} =300 K, the maximum concentration of deuterium present in both states (D atoms and D_2 molecules) in BeO layer is estimated to be about 0.2 D/BeO. This value correlates well with that reported in Ref.[18] where the ratio of D atoms to BeO molecules was found to be 0.24. After the ion implantation at T_{irr} = 700 K, the maximum total concentration of deuterium atoms and molecules in the ion stopping zone was estimated to be $\approx$ 0.07 D/BeO. It is pertinent to note that a similar three-fold decrease of the amount of deuterium in oxidized Be pre-implanted with 1 keV D ions at 300 K to saturation was observed by Möller et al. after sample heating up to 700 K [8].

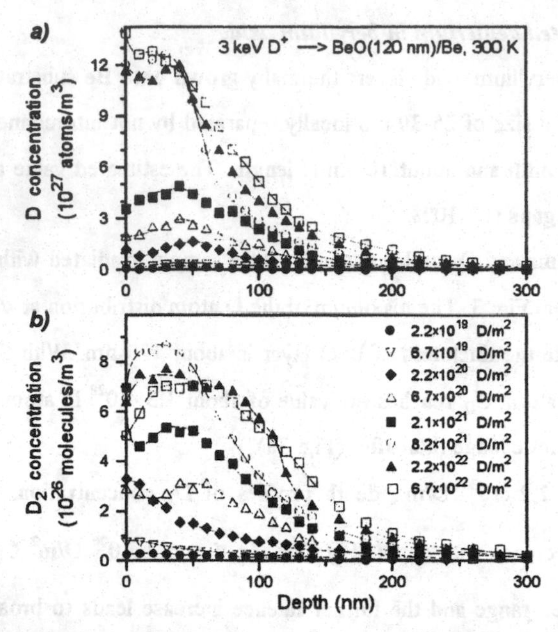

Fig. 3. Concentration profiles of deuterium trapped as D atoms (a) and in the form of D_2 molecules (b) in 120 nm layer of BeO on Be substrate implanted with 3 keV D ions at 300 K.

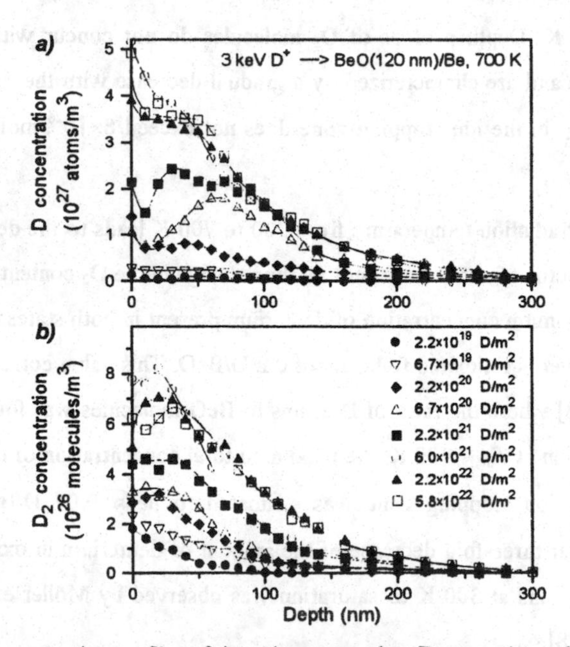

Fig. 4. Concentration profiles of deuterium trapped as D atoms (a) and in the form of D_2 molecules (b) in 120 nm layer of BeO on Be substrate implanted with 3 keV D ions at 700 K.

After D ion irradiation at T_{irr} = 300 K ion-induced defects were revealed within the oxide layer by TEM. Firstly, there is a high volume density of tiny round cavities (mean radius r_b = 0.6 nm, volume density c_b = 5×10^{24} m^{-3}). Taking into consideration the presence of D_2 molecules in D implanted BeO, we have concluded that these cavities were D_2 filled bubbles. Secondly, a slight smearing of BeO diffraction rings on selected area diffraction patterns provide evidences about noticeable modifications on the atomic scale. Their origin can be radiation-induced point defects and their complexes invisible in TEM together with internal stresses within the oxide due to these defects. Thirdly, some changes in the appearance of a gap network take place, namely, curing of a part of the as-grown gaps (under the action of radiation-induced stresses) and widening of some of the remaining gaps (presumably under the action of accumulated deuterium).

The microstructure of BeO layers subjected to D ion implantation at T_{irr} = 700 K, as it is evidenced by TEM, is visually indistinguishable from that of BeO layers irradiated at 300 K. Also at 700 K, very small D_2 bubbles are formed. The parameters of gas bubbles are practically identical to those observed for the case of T_{irr} =300 K. At the same time, one establishes a slight increase of mean radius of the bubbles for 700 K (r_b =0.7 nm) over that for 300 K (r_b =0.6 nm) with a high degree of confidence. No striking differences are found in the appearance of intercrystalline gaps when comparing TEM micrographs of specimens irradiated at T_{irr} = 300 and 700 K.

Whereas the depth profiles of D atoms at T_{irr} = 300 K is roughly described by the distribution of the ion projected range, D_2 profiles at fluences below 2×10^{20} D/m^2 are thought to be consistent with the distribution of damages generated in BeO lattice mainly due to inelastic interaction of moving D particles with electrons of oxygen anions. The anion and cation vacancies (F-centers and V_1-centers, respectively) are formed as a result of the irradiation. The neutral V_1F-centers are probably served as the pore nuclei and the traps for molecular deuterium at low D ion fluences. The fluence increase leads to the deuterium bubble formation at the depth of the ion projected range, and shape of D_2 depth profiles changes - becoming Gaussian-like at a fluence of 6.7×10^{20} D/m^2. Comparison of concentration profiles of atomic and molecular deuterium in BeO allows one to estimate a minimum concentration of separate D atoms in the matrix sufficient for the onset of the appearance of D_2 gas bubbles. It is about 3 at.%. As for metallic Be irradiated with D ions, first D_2 bubbles appear at a D atom concentration of about 1 at.%.

The uniform distribution of deuterium in the BeO film at high fluences cannot be understood from diffusion measurements of hydrogen isotopes in BeO at very low concentrations (below 10^{-4} at.%) [15]. It is the reasonable to suggest that a concentration-induced migration of D atoms takes place out from the lattice saturated with deuterium.

The comparison of the maximum concentrations of D atoms, D_2 molecules and total deuterium present in both states in Be and BeO irradiated with D ions at 300 and 700 K to high fluences is given in Table 1. The study of deuterium retention in beryllium has shown that a high percentage of deuterium is accumulated in the form of D_2 molecules. As for BeO, the majority of implanted deuterium is retained as D atoms both at 300 and 700 K. Attention is drawn to the fact that the value of gas-induced swelling in BeO at T_{irr} = 700 K is two orders of magnitude less then that for Be.

Table 1 Maximum concentrations of D atoms, D_2 molecules and deuterium present in both states in Be and BeO irradiated with D ions at 300 and 700 K

	Be		BeO	
	300 K	700 K	300 K	700 K
D atom concentration	0.02 D/Be	0.01 D/Be	0.18 D/BeO	0.05 D/BeO
D_2 molecule concentration	0.04 D_2/Be	0.01 D_2/Be	0.01 D_2/BeO	0.01 D_2/BeO
Total D concentration	0.10 D/Be	0.03 D/Be	0.20 D/BeO	0.07 D/BeO
Gas swelling, (%)	1.8	50	0.5	0.6

3.3. Deuterium retention in beryllium oxide layer exposed to D atoms

Analysis of TEM micrograph of a film formed onto Be surface exposed to D atoms at 340 and 500 K together with a corresponding selected area pattern diffraction characterized by strong haloes around central electron spots says that the surface film contains at least one distinguishable phase, cph-BeO. According to EPMA, the surface layers contains besides beryllium oxide a lot of carbon. So, this layer can be termed as the "oxide film" very conventionally. Indeed, while the oxygen content here is ~1×10^{21} O/m^2, the content of C atoms is about 5×10^{20} C/m^2.

Depth profiles of deuterium (D atoms and D_2 molecules) and depth distributions of beryllium oxide in Be exposed to D atoms for 1 hour at temperatures of 340 and 500 K are shown in Fig. 5. The beryllium oxide profiles were measured by recording of BeO^- SIMS signal. The comparison of the deuterium and BeO profiles shows that deuterium is mainly accumulated in beryllium oxide layer growing during D atom exposure.

Obviously, deuterium content in beryllium samples is correlated with beryllium oxide thickness. According to sputter measurements the thickness of oxide film at T_{exp} of 340 and 500 K is ~20 nm (Fig. 5) and deuterium contents for these temperatures are approximately the same and equal to $\sim 1\times10^{20}$ D/m^2.

Depth distributions of D atoms and D_2 molecules in Be samples exposed to D atoms for 1 h at 340 K and their evolution during vacuum storage of the exposed samples at room temperature are given in Fig. 6. It has been shown that the majority of deuterium accumulated in beryllium oxide layer is trapped as D atoms. At T_{exp} = 340 and 500 K the D atom concentration in BeO layer reaches the value of (3-4)$\times10^{27}$ D/m^3 (0.04-0.05 D/BeO). The fraction of D_2 molecules in deuterium retention is small and does not exceed 5-10%.

We emphasize that after termination of D atom exposure at T_{exp} of 340 and 500 K the concentration of deuterium atoms in Be samples decreases with storage time at 300 K. D_2 content in the samples exposed at T_{exp} of 340 and 500 K does not practically changed with storage time in vacuum (Fig. 6).

The evolution of D atom content in the Be samples, Q_D, with storage time at room temperature obeys the equation: $Q_D = Q_D^0 \times \exp(-f_D \cdot t)$, where Q_D^0 is initial D atom content in the sample, t - the elapsed time from termination of D atom exposure, f_D - decrease rate constant at 300 K (Fig. 7). The values of f_D for Be samples exposed at different T_{exp} are given in Table 2.

Table 2 The values of D atom decrease rate constant, f_D, at 300 K measured by the use of Be samples exposed with D atoms at different temperatures T_{exp}.

T_{exp} (K)	Decrease rate constant (min^{-1})
340	3.86×10^{-5}
500	3.44×10^{-5}

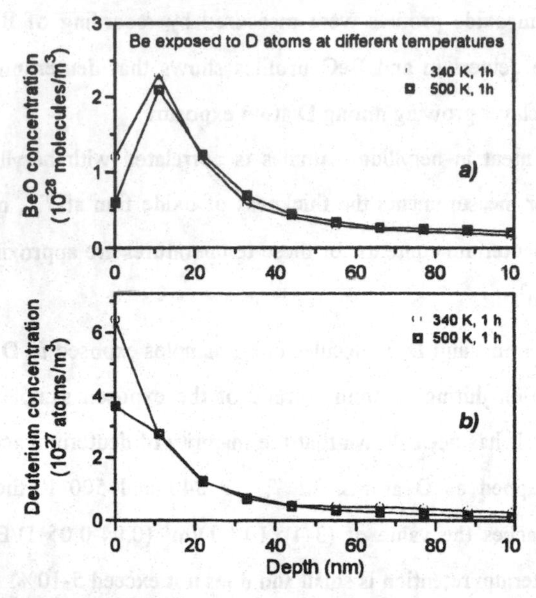

Fig. 5. Depth profiles of BeO (a) and deuterium (b) in Be sample exposed to D atoms for 1 h at 340 and 500 K. Deuterium profiles represent the sum of D atom and D_2 molecule concentrations.

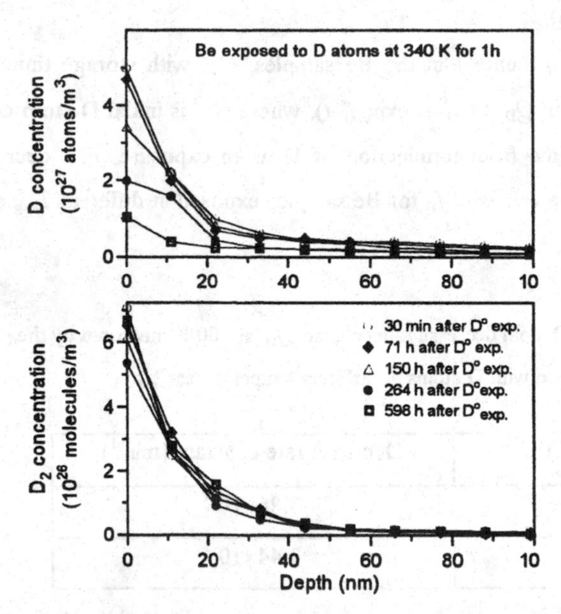

Fig. 6. Evolution of D atom concentration (on the top) and D_2 molecule concentration (on the bottom) in Be sample exposed to D atoms for 1 h at 340 K during its storage in vacuum at room temperature.

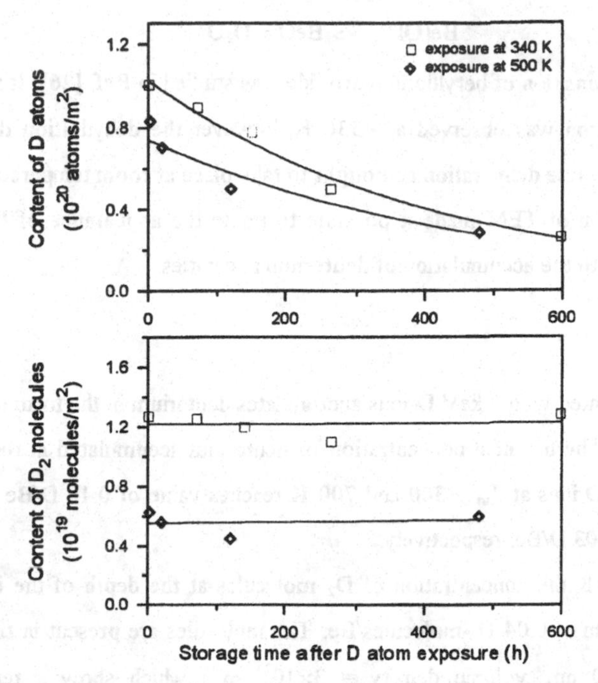

Fig. 7. D atom content (on the top) and D_2 molecule content (on the bottom) in Be samples exposed to D atoms for 1 h at 340 and 500 K as a function of the elapsed time from termination of D atom exposure (the samples were kept in vacuum at room temperature).

One of the probable mechanisms which could be responsible for deuterium retention in growing oxide layer is reaction of D atoms with beryllium oxide and formation of beryllium hydroxide [17]:

$$2BeO + 2D \rightarrow Be(OD)_2 + Be + 0.704\ eV/D \qquad (2)$$

Actually, the presence of $BeOD^-$ line in SIMS spectra confirms the formation of hydroxylic OD^- groups. It should be noted that D atom concentration in the samples exposed at 340-500 K does not exceed 4×10^{27} D/m^3. Therefore, the fraction of hydroxide in BeO layer could be estimated to be less than 8 %. The relatively low concentration of amorphous hydroxide phase in BeO crystalline structure was not detected in TEM study.

The decrease of D atom concentration with the passage of time could be explained by the dehydration of $Be(OD)_2$ according to the reaction:

$$Be(OD)_2 \rightarrow BeO + D_2O \quad (3)$$

The thermal decomposition of beryllium hydroxide was studied in Ref. [26]. It was shown that significant dehydration was observed at ~530 K, however the dehydration did not proceed completely. The low rate dehydration is thought to take place at room temperature also.

The application of TEM made it possible to relate the appearance of D_2 molecules in growing BeO layer to the accumulation of deuterium in cavities.

4. Conclusions

(1) Beryllium implanted with 9 keV D-ions accumulates deuterium in the form of both D atoms and D_2 molecules. The maximal concentration of deuterium accumulated in the both states in Be irradiated with D ions at T_{irr} = 300 and 700 K reaches value of 0.10 D/Be (in the units of atomic ratio) and 0.03 D/Be, respectively.

At T_{irr} = 300 K the concentration of D_2 molecules at the depth of the ion mean range reaches its maximum of 0.04 D_2-molecules/Be. The molecules are present in tiny bubbles (the mean radius = 1.0 nm, volume density = 3×10^{24} m^{-3}) which show a tendency toward interconnection at higher fluences. At T_{irr} = 700 K, along with relatively small facetted bubbles (near the very surface), large oblate gas-filled cavities and channels forming extended labyrinths appear and they accumulate most of the injected gas. The maximal D_2 concentration in the latter case is 0.01 D_2-molecules/Be.

The high concentration of D atoms in the ion stopping zone of Be matrix after implantation to saturated fluences at T_{irr} = 300 and 700 K (about 0.02 and 0.01 D-atoms/Be, respectively) is attributed to deuterium (i) trapped in radiation vacancies, (ii) adsorbed on the walls of bubbles and channels and (iii) bonded to BeO present in the form of metallurgical inclusions in the bulk.

(2) BeO films about 120 nm in the thickness thermally grown onto Be substrates (by annealing of electropolished Be plates at 1023 K for 60 minutes in air atmosphere at a pressure of 0.5 kPa) consist of small close-packed hexagonal microcrystals of size 25-30 nm. Deuterium implanted as 3 keV D-ions is found to be retained in BeO film in the form of both D atoms and D_2 molecules. The maximal concentration of deuterium accumulated in the both states in BeO at T_{irr} = 300 and 700 K reaches value of 0.20 D/BeO and 0.08 D/BeO, respectively.

Irradiation at 300 and 700 K leads to the formation of tiny D_2 bubbles of 0.6-0.7 nm in radius and of high volume density $\cong 4.5\times10^{24}$ m^{-3}. These bubbles together with the intercrystalline gaps are responsible for the accumulation of a molecular fraction of the implanted deuterium. At both irradiation temperatures the D_2 concentration reaches its maximum of 0.01 D_2-molecules/BeO.

At T_{irr} = 300 and 700 K the major part of implanted deuterium is present in BeO film in the form of D atoms, probably, chemically bound to O atoms. The maximal D atom concentration is 0.18 D-atoms/BeO for T_{irr} = 300 K and 0.05 D-atoms/Bo for T_{irr} = 700 K.

(3) Due to the presence of oxygen containing molecules as trace impurities, the beryllium oxide layer is formed on the surface of Be sample exposed to D atoms at elevated temperatures. It has been found that the deuterium atom exposure leads to the deuterium retention in this layer. The majority of deuterium (> 90%) retains as D atoms, the other part is accumulated in the form of D_2 molecules. After termination of D atom exposure D concentration in BeO layer decreases with time at room temperature. It is supposed that the formation of beryllium hydroxide $Be(OD)_2$ under D atoms exposure of growing BeO layer takes place. The decrease of deuterium concentration in time is explained by dehydration of $Be(OD)_2$.

Acknowledgments

This research was sponsored by the United States Department of Energy, under Contracts LC-8102 and LF-7292 with Sandia National Laboratories.

References

1. Technical Basis for the ITER Interim Design Report, Cost Review and Safety Analysis. ITER EDA Documentation Series, No. 7. IAEA, Vienna (1995).
2. A.P. Martinelli, R. Behrisch, A.T. Peacock, J. Nucl. Mater. 212-215 (1994) 1245.
3. R. Behrisch, A.P. Martinelli, S. Grigull, R. Grötzschel, U. Kreissig, D. Hildebrandt, W. Schneider, J. Nucl. Mater. 220-222 (1995) 590.
4. M. Mayer, R. Behrisch, H. Plank, J. Roth, G. Dollinger, C.M. Frey, J. Nucl. Mater. 230 (1996) 67.
5. K.L. Wilson, R.A. Causey, W.L. Hsu, B.E. Mills, M.F. Smith and J.B. Whitley, J. Vac. Sci. Technol. A8 (1990) 1750.
6. W.R. Wampler, J. Nucl. Mater. 122 & 123 (1984) 1598.
7. W.R. Wampler, J. Nucl. Mater. 196-198 (1992) 981.
8. W. Möller, B.M.U. Scherzer and J. Bohdansky, Retention and Release of Deuterium Implanted into Beryllium, IPP-JET Report No. 26, Max-Planck-Institut für Plasmaphysik,

Garching, 1985.

9. H. Kawamura, E. Ishituka, A. Sagara, K. Kamada, H. Nakata, M. Saito and Y. Hutamura, J. Nucl. Mater. 176 & 177 (1990) 661.
10. R.A. Langley, J. Nucl. Mater. 85 & 86 (1979) 1123.
11. W.A. Swansiger, J. Vac. Sci. Technol. A4 (1986) 1216.
12. V.I. Shapovalov and Yu.M. Dukelskii, Russ. Metal. 5 (1988) 210.
13. J.P. Pemsler and E.J. Rapperport, Trans. Metall. Soc. AIME 230 (1964) 90.
14. H. Verbeek and W. Eckstein, in: Application of Ion Beams to Metals (Plenum Press, New York, 1974) p. 607.
15. J. D. Fowler, D. Chandra, T.S. Elleman, A.W. Payne, K. Verghese, J. Am. Ceram. Soc. 60 (1977) 155.
16. R.G. Macaulay-Newcombe, D.A. Thompson, J. Nucl. Mater. 212-215 (1992) 942.
17. M.C. Billone, M. Dalle Donne, R.G. Macaulay-Newcombe, Fusion Eng. Des. 27 (1995) 179.
18. R. Behrisch, R.S. Blewer, J. Borders, R. Langley, J. Roth, B.M.U. Scherzer, R. Schulz, Radiat. Eff. 48 (1980) 221.
19. A.I. Kanaev, V.M. Sharapov, A.P. Zakharov, Atomnaya Energiya 76(2) (1994) 145 (in Russian).
20. V.Kh. Alimov, A.E. Gorodetsky and A.P. Zakharov, J. Nucl. Mater. 186 (1991) 27.
21. V.Kh. Alimov, V.N. Chernikov, A.P. Zakharov, J. Nucl. Mater. 241-243 (1997) 1047.
22. I.I. Papirov, Oxidation and Protection of Beryllium (Moscow, Metallurgiya, 1968) (in Russian).
23. V.N. Chernikov, V.Kh. Alimov, A.V. Markin and A.P. Zakharov, J. Nucl. Mater. 228 (1996) 47.
24. K. Ashida, M. Matsuyama, K. Watanabe, H. Kawamura, E. Ishitsuka, J. Nucl. Mater. 210 (1994) 233.
25. A.V. Markin, V.N. Chernikov, S.Yu. Rybakov and A.P. Zakharov, in: Proc. 2nd IEA Int. Workshop on Beryllium Technology for Fusion, 6-8 September 1995, Jackson Lake Lodge, Wyoming (Lockheed Martin Idaho Technologies, Idaho Falls, 1995) pp. 332-347.
26. J.F. Quirk, N.B. Mosley, and W.H. Duckworth, J. Am. Ceram. Soc. 40 (1957) 416.

THE EFFECT OF RADIATION DAMAGE AND HELIUM ON HYDROGEN TRAPPING IN BERYLLIUM

M.I.GUSEVA[1], V.M. GUREEV[1], L.S.DANELYAN[1],
S.N.KORSHUNOV[1], V.S.KULIKAUSKAS[2]
Yu.V.MARTYNENKO[1], P.G.MOSKOVKIN[1],
V.V. ZATEKIN[2]
1. RRC «Kurchatov Institute»,sq.Kurchatov 1, Moscow,123182, Russia
2. Lomonosov University, Moscow, Russia

The beryllium is a candidate material for the ITER- first wall. At neutron bombardment radiation defects and helium retention, as a result of nuclear reaction $^{9}Be(n,2n) \rightarrow 2He^{4} + 94$ keV, will take place. The investigation of radiation defects and helium affect on hydrogen retention in beryllium were carried out. The beryllium samples were irradiated by 2.8 MeV He ions in Van de Graaf accelerator at the temperature 773 K to simulate neutron irradiation. Helium irradiation fluences were lower than blister formation critical dose and were equal 10^{20}, $5 \cdot 10^{20}$ and 10^{21} He^{+}/m^{2}. After that 3-keV H^{+} ions were implanted into the samples at temperature 573 K. Hydrogen irradiation doses were $5 \cdot 10^{22}$ H^{+}/m^{2} and 10^{23} H^{+}/m^{2}. The microstructure of the samples was investigated after He^{+} and after He^{+} and H^{+} irradiation. Depth distribution profiles was measured using the elastic by recoil detection technique. The integral hydrogen concentration in beryllium with He^{+} ions induced defects was 2.6-2.8 times large then that in beryllium non-irradiated by He^{+} ions. For specimens depth previously irradiated by He^{+} ions peak in the hydrogen distribution was at about 0.15 μm in comparison with to the place ~ 0.03 μm for specimen implanted only with 3 keV H^{+} ions simultaneously. The joint irradiation by He^{+} and H^{+} ions results in small blisters formation. At maximal He^{+} fluence (10^{21} He/m^{2}) the blister caps are broken, hydrogen releases and integral hydrogen content decreases. The observed effects are explained by vacancies and He atoms diffusion and hydrogen trapping by vacancy-helium traps.

1. Introduction

The beryllium is a candidate material for the ITER- first wall due to a low atomic number and due to good mechanical and thermal properties. At the same time, this material is highly sensitive to damage produced in the beryllium lattice by high energy neutrons. The damage in beryllium is caused mainly by nuclear reactions:

C.H. Wu (ed.), Hydrogen Recycling at Plasma Facing Materials, 265–272.

$$^9Be + n \rightarrow {}^8Be + 2n$$
$$^8Be \rightarrow 2\,{}^4He + 94\ keV$$

The presence of interstitial helium atoms in the beryllium lattice provides its distortion and, in the presence of temperature, their diffusing in the material, migration to the external surface and to microcavities, gaseous bubble production [1,2]. The helium bubbles present in the beryllium appear to be the main causes of tritium retention in beryllium.

TABLE I. Operating conditions for irradiating beryllium by 2,8 Mev He^+- ions of (T_{irr} = 773K) and by 3 keV H^+- ions of (T_{irr} = 573K)

№ of sample	Ion	Dose, He^+/m^2	Dose, H^+/m^2	t, s He^+
1	H^+	-	$5 \cdot 10^{22}$	-
2	$He^+ + H^+$	10^{20}	$5 \cdot 10^{22}$	1000
3	$He + H^+$	10^{20}	10^{23}	1000
4	$He + H^+$	$5 \cdot 10^{20}$	$5 \cdot 10^{22}$	5000
5	$He + H^+$	$5 \cdot 10^{20}$	10^{23}	5000
6	$He + H^+$	10^{21}	$5 \cdot 10^{22}$	10000
7	$He + H^+$	10^{21}	10^{23}	10000

In this connection, it is important along with production of displaced lattice atoms, to introduce helium atoms into beryllium to simulate an effect of radiation defects on the hydrogen trapping in beryllium.

Most of hydrogen trapping data in the presents of helium atoms were obtained in the beryllium irradiation experiments carried out in material test reactors, in which beryllium was used as reflector [2-4]. It is now, that the mobility of helium in the beryllium lattice at 725K is low, and the coalescence of helium bubbles does not occur [2-4]. In a given study the Be-samples were preliminarily irradiated by 2,8 MeV helium ions to study an effect of radiation defects and helium on the hydrogen trapping in beryllium. The irradiation of beryllium by 2.8 MeV He^+- ions was done at 773K, when the rapid diffusion of helium, helium bubbles production and their coalescence [2] occur. According to the data of our previous studies [5], the presence of helium in the crystalline lattice makes an essential effect on the hydrogen blistering.

2. Experimental technique

The samples of industrial pressed beryllium, TShP-56- type, polished first with a diamond paste, 14th class, and, then electrolytically, were used in the experiments. The Be- samples were irradiated by He^+- ions with the energy of 2.8 MeV, at the temperature 773K, in the Van-de-Graaf accelerator. At this energy helium ions penetrate about 10 μ m deep that is shorter than the grain size of the beryllium under study. The He^+-ion beam, defocused across the whole target surface, was implanted by three irradiation fluences - 10^{20}, $5 \cdot 10^{20}$, 10^{21} He^+ /m^2 - which were lower than the critical doses of blistering formation. After the preliminary irradiation by He^+- ions, all the targets were simultaneously implanted by H^+ - monoenergetic ions with the energy of 3 keV (9 keV H_3^+ ions) to a doses of $5 \cdot 10^{22}$ H^+/ m^2 and 10^{23} H^+/m^2. The H^+- ion beam intensity was equal to $6 \cdot 10^{19}$ H^+/m^2 s and $1.2 \cdot 10^{20}$ H^+/m^2s for specimen 5, see Table I. The temperature was sustained at the level of 573 K in the process of hydrogen ion implantation.

After implanting the helium and hydrogen ions, as well as after joint irradiation by helium and hydrogen ions, the target surface microstructures were studied with a scanning electron microscopes, JEOL. The component composition of surface layers in the initial beryllium - beryllium irradiated by He^+- and H^+- ions - was determined by the Rutherford backscattering technique. For this purpose the energy spectra of helium ions with the initial energy of 1.5 MeV backscattered from the Be-targets at the angle of 170^0 to the surface were measured in the Van-de-Graaf accelerator. The elastic recoil detection technique was used for studying the hydrogen concentration distribution profiles in beryllium. In those experiments the He^+-ion beam with the energy of 2.2 MeV was incident upon the sample under study at the angle of 15^0 to its surface, the recoil atoms were registered at the angle of 30^0 to the initial He^+- ion incidence direction. The measurement of the energy spectra from the standard calibration mylar foil were done to produce absolute values of the hydrogen atom concentration.

The operating conditions for irradiating Be-targets by He^+- and H^+-ions are summarized in Table I.

3. Experimental results

The Fig. 1 shows a comparison between the experimental (1) and calculated by TRIM [6] (2) 3 keV hydrogen distributions in beryllium after implanting up to $5 \cdot 10^{22}$ H/m^2 at 573 K. The broadening of depth distribution profile 1 is caused by hydrogen and vacancy diffusion at the temperature 573 K. RBS analyssis shown that oxygen concentration after irradiation does not exceed 10 at%.

TABLE II. Effect of the preliminary irradiation fluency by 2.8 MeV He^+- ions at 773 K on the hydrogen retention in beryllium after 3 keV H^+- ions implantation at 573 K

№ of specimen	$D_{He}{}^+$/ m^2	$D_H{}^+$/ m^2	C /m^2	C/D, %
1	-	$5 \cdot 10^{22}$	$1,5 \cdot 10^{21}$	3
2	10^{20}	$5 \cdot 10^{22}$	$3,9 \cdot 10^{21}$	7,8
4	$5 \cdot 10^{20}$	$5 \cdot 10^{22}$	$4,2 \cdot 10^{21}$	8,4
6	10^{21}	$5 \cdot 10^{22}$	$4,0 \cdot 10^{21}$	8,0

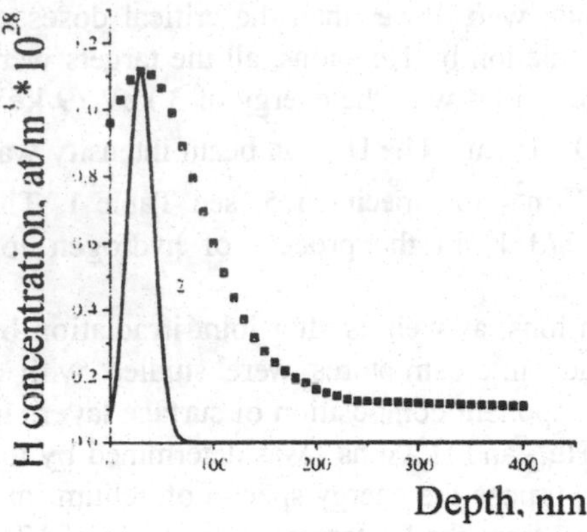

Figure 1. Depth distribution of hydrogen in beryllium after 3 keV H^+ ion implantation at the dose $5*10^{22}$ H/m^2:1 - experimental hydrogen profile after implantation at 573 K; 2 - TRIM simulation.

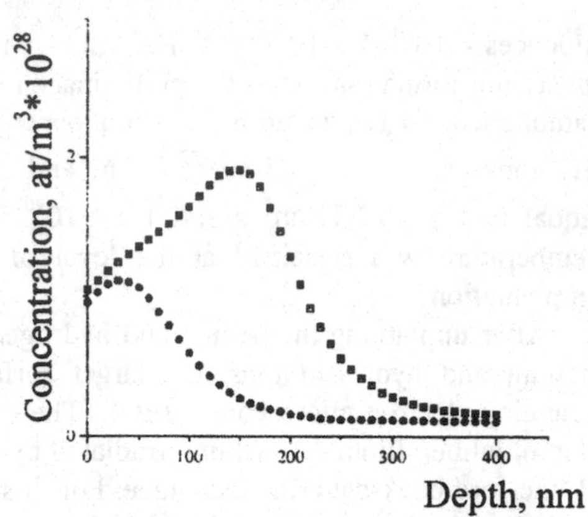

Fig. 2 Distribution profiles of the hydrogen ions implanted into beryllium with the energy of 3 keV at T_{irr}=573K ($D=5*10^{22}m^{-2}$): Curve 1 - H^+; Curve 2 - preliminary irradiation by 2.8 MeV He^+ ions, $D=10^{20}$ He/m^2, T_{irr}=773K.

The profiles of the hydrogen in beryllium after the different irradiation conditions are given in Fig.2 (specimen 1 and 2, see Table I). Specimen 1 (Curve 1) was implanted with hydrogen ions only. The specimen 2 was damaged with a fluence of 10^{20} He^+/m^2 at the temperature of 773 K prior to implantation at the $5 \cdot 10^{22}$ H^+/m^2 dose. Under the same irradiation conditions by H^+- ions in case of a preliminary bombardment - by 2.8 MeV He^+ ions (Curve 2) the hydrogen distribution maximum is shifted deep into the

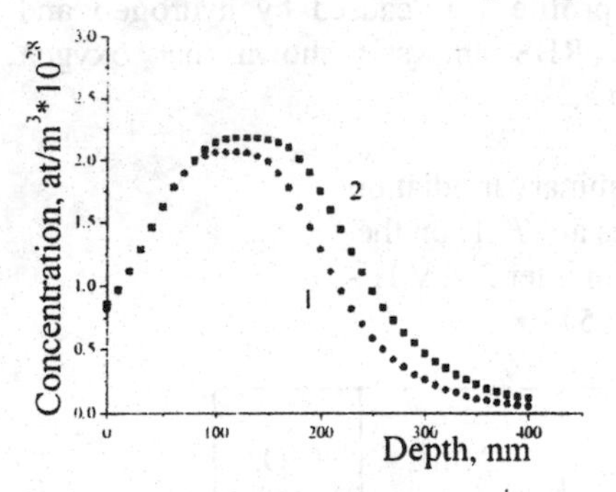

Fig.3. Distribution profiles of the 3 keV hydrogen ions implanted into beryllium at T_{irr}=573 K. Beryllium was preliminary irradiated by 2.8 MeV He+ ions at T_{irr}=773 K: Curve 1 - H^+, $D=5*10^{22}m^{-2}$, Curve 2 - H^+, $D=10^{23}$ m^{-2}.

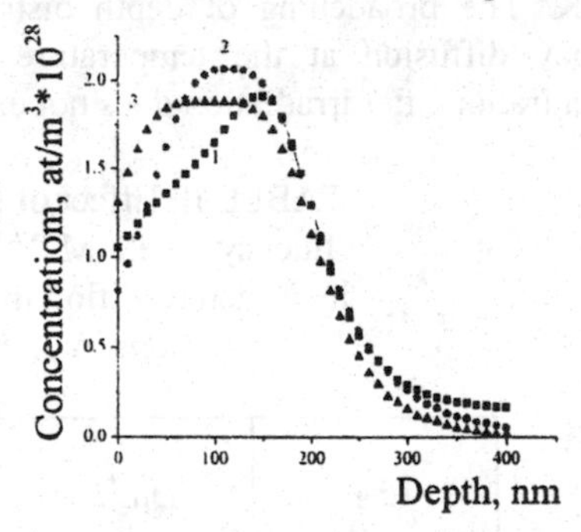

Fig.4. Distribution profiles of the 3 keV hydrogen ions implanted into beryllium at T_{irr}=573 K,($D=5*10^{22}$ m^{-2}):
Curve 1 - He^+, 2.8 MeV,$D=10^{20}m^{-2}$+H^+,
Curve 2 - He^+, 2.8 MeV, $D=5*10^{20}m^{-2}$+H^+,
Curve 3- He^+, 2.8 MeV, $D=10^{21}$ m^{-2} + H^+.

target (specimen 2). For this specimen the peak in the hydrogen distribution is at about 0.15 μm in comparison with the peak at 0.03 μm for specimen 2. The integral concentration of hydrogen is increased by a factor 2.6 from $1.5 \cdot 10^{21}$ H/m^2 (specimen 1) to $3.9 \cdot 10^{21}$ H/m^2 (specimen 2). As a result of previous He-ion irradiation the fractional retention of the hydrogen rises from 3% (specimen 1) up to 7.8% (specimen 2) of the incident particles.

The effect of hydrogen dose implantations on hydrogen depths distributions in beryllium which was previously irradiated at the fluence of $5 \cdot 10^{20}$ He/ m^2 are compared

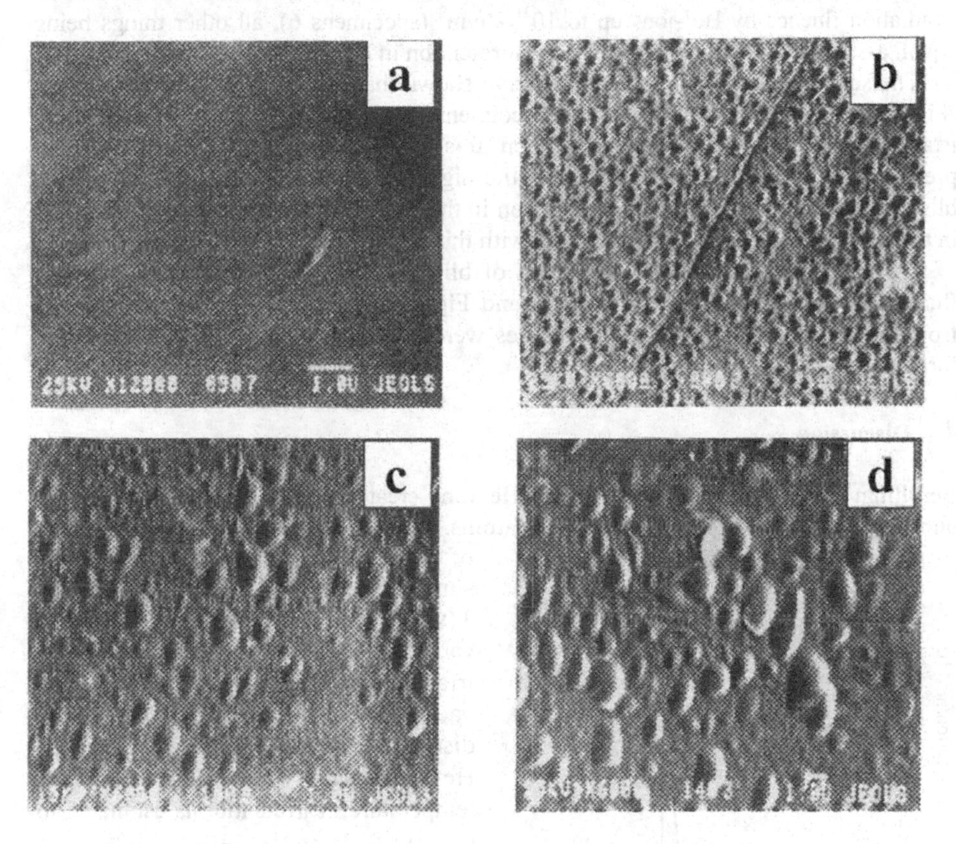

Fig. 5. Microstructure of a beryllium surface after irradiation by 2.8 MeV He^+-ions at 773K and after implantation of 3 KeV H^+-ions:
a-He^+, D= 10^{21} He/m^2; b-H^+, D= 10^{23} H/m^2
c-He^+, D= $5 \cdot 10^{20}$ He/m^2 + H^+, D= 10^{21} H/m^2
d-He^+, D= 10^{21} He/m^2 + H^+, D= $5 \cdot 10^{20}$ H/m^2

The increase of the hydrogen dose implantation from $5 \cdot 10^{22}$ H/m^2 (Curve 1) to $5 \cdot 10^{23}$ H/m^2 (Curve 2) at constant 2.8 MeV He fluence results in the further increasing the integral hydrogen concentration up to $4.5 \cdot 10^{21}$ H/m^2. The hydrogen depth distribution for higher dose implantation (Curve 2) is shifted deeper into the target.

The change of the hydrogen profiles in beryllium as a function of prior incident helium fluence is shown in Fig. 4. One can see that with a rise in the fluence of helium ions the profiles of the hydrogen distribution are shifted towards the surface. The experimental data on the hydrogen accumulation under various irradiation conditions of beryllium by He^+- and H^+- ions are summarized in Table II.

From an analysis of Table II one can conclude that amount of hydrogen and fractional retention of the hydrogen by beryllium to the incident H^+ ions under the same irradiation conditions of hydrogen ions are maximal for the specimen 4 which previously was irradiated with a fluence of $5\cdot 10^{20}$ He/m^2. The further increase in the Be-irradiation fluence by He^+-ions up to 10^{21} He/m^2 (specimens 6), all other things being equal, assists in a reduction in the hydrogen retention in beryllium.

The studies of a surface topography have shown that blisters with undamaged caps (Fig 5 b, c) are produced upon the specimens (1-5) surfaces (See Table I) under irradiation conditions. In difference from this upon the specimens surfaces (6,7), preliminarily - irradiated by He^+- ions, at the highest fluency, 10^{21} He/ m^2 a part of the blister caps are broken (Fig.5d). A reduction in the amount of the accumulated hydrogen in these samples (See, Table II) is related with this effect.

One should note that the exfoliation of blister caps depends only from He^+-ions fluence (compare Fig.5c, specimen N 5, and Fig.5c, specimens N 2). As one can see from the Fig. 5a helium irradiation fluences were lower than critical dose of blistering formation.

4. Discussion

Beryllium preliminary bombardment by He ions creates vacancy clusters and/or He bubbles which work as traps for hydrogen atoms. We do not investigate here the kinetic of the traps formation, but believe for simplicity that traps depth distribution T(x) is like the depth distribution of vacancies during the 2.8-MeV He ion irradiation. The primary created vacancies calculated by TRIM have distribution shown in Fig. 6 [6].Since the He diffusion coefficient at the temperature of irradiation is estimated to be $D \cong 3\cdot 10^{-13}$ m^2/s, a stationary distribution with a linear dependence n(x) $= n_{max} \cdot x/R_p$ at $x < R_p \cong 10.9$ μm is reached in irradiation time $t \cong 10^2$ s (fluency 10^{20} m^{-2}). Therefore we assume that the hydrogen atom kinetic is described by the equation

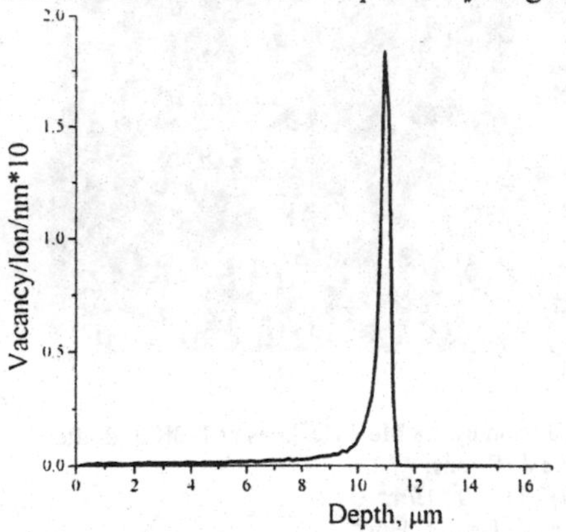

Fig.6. Depth distribution of vacancies in Beryllium after 2.8 MeV He ions irradiation calulated by TRIM.

$$\partial C/\partial t = D_H \cdot \partial^2 C/\partial x - \alpha \cdot D_H \cdot (T(x) - (a/r)^3 \cdot C_t) \cdot C + G(x),$$

$$\partial C_t/\partial t = \alpha \cdot D_H \cdot (T(x) - (a/r)^3 \cdot C_t) \cdot C \qquad (1)$$

where C is the depth distribution of mobile hydrogen atoms, C_t is the depth distribution of trapped hydrogen atoms, $G(x) = j/\Delta R \cdot \exp(-(x-r_p)^2/2\Delta r^2$ is the source function describing depth distribution of implanted hydrogen, r_p and Δr are the mean projected range and straggling of 3-kev hydrogen ions, $\alpha = 4\pi r$, r=5 nm is the radius of the trap, and $T(x) = k \cdot x + T_0$, $k = T_{max}/R_p = 7 \cdot 10^{30}$ m^{-2}, $T_0 = 10^{28}$ m^{-3} describes the traps created by hydrogen ions. The second term in the first equation describes mobile hydrogen atoms trapping with saturation of the traps, a being equal hydrogen atom radius in a trap. The simple boundary condition C(0)=0 is chosen.

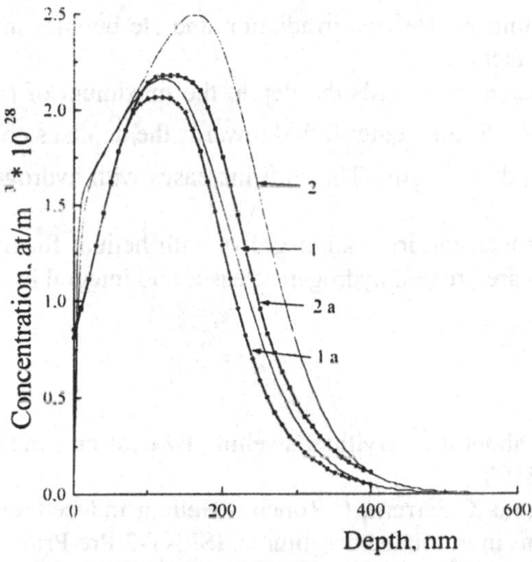

Fig.7. Calculated (1, 2) and experimental (1a, 2a) depth distributions of 3 keV hydrogen implanted in beryllium, which was preliminary irradiated by 2.8 MeV He^+ ions.
1, 1a - He(5*10^{20} m^{-2}) + H(5*10^{22} m^{-2});
2, 2a - He(5*10^{20} m^{-2}) + H(10^{23} m^{-2}).

The Fig 7 shows a comparison between calculated (1, 2) and exprimental (1 a, 2 a)depth distributions for hydrogen in He irradiated Beryllium. The simple model shows good agreement of calculated hydrogen distribution maximum (curve 1) with that of experimental (curve 1 a). The trapped hydrogen concentration agrees also with experimental.

The increase of hydrogen dose results in an increase of the trapped hydrogen concentration and in a shift of depth distribution maximum, calculated concentration and shift (curve 2) being to be larger than those of experimental (curve 2a).

The calculation confirms the main role of the traps in hydrogen retention.

The hydrogen atoms distribution profile measurement shows that at higher doses of He irradiation the depth distribution broadens towards the surface. This broadening can be explained by the blister appearance after hydrogen implantation, which was not taken into account at elastic recoil method of hydrogen profile measurement. Really one assumes that the recoil energy loss is

$$\Delta E = (dE/dz) \cdot (h/\sin\alpha) \qquad (2)$$

where h is the depth where recoil atom was knock on, α is the angle between recoil observation direction and target surface. But if hydrogen atoms are situated in a blister with cap having an angle θ with respect to the surface.

$$\Delta E = (dE/dz)\cdot(h/\sin(\alpha+\theta)). \quad (3)$$

Taking into account that in measurement $\alpha = 15^0$ and θ is approximately the same, one can see that visible depth is reduced about two times. The convexity of the blister grows with He bombardment dose increase and the profile broadening increases also.

5. Conclusions

1. The investigation of preliminary He^+ irradiation fluence on hydrogen retention in beryllium were carried out.
2. Vacancy clusters created at preliminary He^+ ion irradiation and He bubbles increase hydrogen retention in Beryllium by factor 2.8.
3. Since the traps concentration increases towards the depth, the maximum of trapped hydrogen atoms concentration is at 0.15 μm, being shifted towards the depth as compare with hydrogen atom penetration depth, 0.03 μm. The shift increases with hydrogen ion dose.
4. Blister formation rate after hydrogen ion irradiation grows with helium fluency. At high helium ion fluency blister caps are broken, hydrogen releases and integral hydrogen retention decreases.

6. References

1. C.Nardi «Status of knowledge about the beryllium swelling by neutron irradiation» PT /Nucl./91/ 24 ISSN/ 1120-5598.
2. F.Scaffidi-Argentina, M.D.Donne, C.Ferrero, C.Ronchi «Helium induced swelling and tritium trapping mechanisms in irradiated beryllium», ISFNT-3 Pre-Print.
3. J.M.Beeston «Gas Retention In Irradiated Beryium» EGG Idna O Report, EGG-FSP-9125 (1950).
4. Sannen et.al. «Helium content and Swelling of Low Temperature Irradiated/ Post Irradiated Annealed Beryllium/
5. M.I.Guseva, Yu.V.Martynenko, The Blistering of Stainless Steel (0Cr16Ni15Mo3Nb) Under Simultaneous D^+ and He^+ Ions Irradiation., J. Nucl. Mater. 93-94, 1980, 734-738.
6. J.F.Ziegler, J.P.Biersack «The Stopping and Range of Solids», (Pergaman Press, New York, 1985).

ATOMIC HYDROGEN-GRAPHITE INTERACTION

E.A.DENISOV, T.N.KOMPANIETS, A.A.KURDYUMOV
Scientific Research Institute of Physics, St.-Petersburg University
198904 Ulianovskaja 1, St.-Petersburg, Russia

The plasma facing materials in future fusion devices are subjected not only to ion bombardment but interact with neutrals (hydrogen atoms and molecules) also. Present work is devoted to investigation of hydrogen desorption kinetics after exposure of graphite samples to the flux of hydrogen atoms. Three types of graphite (pyrolitic (PG), quasimonocristalline (QM) and RG-Ti graphite) were studied. It is shown that sorption of atomic hydrogen occurs in a similar way for all three types of graphite. Two distinct maximums have been observed in thermal desorption spectra obtained during linear heating of the sample after its exposure to hydrogen atoms. While stopping a linear heating at various temperatures some special features of hydrogen desorption kinetics were observed. The rate of desorption decreases at any fixed temperature very fast and its dependence on temperature does not follow neither first nor second order desorption kinetics. Taking into account these peculiarities of desorption process the mathematical model of hydrogen - graphite interaction is proposed. According to this model most of the sorbed hydrogen is situated inside the bulk of the sample and two types of traps with binding energy of 2.4 and 4.1 eV are present there. We believe that both types of traps are due to hydrogen interaction with edge bonds of graphite layers (or edge dislocations). The distinction between trapping energy for these two types of traps results from different sorption places of hydrogen at edge dislocations.

1. Introduction

In recent years a lot of papers devoted to hydrogen-graphite interaction was published. This is because of the interest to graphite as a plasma facing material for fusion reactors.
Hence most of the papers deals with the interaction of high energy hydrogen ions with technical types of graphite. There are many both experimental and theoretical works in this area [1–12].

Thus far only a few of papers are devoted to the study of atomic hydrogen interaction with graphite [8, 13–18]. Most of authors do no more than describe the experimental results and make some remarks concerning hydrogen trapping at broken bonds of carbon atoms. It is usually suggested that atomic hydrogen - graphite

C.H. Wu (ed.), Hydrogen Recycling at Plasma Facing Materials, 273–280.

interaction takes place at graphite surface only. The possibility of hydrogen dissolution in the bulk of graphite is considered very seldom [17, 18].

The purpose of present work was to study the kinetics of atomic hydrogen interaction with different types of graphite in order to find the characteristics of atomic hydrogen interaction with the bulk and the surface of graphite samples.

2. Samples and methods

RG-Ti graphite (ρ=2.20-2.26g/cm^3) contains about 7.5 wt.% Ti. This reactor type of graphite is produced by hot one-axial pressing from the mixture of carbon and Ti powder [19]. Most of grains in RG-Ti were found to be flatted out relatively to the basal plane of the graphite lattice. The sizes of the disc-shaped grains are in the range from tens to hundreds of microns with the mean thickness of the order of 10 μm.

We have used two types of pyrolitic graphite with different temperatures of the final annealing in the course of graphite fabrication. The first one (with lower annealing temperature) was strictly pyrolytic graphite (PG) with matt rough surface. The second one was quasimonocristalline (QM) graphite. Its surface was more bright and flat indicating a better ordered structure of this graphite. Pyrolitic graphite has a high density (2.2 g/cm^3) and shows a laminated structure with periodicity in the order from 0.5 to 1.0 μm in the direction of the growth (c axis), manifesting its stepwise kinetics.

Experiments were carried out in UHV setup with residual pressure about 10^{-8} torr. Well known method of thermal desorption spectrometry was used. Band graphite samples with character dimension 0.5*1*40 mm were suppressed in copper holders. Heating was performed by passing the current through a sample. Temperature measurements were carried out by W-WRe thermocouple. Experimental setup allowed to regulate sample's temperature according to specified law. Desorbed hydrogen was detected by mass-spectrometer. Before exposure to hydrogen samples were annealed for some hours at the temperature about 1200 C and after that the temperature of the sample was shortly increased up to 1400 C.

Hydrogen atomisation was performed on a tungsten filament placed in front of the sample in such a manner that the flux of hydrogen atoms impacted the basal plane of graphite. The flux of atom irradiation has been calculated taking into account the inlet pressure, the probability of thermal dissociation and relative positions of sample and filament. The flux usually averaged $5 \cdot 10^{13}$H°/cm^{2}s.

3. Results

In contrast to RG-Ti [20] desorption of hydrogen was not detected after exposure of PG and QM to molecular hydrogen. After atomic hydrogen irradiation of pyrolitic graphite there was an essential hydrogen desorption during a linear heating of the sample. The typical thermal desorption spectrum is presented on Fig.1 curve 1. There are two clearly defined desorption maximums at 850 and 1250 C. Nearly the same spectra were obtained for RG-Ti graphite.

No significant difference in sorption kinetics was found between pyrolitic graphite and RG-Ti. Dependencies of the amount of sorbed hydrogen on temperature and fluence of atomic hydrogen demonstrate almost the same behaviour both for pyrolitic and RG-Ti graphite. These results indicate that impurities in graphite and its microstructure have little or no effect on the processes of atomic hydrogen sorption and desorption..

Two desorption maximums may be related to hydrogen release from either surface or bulk traps with different binding energies. Supposing that hydrogen desorption proceeds from the surface we can estimate the activation energy of desorption for the first type of traps. Supposing that hydrogen desorption proceeds from the surface we can estimate the activation energy of desorption for the low temperature maximum. Since the temperature of the first maximum is independent on concentration of sorbed hydrogen, it is evident that the process of desorption from the first state obeys the first order kinetics. Fig.2 presents the dependence of the position of the first maximum on the heating rate. For the first order desorption kinetics a linear dependence should be observed. Estimation of desorption activation energy gives approximately E_{d1}=2.4eV.

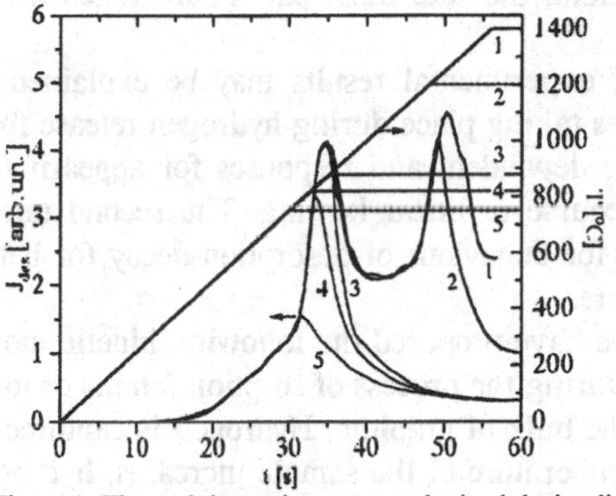

Figure 1. Thermal desorption spectra obtained during linear heating up to various final temperatures after exposure PG to hydrogen atoms.

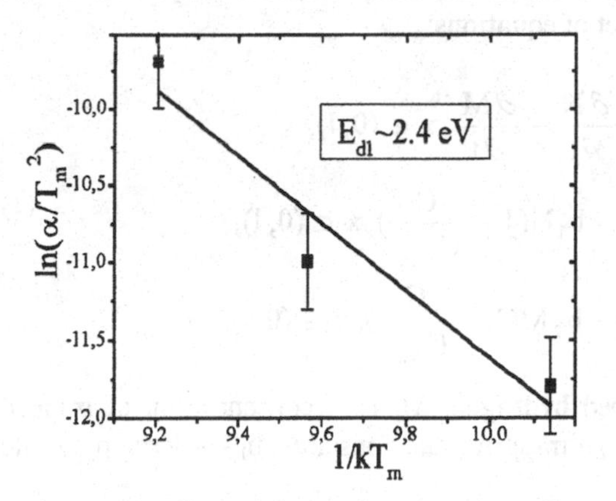

Figure 2. Dependence of first maximum temperature T_m on heting rate α.

Nevertheless we had to correct this model. In the following experiments atomic hydrogen sorption was carried out under the same conditions. During thermal desorption the heating of the sample was performed linearly with the same rate but up to a various final temperature (Fig.1). One can see that at a fixed temperature the rate of hydrogen desorption decreases much more rapidly than in the case of the ordinary desorption. Moreover desorption-decays curves obtained within one desorption maximum do not cross each other.

These peculiarities can not be attributed either to desorption from the surface or to activated diffusion from the bulk of graphite.
It should be noted that similar behaviour of desorption-decay curves was observed after irradiation of graphite with hydrogen ions with the energy about 100eV as well.

4. Discussion

It is a matter of common knowledge that in the course of ion irradiation of graphite hydrogen ions are implanted at some depth inside the bulk of the sample. In this case the processes inside the bulk should play a definite role in the release of the implanted hydrogen.. Similarity of kinetics of hydrogen release after exposure of the sample to hydrogen atoms and ions allows to conclude that the most part of hydrogen atoms enters the bulk of graphite.

We believe that the basic part of experimental results may be explained by supposing of the existence of two processes taking place during hydrogen release from graphite. The first process is temperature dependent and responses for appearing of maximums on the release curve in the course of linear heating. The second one is temperature independent and responsible for behaviour of desorption-decay (or better to say- release) curves at a fixed temperature.

On the basis of this assumptions we have proposed the following kinetic model for the description of degassing process. During the process of sorption (atoms or ions) a saturated hydrogen layer is formed in the bulk of graphite. Hydrogen is captured in this layer by two types of traps. As the temperature of the sample increases, hydrogen releases from traps and begins to migrate inside the bulk, namely between graphite layers. Migration of released hydrogen is accompanied by reversible trapping. Hydrogen that reached graphite surface leaves the sample immediately. This model may by represented by the following set of equations:

$$\frac{\partial C}{\partial t} = D\frac{\partial^2 C}{\partial x^2} - \frac{\partial N}{\partial t} - \frac{\partial M}{\partial t}, x \in (0,1),$$

$$\frac{\partial N}{\partial t} = r_1 C(1 - \frac{N}{N_{max}}) - b_1 N(1 - \frac{C}{C_{max}}), x \in (0,1), \quad (1)$$

$$\frac{\partial M}{\partial t} = r_2 C(1 - \frac{M}{M_{max}}) - b_2 M(1 - \frac{C}{C_{max}}), x \in (0,1),$$

were C– concentration of non-trapped hydrogen; M, N – concentration of trapped hydrogen; D – diffusion coefficient; $r_{1,2}$ – trapping rate constants; $b_{1,2}$ – detrapping rate constants.

We believe that temperature depended process is the release of hydrogen from traps. Temperature independent process is considered to be migration of hydrogen inside the bulk

As a first approximation the activation energy of hydrogen release from the first type of traps was taken to be 2.4 eV. This value has been obtained above under the assumption that sorbed hydrogen is situated on the surface. Additional computations have shown that location of traps does not matter. Location of the traps either on the surface or inside the bulk of graphite gives exactly the same slope of dependencies $\ln(\alpha/T_m^2)$ vs. $1/T_m$ (and thus the same activation energy). Moreover the slope does not depend on diffusion coefficient in the range from 10^{-7} to 10^{-4} cm^2/s. Because of this we took a diffusion coefficient to be $D=10^{-6} cm^2/s$. This value has been obtained earlier for RG-Ti graphite [21].

Computations have shown that under these conditions trapping rate for the first type of traps should be about $25c^{-1}$ with negligible activation energy. The rate constants of trapping and detrapping processes for the second type of traps can be calculated in the similar manner.

TABLE 1. Parameters of hydrogen-graphite interaction

Diffusion coefficient [cm^2/s]	$D_0=10^{-6}$
Activation energy of diffusion [eV]	$E_D=0$
Trapping rate constant for the first type of traps [1/s]	$r_1=25$
Activation energy of trapping [eV]	$E_{r1}=0$
Detrapping rate constant for the first type of traps [1/s]	$b_1=2\cdot10^{11}$
Activation energy of detrapping [eV]	$E_{b1}=2.4$
Trapping rate constant for the second type of traps [1/s]	$r_2=7\cdot10^7$
Activation energy of trapping [eV]	$E_{r2}=2.0$
Detrapping rate constant for the second type traps [1/s]	$b_2=5\cdot10^{13}$
Activation energy of detrapping [eV]	$E_{b2}=4.1$

Fig.3 represents set of thermal desorption spectra computed using the proposed model. These spectra are in reasonably good agreement with the spectra obtained experimentally. Table 1 sets out the parameters of hydrogen-graphite. Some of these parameters are derived from experimental data and others - from computer simulation process.

Although the majority of authors [1–3, 5, 7–11, 17, 22–26] involved hydrogen traps in graphite, only certain of them tried to explain the trapping mechanism.

We believe that hydrogen trapping for both types of traps occurs on the broken bonds of carbon atoms placed at the edge of graphite layers. Difference in binding energies for the two types of traps may be explained by hydrogen sorption on different planes of edge of graphite layers (or edge dislocations). A great amount of edge dislocation is believed to be produced in graphite while in fabrication as a result of crawling of one grown layer over another. As mentioned in [27], binding energy at plane (10-10) exceeds this energy at plane (11-20) by 0.8 eV (see Fig.4). If we suppose that hydrogen migrates in graphite as molecule and this molecule is captured at neighbouring bonds of carbon atoms placed on dislocation hence hydrogen adsorption enthalpy should differ by 1.6eV for these two types of traps. This value is in a good agreement with the data listed in the table 1.

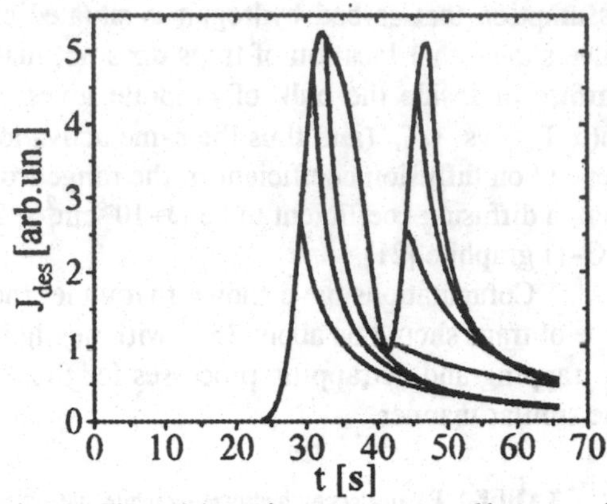

Figure 3. Thermal desorption spectra computed by proposed model.

The probable mechanism of atomic hydrogen sorption in graphite seems to be the following.

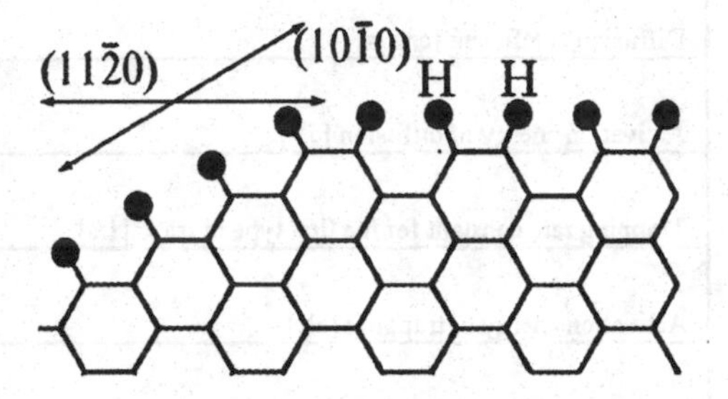

Figure 4. More probable places of hydrogen sorption on edge dislocations in graphite.

When arriving on graphite surface atomic hydrogen penetrates into the bulk of graphite. Due to a small atomic diameter of hydrogen it can probably penetrate through graphite layers (through hexagonal pores or vacancies in a layer). Nevertheless the mobility of hydrogen along the layers should be significantly higher than through them.

As a result a process of graphite saturation with hydrogen occurs layer by layer. Eventually free carbon bonds on dislocations become fully occupied by hydrogen atoms in every saturated layer. Further permeation of hydrogen into the bulk of graphite may be limited by inner defects like cracks and pores.

As temperature of the saturated sample rises the detrapping process starts. Released hydrogen migrates along graphite planes and may be trapped again. Migration process is limited by diffusion and reversible trapping on edge dislocations. In the long run hydrogen reaches a grain boundary and leaves a sample very fast *over* a set of interconnected pores.

5. Conclusions

Atomic hydrogen interaction with pyrolitic, quasimonocristalline and RG-Ti graphite was studied. Two distinct maximums in thermal desorpton spectra and some peculiarities of desorption kinetics at a fixed temperature are explained by the existence of two types of traps with binding energy 2.4 and 4.1 eV in the bulk of graphite. These traps are situated on different planes of edge dislocations. The process of degassing of graphite saturated with hydrogen occurs through diffusion and reversible trapping on edge dislocations.

6. References

1. W.Moller (1989) Hydrogen Trapping and Transport in Carbon, *J.Nucl.Mat.* **162-164**, 138-150.
2. K.L.Wilson, W.L.Hsu (1987) Hydrogen Recycling Properties of Graphite, *J.Nucl.Mat.* **145-147**, 121-130.
3. R.A.Causey (1989) The Interaction of tritium with Graphite and its Impact on Tokamak Operations, *J.Nucl.Mat.* **162-164**, 151-161.
4. P.G.Fisher, R.Hecker, H.D.Rohrig, D.Stover (1977) Zum Verhalten Ion Tritium in Reaktorgraphiten, *J.Nucl.Mat.* **64**, 281-288.
5. K.Nakayama, S.Fukuda, T.Hino, T. Yamashina (1987) Thermal Desorption Process and Surface Roughess of POCO Graphite Irradiated by Hydrogen Ion Beam *J.Nucl.Mat.* **145-147**, 301-304.
6. W.R.Wampler, Ĉ.W.Magee (1981) Depth Resolved Measurements of Hydrogen Isotope Exchange in Carbon, *J.Nucl.Mat.* **103&104**, 509-512.
7. G.Hansali, J.P.Biberian, M.Bienfait (1990) Ion Beam Implantation and Thermal Desorption of Deuterium Ions in Graphite, J.*Nucl.Mat.* **171 (2&3)**, 395-398.
8. V.Phylipps, E.Vietzke, M.Erdweg, K.Flaskamp (1987) Thermal Desorption of Hydrogen and Various Hydrocarbons from Graphite Bombarded with Thermal and Energetic Hydrogen, *J.Nucl.Mat.* **145-147**, 292-296.
9. S.Fukuda, T.Hino, T.Yamashina (1989) Desorpion Processes of Hydrogen and Methane From Clean and Metal-Deposited Graphite Irradiated By Hydrogen Ions, *J.Nucl.Mat.* **162-164**, 997-1003.
10. D.K.Brice (1990) Evidence for Single Shallow Hydrogen Trap in Hydrogen Implanted Graphite, *Nucl.Instr. and Meth.* **B44**, 302-312.
11. Graphite, R.A.Causey, M.I.Baskes, K.L.Wilson (1986) The Retention of Deuterium and Tritium in POCO AXF-5Q, *J.Vac.Sci.Technol.* **A4(3)**, 1189-1192.
12. G.Federici, C.H.Wu (1992) Modeling of Plasma Hydrogen Isotope Behavior in Porous Materials (Graphites / Carbon-Carbon Composites), *J.Nucl.Mat.* **vol.186**, No.2 131-152.
13. P.C.Stangeby, O.Auciello, A.A.Haasz, B.L.Doyle (1984) Trapping of Sub-eV Hydrogen and Deuterium Atoms in Carbon, *J.Nucl.Mat.* **122&123**, 1592-1597.
14. P.Hucks, K.Flaskamp, E.Vietzke (1980) The Trapping of Thermal Atomic Hydrogen on Pyrolytic Graphite, *J.Nucl.Mat.* **93&94**, 558-563.
15. J.W.Davis, A.A.Haasz, P.C.Stangeby (1988) Hydrocarbon Formation due to Combined H^+ Ion and H^o Atom Impact on Pyrolytic Graphite, *J.Nucl.Mat.* **155-157 part A**, 234-240.
16. I.S.Youle, A.A.Haasz (1991) Retention of Sub-eV Atomic Tritium and Protium in Pyrolytic Graphite, *J.Nucl.Mat.* **182**, 107-112.
17. T.Tanabe, Y.Watanabe (1991) Hydrogen Behavior in Graphite at Elevated Temperatures, *J.Nucl.Mat.* **179-181, part A**, 231-234.
18. M.Balooch, D.R.Olander (1975) Reactions of Modulated Molecular Beams with Pyrolytic Graphite.III.Hydrogen, *J.Chem.Phys.* **63**, 4772-4786.
19. Final Report on the Contract N 7/4 between NTC "Sintes" St.Petersburg, Russia and Fusion Centre, Moscow, Russia (1995) Hydrogen Retention and Release From Graphite, principal investigator: A.P.Zakharov, 7-11.
20. E.Denisov, T.Kompaniets, A.Kurdyumov, S.Mazayev (1996) Molecular Hydrogen Interaction with Unirradiated Graphite, *J.Nucl.Mat.* **233-237**, 1218.
21. E.A.Denisov, T.N.Kompaniets, A.A.Kurdyumov, S.N.Mazayev, Yu.G.Prokofiev (1994) Comparison of Hydrogen Inventory and Transport in Beryllium and Graphite Materials, *J.Nucl.Mat.* **212-215**, 1448.

22. E.Hoinkis (1991) The Chemosorption of Hydrogen on Porous Graphites at Low Pressure and Elevated Temperature, *J.Nucl.Mat.* **182**, 93-106.
23. W.R.Wampler, B.L.Doyle, R.A.Causey, K.Wilson (1990) Trapping of Deuterium at Damage in Graphite, *J.Nucl.Mat.* **176&177**, 983-987.
24. J.P.Redmond, P.L.Walker (1960) Hydrogen Sorption on Graphite at Elevated Temperatures, *J.Phys.Chem.* **64(9)**, 1093-1099.
25. S.L.Kanashenko, A.E.Gorodetsky, V.N.Chernikov, A.V.Markin, A.P.Zakharov, B.L.Doyle, W.R.Wampler (1994) Hydrogen Adsorption on and Solubility in Graphites, *J.Nucl.Mat.* **233-237, part B**, 1207-1212.
26. V.N.Chernikov, A.E.Gorodetsky, S.L.Kanashenko, A.P.Zakharov, W.R.Wampler, B.L.Doyle (1994) Trapping of deuterium in boron and titanium modified graphites before and after carbon ion irradiation, *J.Nucl.Mat.* **217**, 250-257.
27. J.P.Chen, R.T.Yang (1989) Chemosorption of Hydrogen on Different Planes of Graphite – a Semiempirical Molecular Orbital Calculation, *Surf.Sci.* **216**, 481-488.

DEUTERIUM RETENTION IN SI DOPED CARBON FILMS

E. VAINONEN-AHLGREN, T. SAJAVAARA, W. RYDMAN,
T. AHLGREN, K. NORDLUND AND J. KEINONEN
Accelerator Laboratory, P. O. Box 43, FIN-00014 University of Helsinki, Finland

J. LIKONEN AND S. LEHTO
Technical Research Centre of Finland, Chemical Technology, P.O. Box 1404, FIN-02044 VTT, Finland

AND

C. H. WU
The NET Team, Max-Planck-Institute für Plasmaphysik, Boltzmannstrasse 2, D-85748 Garching bei München, Germany

Abstract. Deuterium retention, solubility and out-diffusion have been studied in silicon doped carbon films produced by physical vapor deposition. The deuterium concentration profiles were measured by the time-of-flight elastic recoil detection analysis technique and secondary ion mass spectrometry. The D retention and solubility were measured in D implanted carbon samples. The out-diffusion of D was investigated in D co-deposited samples. The solubility of D was shown to increase as a function of Si concentration in the co-deposited samples while in the implanted samples no dependence of the Si content was observed. It was proposed that annealing behavior of deuterium has a trapping-like character.

1. Introduction

In the next-step fusion device ITER, carbon fiber composites are interesting candidates for divertor armour materials. In the presence of plasma, redeposition of sputtered carbon particles, formation of diamondlike carbon films and carbon based composite films will take place. The uptake and release of deuterium and tritium from those films will significantly affect the recycling of D and T fuel as well as tritium retention in the fusion de-

C.H. Wu (ed.), Hydrogen Recycling at Plasma Facing Materials, 281–287.

vice. Therefore, an understanding of the processes which involve trapping and retention of hydrogen isotopes in those films is important.

A decrease of the chemical sputtering by a factor of 2 to 3 [1] in silicon doped carbon compared to pure carbon makes this material attractive for application in a fusion device. In addition to this, Si doping will decrease the baking temperature needed to remove impurities from surfaces as well. Si is also known to be a good oxygen getter and impurity which increases thermal conductivity.

This work continues our studies on the migration of hydrogen isotopes in carbon films [2, 3]. To our knowledge there are no systematic experimental data in the literature on the migration of deuterium in Si doped carbon materials.

2. Experimental arrangements

The carbon films studied were prepared by the company DIARC-Technology Inc. using the arc discharge method. Characterization of the films and the deposition method have been described in detail elsewhere [2]. Silicon doped samples (6, 15 and 33 at.% Si) were produced by mixing pure graphite and silicon powders which were further solidified by the hot isostatic pressing technique.

Two sets of samples were prepared for deuterium migration studies. The first set of films was grown in a D atmosphere under the pressure of 1.0 and 0.2 mPa. In the second set 2 keV D^+ ions were implanted with doses from 1.0×10^{16} to 1.0×10^{18} ions cm^{-2}. The implantations were performed in a vacuum of 5×10^{-7} Pa at room temperature with the 120-keV isotope separator of the Accelerator laboratory.

Annealing was done in a quartz-tube furnace (pressure below 2×10^{-6} Pa) at temperatures from 400 to 1050 °C. The annealing time varied from 1 to 24 h.

The time-of-flight elastic recoil detection analysis (TOF–ERDA) measurements of D and impurity profiles were performed with the 5 MV EGP–10–II tandem accelerator of the University of Helsinki. In the measurements a 53 MeV $^{127}I^{10+}$ beam was used. The detection angle was 40° and the sample was tilted 20° relative to the beam direction. The beam electric current varied between 5 and 12 nA. Energy spectra of the heavy recoils were calculated from the time-of-flight signal and the energy detector was used for mass separation. The energy spectra for hydrogen isotopes were obtained from the energy detector. By using the fact that other recoils than hydrogen isotopes have a detection efficiency of 100%, a total detection efficiency of 100 % could be obtained by adding non-coincident energy detector events and coincident hydrogen events. The atomic concentrations were calculated with known geometry and ZBL-stopping powers [4, 5].

The depth profiling of D atoms was also carried out by secondary ion mass spectrometry (SIMS) at the Technical Research Centre of Finland using a double focusing magnetic sector SIMS (VG Ionex IX70S). The current of the 5-keV O_2^+ primary ions was typically 400 nA during depth profiling and the ion beam was raster-scanned over an area of $270 \times 430\ \mu m^2$. Crater wall effects were avoided by using a 10 % electronic gate and 1 mm optical gate. The pressure inside the analysis chamber was 5×10^{-8} Pa during the analysis. The depth of the craters was measured by a profilometer (Dektak 3030ST). The uncertainty of the crater depth was estimated to be 5 %. Data of TOF-ERDA measurements were used to normalize the D concentration obtained in SIMS experiments.

3. Results and Discussion

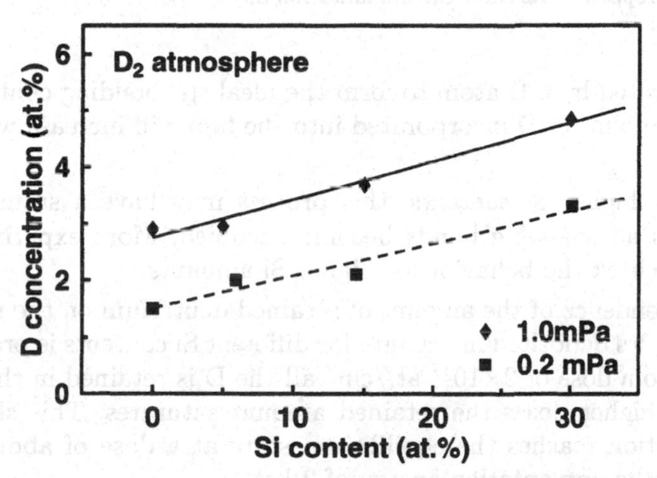

Figure 1. D concentration as a function of Si content for samples deposited in a deuterium atmosphere. The solid and dashed lines are the linear fits to the experimental data.

The dependence of the amount of retained deuterium on the Si concentration in the samples deposited in deuterium atmosphere is presented in Fig. 1. As can be seen, a higher Si content leads to an increase in the retained D amount. The dependence on Si concentration could be explained by the following facts. Carbon has practically the same cohesive energy in the sp^2 and sp^3 configurations, and Si prefers an sp^3 bonding. Thus during the growth process an sp^2 bonded C atom at the surface does not necessarily form a bond with a D atom in the surrounding atmosphere. A Si atom in a three-fold bonded state, however, has a dangling bond, which

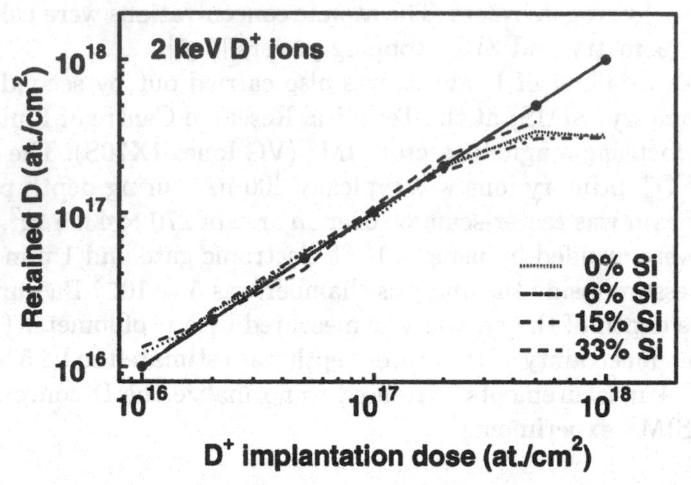

Figure 2. Amount of retained D as a function of implantation dose in samples with different Si concentrations. The solid line presents an ideal case when all implanted deuterium is trapped. Dots show the implantation doses used.

can be saturated by a D atom to form the ideal sp^3 bonding configuration. Hence the amount of D incorporated into the film will increase with the Si content.

For even higher Si contents, this process may have a saturation-like character as all available bonds become occupied. More experiments are required to check the behavior for higher Si amounts.

The dependence of the amount of retained deuterium on the implanted dose in samples deposited in vacuum for different Si contents is presented in Fig. 2. Up to a dose of 2×10^{17} at./cm^2 all the D is retained in the sample, whereas at higher doses the retained amount saturates. This shows that the D retention reaches the equilibrium state at a dose of about 5×10^{17} at./cm^2 for the implantation energy of 2 keV.

The figure also shows that the amount of deuterium does not depend on the Si content of the sample, in strong contrast to the material produced in a D atmosphere. We propose the following reason to be behind this fact. In the implanted sample all Si and C atoms are already bonded in an energetically favorable state (sp^3 for Si and sp^2 or sp^3 for C). To be retained in the sample, an implanted D atom has to form a covalent bond with one of the sample atoms. Since a five-fold coordination is not energetically favorable for either C or Si, the most likely process by which a D atom can form a bond is that it reacts with an sp^2-bonded C atom, which then enters into the sp^3 state. Since no (or very few) Si atoms are in the sp^2 configuration, and since experiments show that the total amount of sp^2-bonded carbon atoms remains roughly the same with increasing Si content

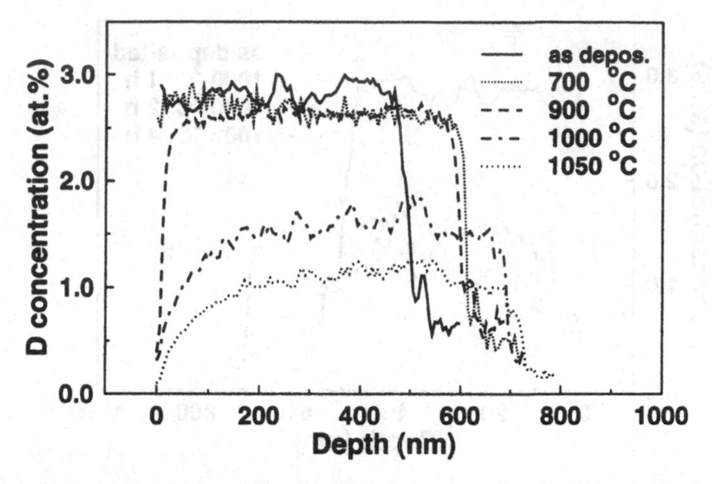

Figure 3. Deuterium concentration profiles after annealing at different temperatures for 1 h as obtained by SIMS for samples not containing Si and deposited in D atmosphere under a pressure of 1.0 mPa.

[7], the D retention in this case will not depend on the Si concentration of the film.

Out-diffusion of deuterium was studied in the samples not containing Si and deposited in a D atmosphere under the pressure of 1.0 mPa. Figure 3 shows concentration profiles measured with SIMS after annealing at different temperatures. The difference in thicknesses is due to the fact that different sample pieces were analyzed. As can be seen the amount of D has not changed significantly at temperatures 700 – 900°C. However, at higher temperatures, out-diffusion was observed near the surface. In addition to this, the amount of D has decreased in the film. Figure 4 shows D concentration profiles for different annealing times at 1000°C. It can be observed that longer annealing times lead to decrease of D content and out-diffusion at the near-surface region. After 1 h annealing about half of D has migrated away. The following annealing for 1 h leads to release of a small additional fraction of deuterium. Further 2 h reduce only little the D amount. The D concentration is after the 4 h annealing at the level of 1 at.% which is about the same as the D content after annealing at 1050°C for 1 h. From the said above it is possible to suggest that deuterium retention within the sample has a trapping-like character.

Figure 5 shows the temperature dependence of the ratio of retained D to the initial D amount. Presented data are for the samples deposited in vacuum without Si. In this case the release point, i.e. the temperature at which the D amount drops by a factor of 2, was found to be 984°C, in agreement with our previous data for H release from the samples deposited

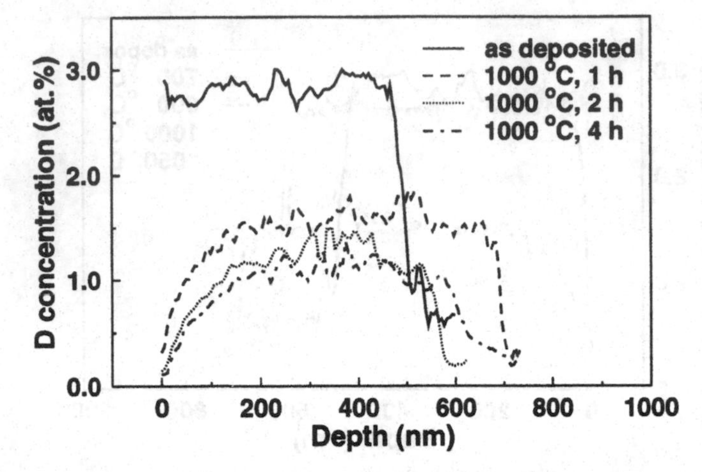

Figure 4. Deuterium concentration profiles obtained by SIMS after annealing at 1000° for 1, 2 and 4 hours for samples deposited in deuterium atmosphere of 1.0 mPa without Si.

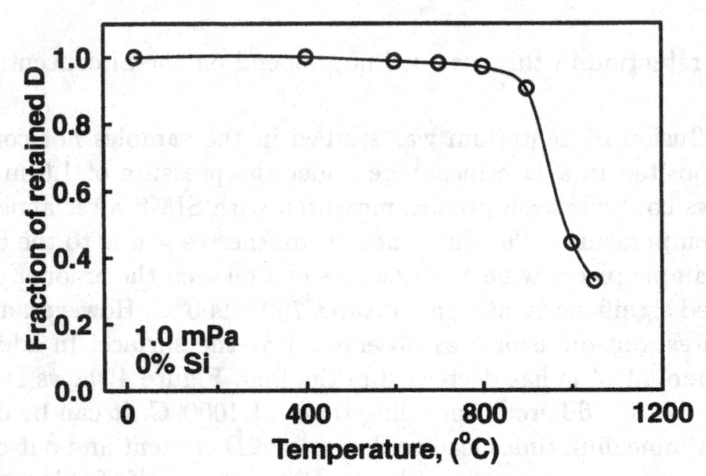

Figure 5. The ratio of retained D amount to the initial D amount as a function of annealing temperature for samples deposited in deuterium atmosphere of 1 mPa without Si.

in hydrogen atmosphere with pressures between 0.06 and 0.6 mPa [2].

4. Conclusions

Retention of deuterium in D^+-ion-implanted and D co-deposited carbon samples with different Si contents was studied. We observed that the D amount depends on Si concentration in the co-deposited but not in implanted samples, and proposed a microscopic mechanism explaining this behavior. Annealing behavior of deuterium in films not containing Si has a trapping-like character in the bulk and out-diffuses in the near-surface region. Temperature of release was found to be 984°C.

Acknowledgements

This work was supported by the Association Euratom-TEKES within the Finnish fusion program (FFUSION–2). Authors want to thank Mr Jukka Kolehmainen and Mr Janne Partanen (DIARC-Technology Inc.) for sample preparation.

References

1. Balden, M., Roth, J., and Wu, C.H.: Thermal stability and chemical erosion of the silicon doped CFC material NS31, *J. Nucl. Mater.*, **258–263** (1998), 740–744.
2. Vainonen, E., Likonen, J., Ahlgren, T., Haussalo, P., Keinonen, J., and Wu, C.H.: Hydrogen migration in diamond-like carbon films, *J. Appl. Phys.*, **82** (1997), 3791–3796.
3. Ahlgren, T., Vainonen, E., Likonen, J., and Keinonen, J.: Concentration-dependent deuterium diffusion in diamondlike carbon films, *Phys. Rev.* **B 57** (1998), 9723–9726.
4. Ziegler, J.F.: SRIM-96 computer code, private communication.
5. Jokinen, J., Keinonen, J., Tikkanen, P., Kuronen, A., Ahlgren, T., and Nordlund, K.: Comparison of TOF-ERDA and nuclear resonance reaction . . . , *Nucl. Instr. and Meth.* **B** 119 (1996), 533–542.
6. Sajavaara, T., Jokinen, J., Arstila, K., and Keinonen, J.: TOF-ERDA spectrometry applied for the analysis of Be migration in (100) GaAs, *Nucl. Instr. and Meth.* **B** 139 (1998), 225–229.
7. *FFUSION yearbook*, edited by Karttunen, S., VTT Energy, Espoo, 1996.

Retention of 100 eV Tritium in Tungsten at High Fluxes

Thomas J. Venhaus* and Rion A. Causey
Sandia National Laboratories
Livermore, CA 94550

Tungsten ideally suited for many fusion applications due to its high threshold for sputtering as well as its very high melting point. If tungsten is to be used as a plasma-facing material in a fusion reactor, it is necessary to know its hydrogen isotope recycling and retention characteristics. The Tritium Plasma Experiment (TPE) has been used in a research program to determine the retention of tritium in 99.95% pure tungsten exposed to high fluxes of 100 eV tritons. Plasma exposures were performed at a temperature of 623 °K. The flux of ions ranged from 10^{21} to 10^{22} ions/m^2/sec. After exposure to the tritium plasma, the samples were transferred to an outgassing system containing an ionization chamber for detection of the released tritium. The retention results closely followed a power law. Modeling was performed on the retention as a function of flux and the outgassing characteristics.

1 Introduction

Tungsten has been selected as a plasma facing material in the next generation of tokomak fusion reactors due to its high melting temperature and low sputter yield [1,2]. It is important to understand the hydrogen recycling characteristics of tungsten, from the standpoint of plasma fueling, tritium retention, and tritium permeation. A number of experiments have been performed to investigate hydrogen recycling and migration in tungsten as a function of temperature [3, 4, 5], surface impurities [6], fluence [7], ion energy [8], neutron damage [9], and material type [3, 10]. To date, the flux dependence of hydrogen recycling has not been extensively investigated. This type of experiment is important, as divertor components in tokomak machines are exposed to very high fluxes of particles. This report investigates the dependence of hydrogen isotope recycling on the flux of incident particles at a given ion energy and target temperature.

2 Experimental Procedures

Plasma exposures were performed in the Tritium Plasma Experiment (TPE) at Los Alamos National Laboratory. The TPE is an arc-reflex plasma source capable of producing high fluxes of hydrogen isotope ions. This device is described in greater

---

* On Assignment at the Tritium System Test Assembly (TSTA) at Los Alamos National Laboratory

C.H. Wu (ed.), Hydrogen Recycling at Plasma Facing Materials, 289–299.

detail elsewhere [11]. The working gas consisted of 97% deuterium and 3% tritium, and was fed to the system at a flow rate of 30 sccm. The primary impurity in the system was water vapor at 1.5×10^{-5} Pa. For these experiments, samples were exposed to particle fluxes ranging from 1×10^{21} (D+T)/m^{2}s to 1×10^{22} (D+T)/m^{2}s. Targets were biased to accelerate the ions from the plasma to ~120 eV. The plasma provided the heating of the samples to 623 °K. A copper disk behind the sample provided an even heat distribution. By varying the thickness and composition of a third disk (the one in contact with a water-cooled holder), the sample temperature could be maintained at 623 °K for the various fluxes (heat loads). Exposure time was 1 hour.

The samples used in these experiments were 50.8 mm in diameter, and 2 mm thick. The samples, provided by Plansee Aktiengesellschaft, were powder metallurgy products. The tungsten was 99.95% pure. After receipt, the samples were polished, then annealed at 1273 °K for one hour in an inert atmosphere.

Following exposure to the plasma, the samples were transferred in air to an outgassing system. The temperature of the sample was ramped at a rate of 20 °K/min to a final temperature of 1473 °K. Gas consisting of 99% argon and 1% hydrogen was swept over the sample and through an ionization chamber where the tritium concentration was monitored. The total retention was calculated from ionization chamber readings. Modeling of the retention data and thermal desorption spectra was performed with TMAP4, a finite difference computer code [12]. Results were also compare to analytical functions found in literature.

3 Results

The results of the tritium retention for the different ion fluxes are shown in Figure 1. The thermal desorption data is shown in Figure 2. A major peak was observed in the TDS spectra at approximately 850 °K for the flux values of 2.3×10^{21} and 4.6×10^{21} ions/m^2/sec, and two peaks at 750 °K and 850 °K for the flux of 1.3 x 10^{22} ions/m^2/sec. The spectra also show a broad shoulder centered around 1000 K.

4 Modeling

4.1 Analytical Expressions

A number of analytical expressions have been developed to predict the so-called recycling coefficient and the total expected retention in a PFC [13, 14]. The following section considers these expressions in view of the current data. It will be assumed that the trapped hydrogen in the tungsten target will be the dominant contributor to the total retention [18]. This assumption is considered valid due to the high trapping energies in tungsten and the low target temperatures used in this experiment. The material constants used in the modeling efforts are shown in Table 1.

Constant	Value	Units	Reference
Diffusivity	$4.1 \times 10^{-7}\exp(-0.39/kT)$	m^2/s	Frauenfelder[15]
Recombination	$6.0 \times 10^{-20}\exp(-0.81/kT)$	m^4/s	Anderl [16]
Trap Energy	1.42	eV	
Trap Density	26	ppm	
Reflection Coefficient	0.39	per ion	TRIM Code [17]

Table 1 Values used in analytical expressions and TMAP simulations

4.1.1 Boundary Conditions

The rate-limiting step for hydrogen release from the surface can be either diffusion or recombination. For the high fluxes used in these experiments, it is assumed that there is sufficient hydrogen from the plasma to allow immediate recombination. Doyle and Brice [14] developed a dimensionless transport parameter W that determines the rate-limiting step at the surface for a set of conditions. The parameter is given as:

$$W = \left(\frac{\phi \cdot R}{D}\right)\left(\frac{k_r}{\phi}\right)^{1/r}$$

where ϕ is the non-reflected incident flux, R is the average depth of implantation, D is the diffusivity, k_r is the recombination coefficient, and r is the reaction order (i.e., r=2 for recombination kinetics of H). Using the material constants shown in Table 1, an average implantation depth of 3 nm as given by TRIM code[17], and the experimental fluxes used in this report, $w(x)$ on the plasma facing side is greater than one. This parameter distinguishes the rate-limiting step on the surface as diffusion for the entire flux range.

4.1.2 Model and Results

Trapping in tungsten has been shown to significantly effect the diffusion of hydrogen through the material [16, 18]. In a material with active traps, the diffusion profile is nearly constant for a characteristic depth from the surface before decreasing rapidly [19]. This is in contrast to the expected profile in a trap-free material, that is, one that follows the equation $C(x,t) = C_o erfc\left(x/\sqrt{4Dt}\right)$. Federici et al. [20] developed a model for trap front movement as a function of time in the presence of sputter erosion. The case for zero erosion was also considered, and the trap front position is given as:

1) $$x_f = \sqrt{2D\frac{C_o}{C_T}t}$$

where x_f is the position of the trap front as measured from the plasma-facing surface, D is the diffusivity, C_o is the surface concentration, C_T is the fraction of trapped gas atoms, and t is time. Assuming that the flux of atoms penetrating the bulk is small when compared to the recombination flux, the concentration at the surface is given by:

2) $$C_o \approx \sqrt{\frac{\phi_i}{2k_r}} + \frac{\phi_i}{D}\delta$$

Where ϕ_i is the non-reflected flux, and δ is the average implantation depth. In a material such as tungsten where trapping energy is high, the trapped hydrogen fraction dominates at the temperature used in this study. If all traps are filled behind the trap front, the area density N (atoms/m^2) of the trapped hydrogen can be computed from equ. [2] and [3] by the equation:

3) $$N = x_f C_T$$

Figure 3 shows the results of the analytic expressions along with the experimentally determined values. The analytical expression predicts the experimental values quite well, especially in the light of the fact that no fitting parameters were used.

4.2 TMAP Modeling

An attempt was made to model the hydrogen isotope retention with TMAP4, a finite difference calculation code developed at the Idaho National Engineering Laboratory[12]. The following section presents the model development and the results from the simulations.

4.2.1 Model Parameters

An entry is made in the TMAP input file that defines the rate at which the solute atoms are introduced (e.g., by implantation), as well as the spatial distribution of the implantation profile. Implantation profiles and backscatter coefficients were taken from TRIM simulations.

The material constants used in the code were derived from literature, and are the same as those used in the analytic expressions. TMAP allows for a single trap energy and concentration. This concentration will be assumed to be uniform throughout the sample. It will also be assumed, for simplicity, that the 2 mm thick samples will have a uniform temperature distribution.

4.2.2 Simulation Results – Retention

The results for the TMAP simulations are shown Figure 3. The retention values are slightly lower than those from the analytical expression, but still match the experimental data well. The agreement between the TMAP results and the analytical expression give confidence to the assumptions made.

4.2.3 Simulation Results – Thermal Desorption

The desorption of the hydrogen isotopes was simulated with TMAP4. After the simulation of the plasma exposure, the sample temperature was ramped at the same rate as in the experiment. The surface flux from the plasma facing side as the hydrogen diffuse out of the sample was summed with the flux from the back side of the sample, as both sides of the sample would contribute to the TDS spectrum. The results are shown in Figure 4. A minor second peak is observed in the simulated spectra at approximately 1000 °K, and is a result of hydrogen flux from the back side. The broad shoulder in the experimental data may be a result of diffusion from the back side of the sample.

5 Discussion

One hundred eV hydrogen ion energy is below the sputtering threshold for tungsten [21, 22], so surface erosion was not considered in the modeling. Work by Alimov et al.[23] showed that the retention of 6 kV deuterium atoms (above the damage threshold) into single crystal tungsten at 480 °K was not measurable. This indicates that the traps formed in tungsten by ion bombardment damage are low in energy. The thermal desorption spectrum from the sample exposed to 1.3 x 10^{22} ions/m^2/sec shows the formation of a possible second peak lower in energy. This may indicate the formation of some type of surface damage at the high flux levels. Certainly this trap would need to be active at 623 °K to appear in the spectrum.

As 100 eV deuterium ions are below the sputtering threshold, damage may evolve from some type of bubble formation and eruption of hydrogen-filled voids from the surface. Bubbles form as atomic hydrogen diffuses through the lattice and becomes trapped at a defect. Another diffusing atom will combine with the trapped atom and form a hydrogen molecule. Once formed, molecular hydrogen is very difficult to remove from the lattice. So, the thermal desorption may not give any indication of molecular hydrogen filled viods. Condon and Schober examined the conditions necessary for hydrogen bubble formation in metals [24]. In their review article, vacancy clustering was given as a possible mechanism for cavity growth, based on work by Van Veen et al. [25] and Katrich and Budnikov [26]. In the work by Van Veen, the voids were created by 6 MeV proton and electron bombardment. Katrich and

Budnikov observed blister growth in tungsten exposed to 30 keV hydrogen ions. Sze et al. [27] had reported the formation of bubbles on the surface of pure tungsten foil samples exposed to similar conditions in the PISCES-B facility. The PISCES-B facility has, in the past, operated with graphite. As a result, the hydrocarbon content in the plasma was reported to be 0.25%. Carbon targets have never been exposed in TPE, and hydrocarbon partial pressure in the TPE plasma is less than 10^{-6} Pascal. The plasmas in the two facilities differ in this respect and the role of the carbon, and if it indeed plays a role, is not clear. Bubble formation was not observed in the present experiments, and future experiments are planned to investigate bubble formation from targets exposed to the TPE plasma.

6 Conclusion

Tritium retention measurements have been performed on powder metallurgy samples of relatively pure tungsten. The samples were exposed to 100 eV tritium and deuterium particles at a temperature of 623 °K. The flux values were 2.3×10^{21}, 4.6×10^{21}, and 1.3×10^{22} ions/m^2/sec. The pure tungsten showed a retention behavior predictable with TMAP4 modeling and simple analytical equations based on trap front movement. The TDS spectrum gave indications of the formation of slightly lower energy traps at the high flux value, perhaps caused by some type of surface damage. Future experiments will involve higher fluxes of particles, and samples doped with 1% lanthanum oxide added to increase machinability. The surface will be analyzed for the formation of hydrogen bubbles.

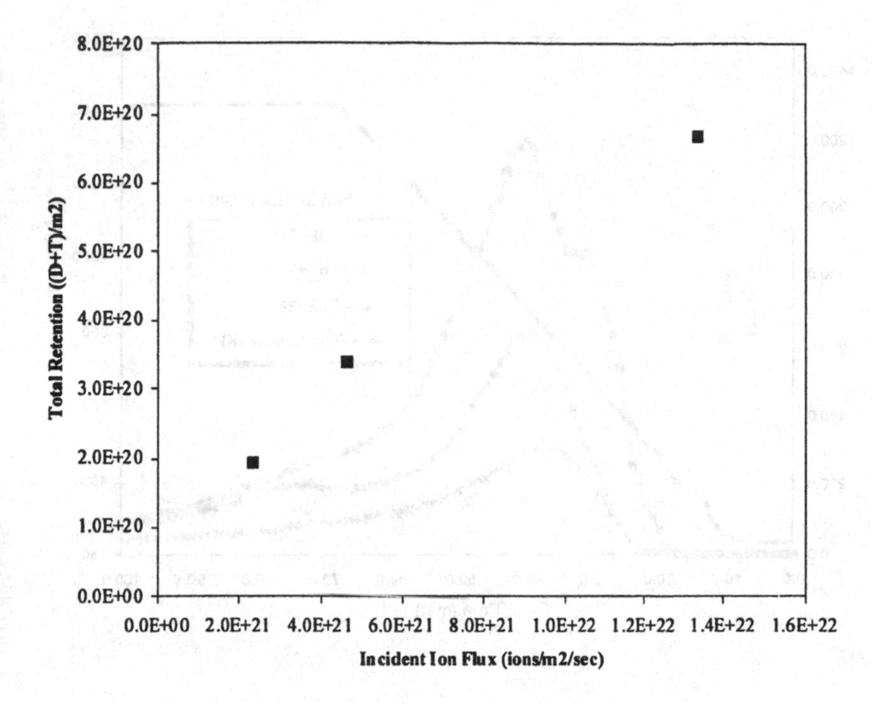

Figure 1. Total hydrogen isotope retention for the pure tungsten samples. Sample temperature was 623 °K. Exposure time was one hour.

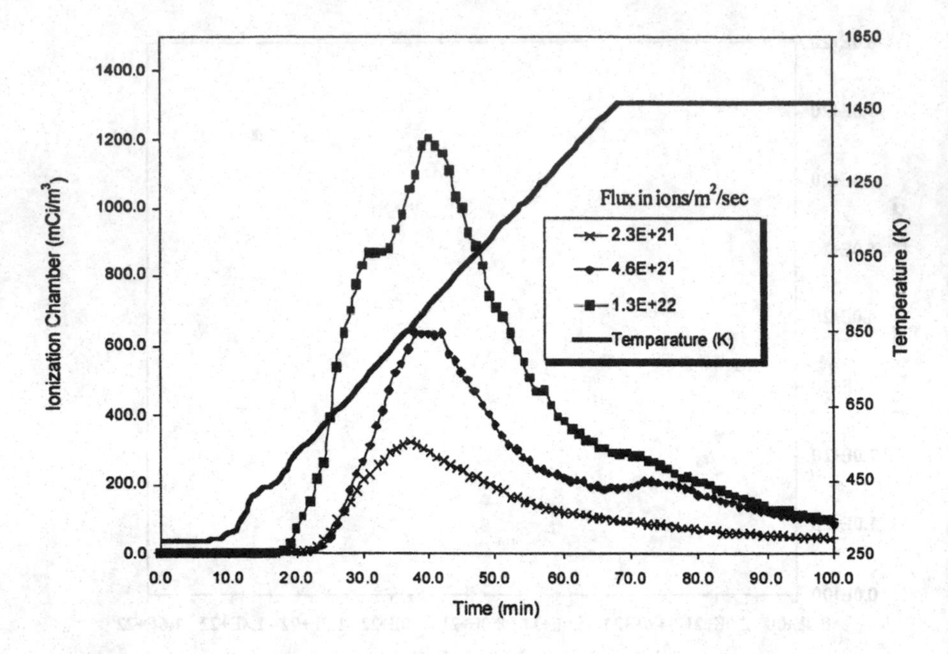

Figure 2. Thermal desorption spectra from tungsten

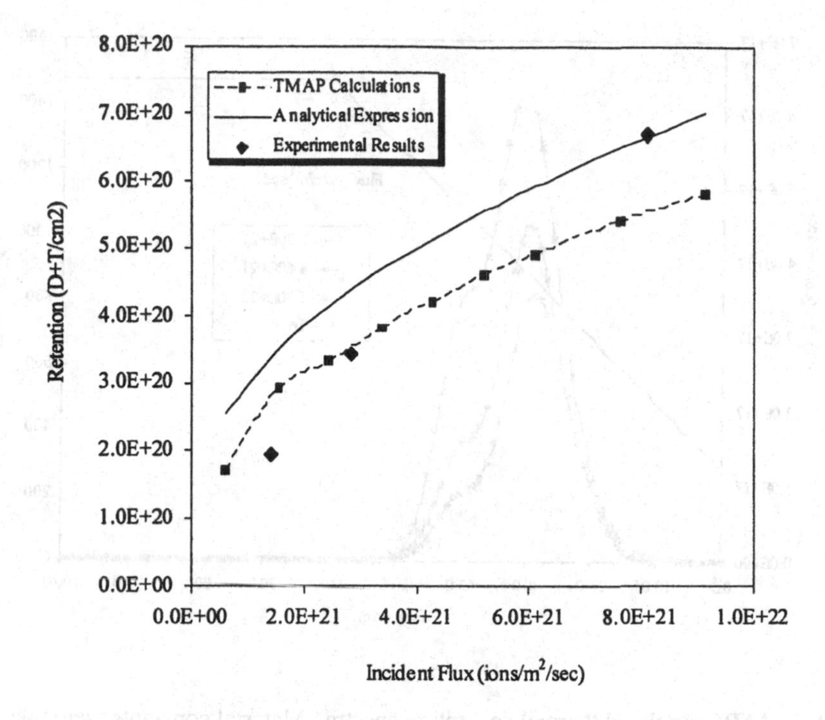

Figure 3. TMAP4 simulated total retention values, results from equation 4, and experimental results for the three tungsten samples. Results are plotted against non-reflected flux.

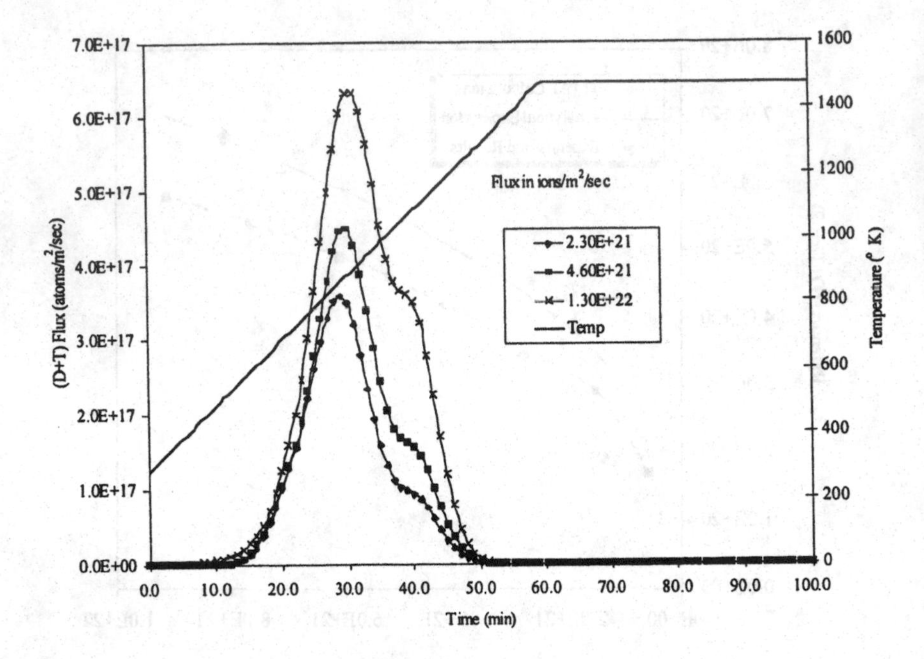

Figure 4. TMAP4 simulated thermal desorption spectra. Material constants were taken from literature.

7 References

[1] K. Tomabechi for the ITER Team, J. Nucl. Mater. 179-181 (1991) 1173
[2] G. M. Kalanin, J. Nucl. Mater 179-181 (1991) 1193
[3] V. Kh. Alimov, B. M. U. Scherzer, J. Nucl. Mater. 240 (1996) 75
[4] J. W. Davis, A. A. Haasz, J. Nucl Mater. 223 (1995) 312
[5] R. Causey, K. Wilson, T. Venhaus, W. R. Wampler, J. Nucl. Mater. 266-269 (1999) 467
[6] H. Eleveld and A. van Veen, J. Nucl. Mater. 191-194 (1992) 433
[7] A. A. Pisarev, A. V. Varava, S. K. Zhdanov, J. Nucl. Mater. 220-222 (1995) 926
[8] V. V. Bandurko and V. A. Kurnaev, Vacuum, 44 (1993) 937
[9] A. Van Veen, H. A. Filius, J. De Vries, K. R. Bijkerk, G. J. Rozing, J. Nucl. Mater., 155-157 (1998) 1113
[10] C. Garcia-Rosales, P.
[11] R. A. Causey, D. Buchenauer, W. Harbin, D. Taylor, and R. Anderl, Fusion Technology, 28 (3) (1995) p.1114
[12] B. J. Merrill at al., TMAP4 Tritium Migration Analysis Program Code, Description and User's Manual, INEL report, EG&G Idaho, Inc., EGG-FSP-10315
[13] A. A. Pisarev, S. K. Zhdanov, and O. V. Ogorodnikova, J. Nucl. Mater. 211 (1994) 127
[14] B. L. Doyle and D. K. Brice, Radiation Effects, 89 (1985) 21
[15] R. Frauenfelder, "Solution and Diffusion of Hydrogen in Tungsten," Journal of Vacuum Science and Technology 6 (1969) 388
[16] R. A. Anderl, D. F. Holland, and G. R. Longhurst, J. Nucl. Mat. 176 & 177 (1990) 683
[17] *The Stopping and Range of Ions in Solids*", by J. F. Ziegler, J. P. Biersack and U. Littmark, Pergamon Press, New York, 1985
[18] G. R. Longhurst, R. A. Anderl, D. F. Holland, "Evidences of hydrogen trapping in tungsten an implications for plasma-facing components", Fusion Technology 19 (1991) p. 1799
[19] X. W. Wu and B. Y. Tong, Philos. Mag. Lett. 61 (1990) 147
[20] G. Federici, D. F. Holland, B. Esser, J. Nucl. Mater. 227 (1996) 170
[21] R. Sakamoto, T. Muroga, N. Yoshida, J. Nucl. Mater., 220-222 (1995) 819
[22] Yoshida, N. J. Nucl. Mater. 266-269 (1999) 197-206
[23] V. Kh. Alimov, I. I. Arkhipov, A. E. Gorodetsky, A. V. Markin, A. P. Zakharov, R. Kh. Zalavutdinov, Progress Report on Task 2 of Contract LF-9196, April, 1999
[24] J. B. Condon and T. Schober, J. Nucl. Mater. 307 (1993) 1-24
[25] A. Van Veen, H. A. Filius, J. De Vries, K. R. Bijkerk, G. J. Rozing, and D. Segers, J. Nucl. Mater., 155-157 (1988) 1113
[26] N. P. Katrich and A. T. Budnikov, Zhurnal Tekhnicheskoi Fiziki, 52 (1982) 1236
[27] F. C. Sze, R. P. Doerner, and S. Luckhardt, J. Nucl. Mater. 264 (1999) 89-98

HYDROGEN ABSORPTION AND DESORPTION BEHAVIOR WITH A BORONIZED WALL

N. NODA, K. TSUZUKI, A. SAGARA
National Institute for Fusion Science, Toki-shi 509-5292, JAPAN

Abstract

A systematic study on hydrogen behavior with boronized walls has been carried out. . Hydrogen in boron films can be completely removed below 400°C in a pure boron film, This is not significantly changed with a boron film fully contaminated by oxygen. Strong absorption of hydrogen by a hydrogen-depleted boron film is seen just after start of hydrogen/glow discharge. Absorption during the discharge is not completely saturated with time, even after 10 hours. Behavior of hydrogen pressure just after the termination of the glow discharge suggests so called "dynamic retention". It is comparable to saturated retention. The dynamic behavior can be reproduced with a simple model, in which trapping, detrapping, recombination processes are included.

1. Introduction

Boronization [1] is one of the wall conditioning techniques that have been established in many fusion experimental devices. Roles of the boron walls at present are (1) reduction in oxygen, (2) reduction in hydrogen recycling compared to carbon walls, (3) reduction in first wall materials such as iron [2]. In future, first wall protection will become another important role. Thus, understanding of hydrogen behaviors is important for the boronized wall.

A small experimental device named SUT (SUrface modification Teststand) has been used for studies on the boron film [3,4]. This has a replaceable liner, which can be heated up to 600°C, an *in-situ* surface analysis station with AES, and a good vacuum condition. It gives us a chance to make a systematic study on hydrogen behavior with a

C.H. Wu (ed.), Hydrogen Recycling at Plasma Facing Materials, 301–306.

well-defined surface condition. Detail description on the SUT device is seen in ref. [4]. This paper gives an overview of the results obtained in the SUT experiments on boronized wall.

2. Experimental Procedure

A boron film are deposited by a glow discharge in a gas-mixture of helium and diborane (B_2H_6). Typical concentration of diborane is 5 %. It has been confirmed that most of hydrogen in the original boron film can be desorbed below 400°C. Usual procedure of the experiment is as follows:

i) flesh boron coating on the SS liner with B_2H_6/He,

ii) removal of hydrogen in the film by heating the liner up to 500°C,

iii) a series of experiments with a particular aim,

iv) out-gassing of the boron film by heating up to 500°C,

(almost identical to the procedure ii, to reset the boron film condition).

A glow discharge in pure hydrogen is applied in the procedure iii for studying the simple H absorption behavior into the boron film. Repetition of alternate discharge in hydrogen and in helium is applied to study the H desorption behavior with helium. In any case, the procedure iv brings back to the boron film to the condition similar to that after procedure ii.

3. Experimental Results

In Fig. 1, typical behavior of total pressure is shown for pure hydrogen discharge in the procedure iii. The pressure drops rapidly just after the ignition of the discharge (phase I). This is due to the absorption of hydrogen ions by the boron film. The pressure re-increases quickly (phase II) but not reaches to the original level. Then the increasing rate becomes much slower (phase III). After the discharge stops, sudden but a little increase (overshoot) in the pressure (phase IV) is seen as in an enlarged picture Fig. 2. It is followed by phase V, in which the pressure returns to the original level before the start of the discharge.

It was confirmed that slow increase in Phase III continues for at least 10 hours. It indicates that the film surface is not completely saturated even for such a long time. It could be explained by gradual penetration of hydrogen atoms deeper into the film. It may be caused by continuous impact of hydrogen ions at the surface.

A glow discharge in helium is applied to a hydrogen saturated boron wall. Reemission of hydrogen is observed. Emitted amount is 10 – 20 % of the absorbed one. After the helium discharge, hydrogen absorption behavior similar to phase I appears again. Integrated hydrogen amount is larger than that of reemission during the preceding He discharge. It means that repetitive application of the alternate discharges in helium and hydrogen causes hydrogen accumulation, too. It suggests that a fraction of hydrogen atoms in the top surface is pushed deeper into the film by helium ion impact.

Total amount of hydrogen absorbed in the procedure iii is calculated by integrating the decrease in the pressure as is seen in Fig. 1. It is typically around 1 x $10^{17}/cm^2$. The desorbed hydrogen amount is obtained by integrating the pressure during the heating procedure of iv. It is always a little more than absorbed one in the procedure iii. It means that at least newly implanted hydrogen is completely removed by heating up to 500°C. The peak of the thermal desorption curve is located between 350°C and 400°C. A temperature below this level is sufficient to reduce hydrogen-isotope inventory.

This result is obtained for pure boron film. It is reported that the peak is shifted to higher temperature in a carbon-boron compound film [1]. A question was whether oxygen contamination impacts hydrogen behavior or not. Using a glow discharge in oxygen/helium mixture, the boron film was saturated by oxygen. After oxidization, hydrogen absorption/desorption experiment was repeated. It has been found that amount of absorption is reduced after oxidization and the peak temperature in thermal desorption does not change significantly [5].

The pressure behavior just after the shutdown of the discharge indicates effect of so called "dynamic retention" [6]. The surface is over-saturated during the discharge by impact of energetic ions. Detail is discussed in the next section.

4. Discussion

The behavior at the end of the discharge is interpreted based on a simple model as follows. During the discharge, a significant fraction of hydrogen atoms are in "solute" situation due to detrapping by ion impact. Solute hydrogen is spontaneously trapped again and trapping and detrapping are balanced. A fraction of these solute atoms are recombined and reemitted from the surface. The reemission is caused through recombination between two solute atoms but not between solute and trapped atoms. This is verified by careful comparison of the decay in phase V [3].

If we neglect the recombination between solute and trapped atoms, the situation during the discharge is written by the following equation,

$$dc_s/dt = S_r - 2k_{ss}c_s^2 - k_{st}c_s(c_T - c_t) + k_{st}\phi_i c_t \quad (1),$$

$$dc_t/dt = k_{st}c_s(c_T - c_t) - \sigma_d\phi_i c_t \quad (2),$$

where c_s and c_t are the solute and trapped hydrogen density, respectively, c_T the density of the trap site, S_r the source term due to hydrogen implantation, k_{ss} the recombination coefficient, k_{st} the trapping rate coefficient, σ_d the ion impact detrapping cross section, and ϕ_i the ion flux at the surface. Since the pressure increase in phase IV is much smaller than implanted flux, the third and fourth terms are balanced each other and much larger than first two terms, which corresponds to large k_{st} and ϕ_i. It indicates that if we know the value of c_s , then we can obtain k_{ss}. The last two terms in Eq. (1) quickly become zero, and the decay in Fig. 2 is represented by the equation

$$dc_s/dt = - 2k_{ss}c_s^2 \quad (3)$$

after the end of the discharge. Then the reemission flux ϕ_r is written as

$$\phi_r = R_i/2k_{ss}\,(t + 1/\,k_{ss}c_0)^2 \quad (4),$$

where R_i is implanted depth and R_ic_0 the total amount of over saturated hydrogen atoms, which can be estimated to be $1 \times 10^{15}/cm^2$ by integrating the curve in Fig. 2. The depth R_i is estimated to be 6.5 nm based on TRIM code calculation. Then, the k_{ss} is estimated to be $7.8 \times 10^{-24} cm^3/s$.

The last phase of the discharge (phase III) can be regarded as a quasi-steady state. Then Eq. (1) gives $S_r = 2k_{ss}c_s^2$, from which c_s is estimated to be $8.8 \times 10^{21}/cm^3$. If you assume the trap density as $c_T = 4 \times 10^{22}/cm^3$, c_t is obtained as $3.2 \times 10^{22}/cm^3$. Thus in the qausi-steady state of phase III, 22% of hydrogen is in solute state.

5. Summary

Study on hydrogen behavior with boron film has been investigated with an experimental device SUT.

Important results are, (1) hydrogen in boron films can be completely removed below 400°C in a pure boron film, (2) this is not significantly changed with a boron film fully contaminated by oxygen, (3) strong absorption of hydrogen by a hydrogen-depleted boron film is seen just after start of hydrogen/glow discharge, (4) absorption during the discharge is not completely saturated with time, even after 10 hours, (5) dynamic balance in hydrogen retention is confirmed during the glow discharge, (6) the dynamic retention is comparable to static retention, (7) the dynamic behavior can be interpreted with a simple model, in which trapping, detrapping, recombination processes are included.

References

[1] J. Winter et al., J. Nucl. Mater. 145-147 (1987) 131.

[2] N. Noda et al., J. Nucl. Mater. 266-269 (1999) 234.

[3] N. Noda et al., J. Nucl. Mater. 220-222 (1995) 623.

[4] K. Tsuzuki et al., J. Nucl. Mater. 256 (1998) 166.

[5] K. Tsuzuki et al., J. Nucl. Mater. 266-269 (1999)

[6] K. Morita et al., J. Nucl. Mater. 145-147 (1987).

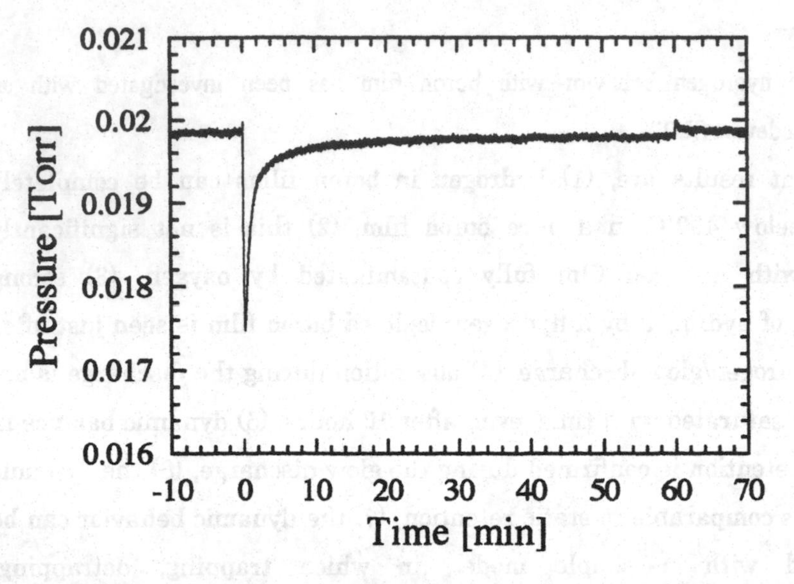

Fig. 1 **Time behavior of hydrogen pressure. A glow discharge is started at 0 min. and ended at 60 min.**

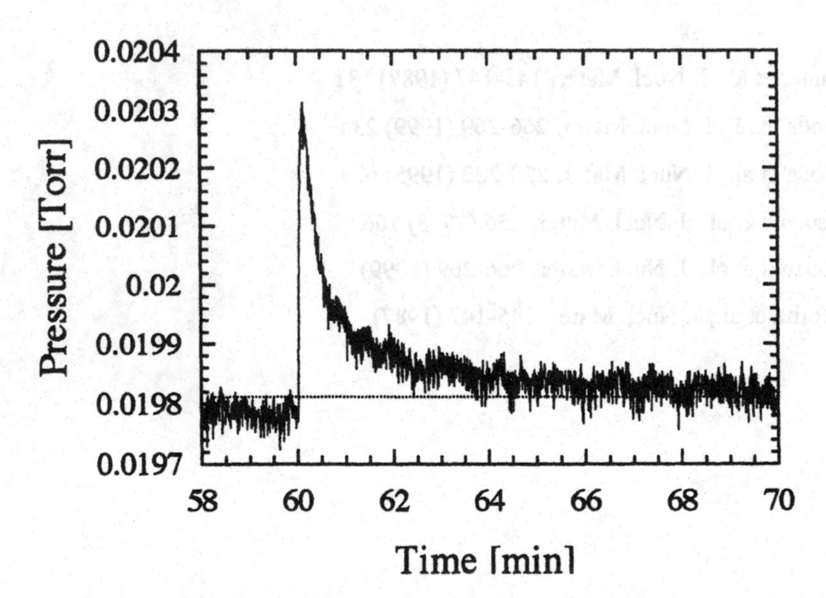

Fig. 2 **Enlarged picture of Fig. 1 around 60 min., end of the discharge. The dotted line indicates the initial level of the pressure before start of the discharge**

OUT OF PILE EXPERIMENTS ON THE INVESTIGATION OF HYDROGEN INTERACTION WITH REDUCED ACTIVATION FERRITIC-MARTENSITIC STEEL F82H.

A.Kh.Klepikov, T.V.Kulsartov, O.G.Romanenko, Y.V.Chikhray, V.P.Shestakov, I.L.Tazhibaeva

Science Research Institute of Experimental and Theoretical Physics of Kazakh State University, Tole bi Str., 96a, Almaty, 480012, Kazakhstan, Fax (7-3272)-50-62-88. E-mail: klepikov@ietp.alma-ata.su

1. Introduction

One of the main concerns of fission and fusion reactors development is the ecological aspect of their operation. If we consider fusion reactor operation with high energy and density neutron fluxes, the problem of activated materials management will be very important. To reduce the level of induced radioactivity it is supposed to use reduced-activation steels, for example F82H steel. Nevertheless the main reason of these steels application is their good stability against swelling and embrittlement. Because of possible application of this steel in the design of DEMO liquid metal blanket it is very important to know the parameters of its interaction with hydrogen isotopes for evaluation of tritium uptake and permeation flux in structural elements made of this material.

2. Experimental

The samples of F-82H reduced activation steel were used in the experiments. The chemical composition of the samples is shown in Table 1. The steel was heat treated before the sample machining. The heat treatment process consisted of normalizing at the temperature 1313 K for 30 min with the following tempering at 1013 K for 2 hours.

Table 1 Chemical composition of F82H steel samples, used in the experiments

C	Cr	Ni	Mo	V	Nb	Si	Mn	S
0.09	7.8	0.04	<0.01	0.16	<0.01	0.13	0.18	0.003
P	**B**	**N**	**Co**	**Ta**	**Al**	**Cu**	**W**	**Ti**
0.004	<0.001	0.06	-	0.02	<0.01	<0.007	2	<<0.02

Two types of experimental techniques were used to determine hydrogen interaction parameters for this steel, hydrogen permeation techniques (HP) and equilibrium loading from molecular hydrogen, followed by thermodesorption spectroscopy (TDS).

C.H. Wu (ed.), Hydrogen Recycling at Plasma Facing Materials, 307–312.

3. Results and discussion

For HP experiments the membrane with the thickness 1 mm ant the diameter 16 mm was used. Permeation curves were obtained in the temperature range 573–923 K for input hydrogen pressures 0.7×10^2 - 5×10^3 Pa. Several sets of permeation experiments revealed instability of the obtained permeation parameters. Hydrogen diffusivity and permeability were increasing in the course of permeation experiments, Fig. 1, and hydrogen permeation parameters, determined from the data obtained, were different from the literary data on this steel [1]. It was attributed to the changes of surface composition and the sample was oxidized at 973 K (O_2 pressure 133 Pa) for 1 hour with the following disoxidation in hydrogen (P=1330 Pa) for 3 hours for surface cleaning. This treatment did not result in significant changes in the behavior of hydrogen permeation parameters. The situation changed only after heat treatment at 973 K for 20 hours. Permeation curves became reproducible and the calculated values of permeation parameters became agree with the data of other authors (Fig. 2 and 3).

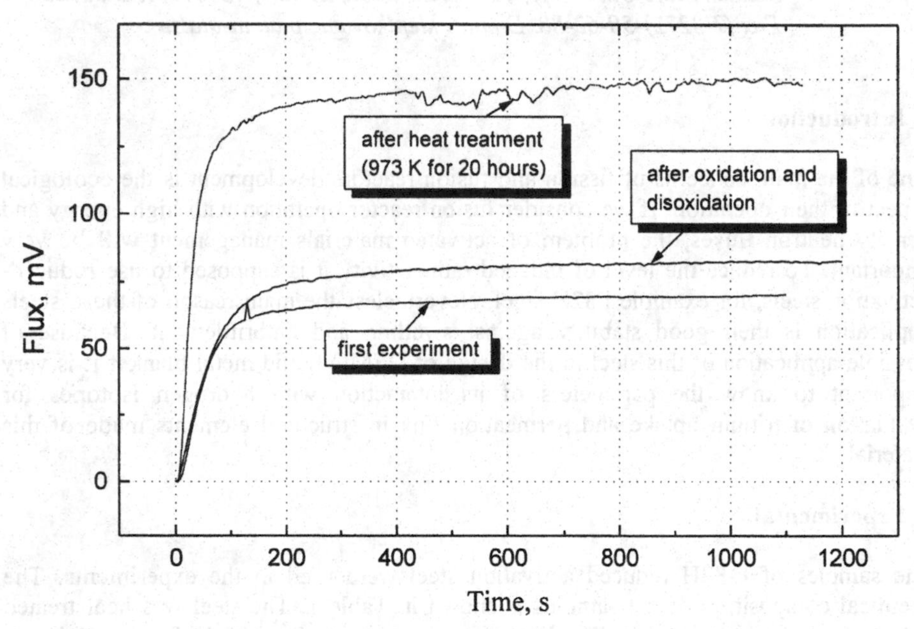

Fig 1. Hydrogen permeation curves at the different stages of experiment (T=600°C, P=1330 Pa)

Hydrogen permeability and diffusivity, determined in the experiments are shown in Fig. 2 and 3 (see also Table 2).

The dependence of H_2 permeation flux on the upstream pressure (P^n) does not follow the square root law. The n value was in the range $0.5<n<1$ and all the treatment does not result in the change of permeation flux dependence to bulk diffusion limited case.

Table 2. Hydrogen permeation parameters of F82H sample for different stages of experiment.

	P_0 mole/m*s*Pa$^{1/2}$	Ep, kJ/mole	D_0 m^2/s	Ed, kJ/mole
First experimental set	3.38×10^{-8}	47.8	2.29×10^{-8}	15.1
After annealing for 20 hours at 973 K	1.0×10^{-7}	47.3	4.7×10^{-8}	10.7

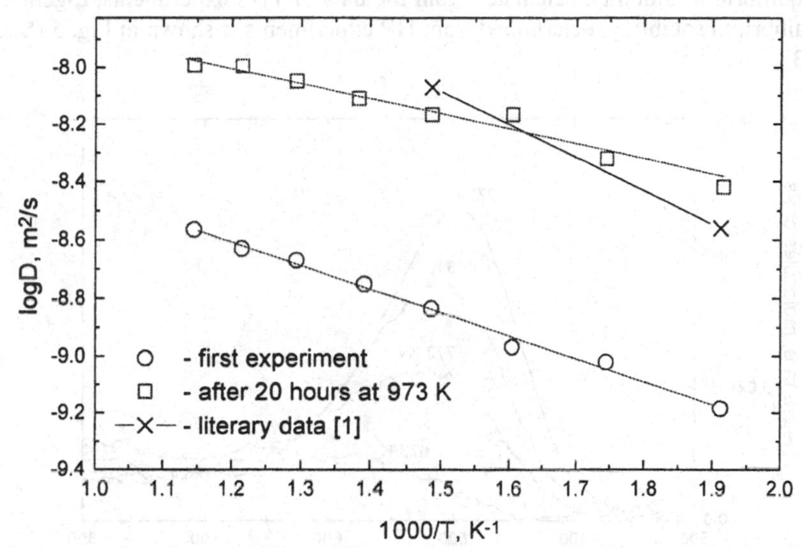

Fig 2. Temperature dependence of hydrogen diffusivity in F82H steel on the different stages of experiment

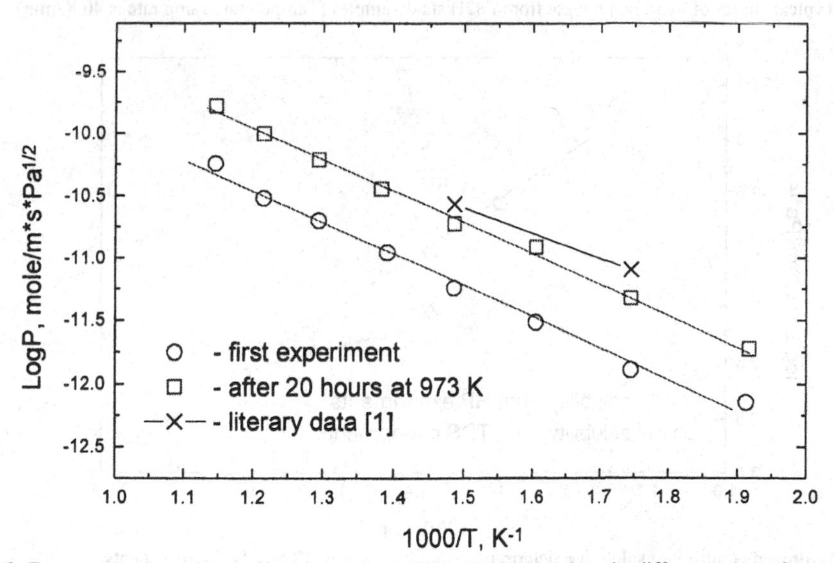

Fig 3. Temperature dependence of hydrogen permeability in F82H steel on the different stages of experiment

For TDS experiments the samples of F82H steel with the following dimensions were used: diameter – 3 mm, length 55 mm. The samples were loaded with hydrogen at different temperatures (573-973 K) and then degassed in a high vacuum to determine equilibrium solubility.

Typical curves of hydrogen gas release are shown in Fig. 4. It is clearly seen that TDS curves demonstrate the second order desorption mechanism, since the maximum of hydrogen release shifts to lower temperatures with the increase of concentration. This fact is also the evidence of surface influence on H_2 release from the sample.

Equilibrium solubility, calculated from the data of TDS experiments, together with the equilibrium solubility, determined from HP experiments is shown in Fig. 5 (See also Table 3).

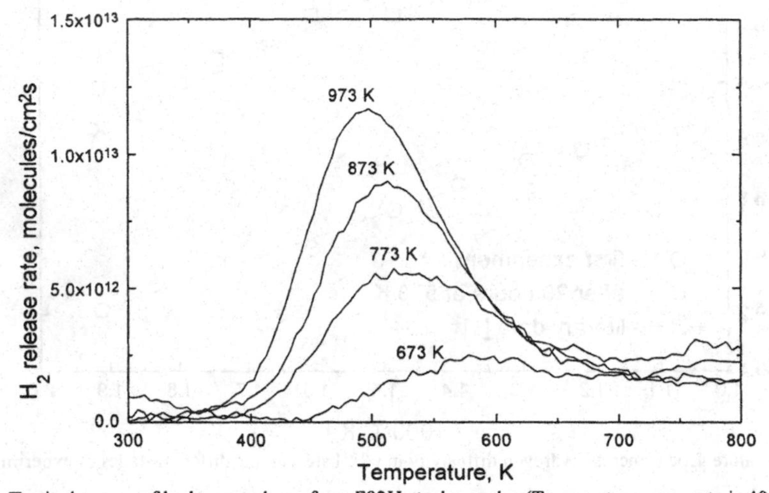

Fig. 4. Typical curves of hydrogen release from F82H steel samples (Temperature ramp rate is 40 K/min).

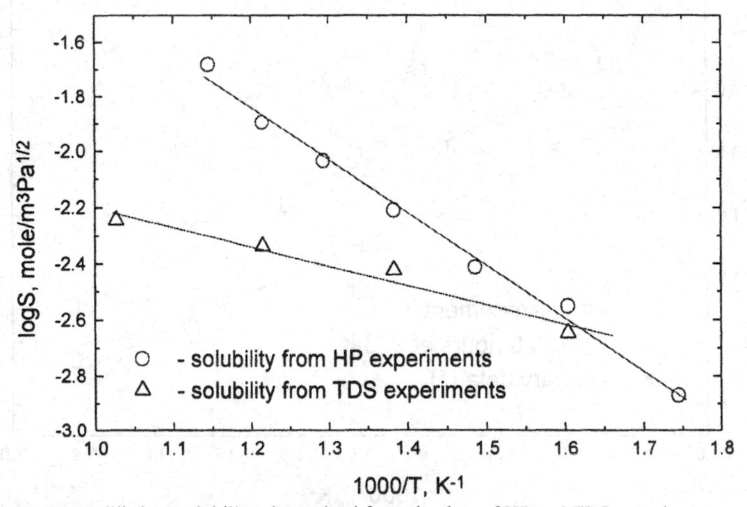

Fig. 5. Hydrogen equilibrium solubility, determined from the data of HP and TDS experiments.

Table 3. Equilibrium solubility of F82H steel, calculated from the data of HP and TDS experiments

	S_0, mole/m^{3}Pa$^{1/2}$	Es, kJ/mole
HP experiment	2.65	36.1
TDS experiment	0.031	13.2

The data of Fig. 5 and Table 3 also demonstrate the influence of surface on hydrogen penetration, since the value of solubility calculated from HP experiments (surface sensitive technique) differs from the equilibrium solubility determined by equilibrium loading technique (surface independent technique).

4. Further prospective

The experiments carried out are the first stage of the experiments on determination of hydrogen isotopes permeation parameters through the structural materials that are planned to be used in lithium blanket design. The experiments on tritium permeation through F82H steel from Li-17Pb eutectic alloy will be carried out in IVG.1M reactor of Kazakhstan National Nuclear Center. A sketch of the experimental assembly is shown in Fig. 6.

The assembly will be initially pumped out, annealed, and after the experiments with molecular hydrogen will be filled with Li-17Pb eutectic alloy.

Then the assembly will be positioned in the reactor channel and permeation of tritium produced in the result of $(Li^6,n)\rightarrow(T,\alpha)$ nuclear reaction will be measured.

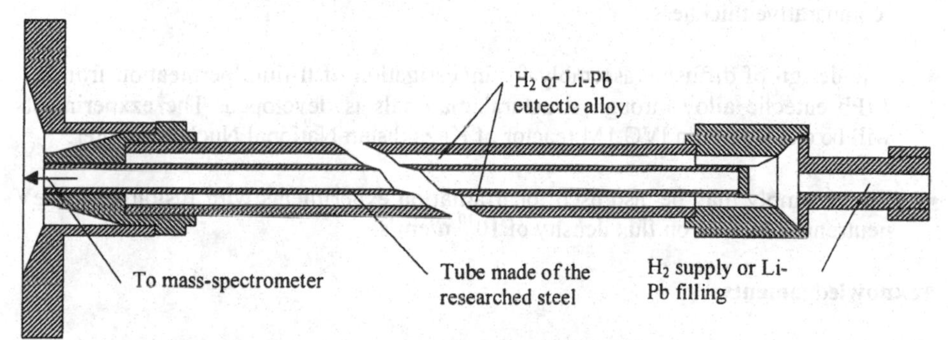

Fig. 6. Experimental assembly for experiments with Li-17Pb eutectic alloy.

The rate of tritium production in the assembly due to reaction:

$$(Li^6,n)\rightarrow(T,\alpha),$$

for IVG.1M reactor at the power level of 6 MW is 2.3×10^{-8} mole/s

If the assembly is filled with LiD it is also possible to use such experimental assembly for fast neutron irradiation tests due to fusion reaction:

$${}_1^2D+{}_1^3T\rightarrow{}_2^4\alpha(3.5MeV)+{}_0^1n(14.1MeV).$$

According to the literary data [2], conversion factor of thermal neutrons into fast neutrons is $1\div3\times10^{-4}$. If WWRK reactor of Kazakhstan National Nuclear Center will be used it allow to obtain samples irradiated up to the fluence of 1×10^{17} of fusion 14 MeV neutrons for 500 hours.

5. Summary

- The parameters of hydrogen permeation through F82H reduced activation steel are determined by hydrogen permeation techniques (apparent diffusion coefficient and permeation constant):

$$D = 4.7\times10^{-8}\exp\left(\frac{10.7kJ/mole}{RT}\right),$$

$$P = 1.0\times10^{-7}\exp\left(\frac{47.3kJ/mole}{RT}\right).$$

  and by the technique of equilibrium loading from molecular hydrogen, followed by thermodesorption spectroscopy (equilibrium solubility):

$$S = 0.031\times10^{-8}\exp\left(\frac{13.2kJ/mole}{RT}\right).$$

  The experiments have shown that surface has a strong influence upon hydrogen permeation process and the obtained diffusivity and permeability values may be used for estimations of hydrogen isotope permeation through the samples with comparative thickness.

- The design of diffusion assembly for investigation of tritium permeation from Li-17Pb eutectic alloy through structural materials is developed. The ezxperiments will be carried out in IVG.1M reactor of Kazakhstan National Nuclear Cemter.

- The assembly may be also used for irradiation experiments with fusion 14.1 MeV neutrons with neutron flux density of 10^{10} n/cm^{2}s.

Acknowledgements

The work was supported by ISTC K39 project.

References

1. E.Serra, A.Perujo, G.Benamati. J.Nucl.Mater. 245(1997) p.108-114
2. W.Miller, W.Law, R.Brugger. Thermal neutron driven 14,1 MeV neutron generators. Nucl. Instr. Meth., 216(1983), p.219-226.

STRUCTURE AND RELIEF MODIFICATION OF TITANIUM AND BORON MODIFIED GRAPHITE UNDER LIGHT ION IRRADIATION.

L.B. BEGRAMBEKOV[1], O.I. BUZHINSKIJ[2], A.L. LEGOTIN[1], S.V. VERGAZOV[1].
[1]Moscow State Engineering & Physics Institute
31, Kashirskoe sh., Moscow, 115409, Russia
[2]Troitsk Institute of Innovation and Fusion Research
Troitsk, Moscow Region, 142092, Russia

In the paper influence of both the titanium and the boron dopants on the graphite surface modification under helium ion bombardment have been studied experimentally. Incorporating of 3÷5% of titanium in the fine grain graphite was shown to stimulate the ion induced surface stresses, when temperature of the samples was 500 C. The stresses initiated development of cone like structures on the irradiated surface. Analogous quantity of boron did not stimulate the development of remarkable surface stresses under ion bombardment. Accordingly, the surface relief modification was much less, than on the titanium modified graphite.

C.H. Wu (ed.), Hydrogen Recycling at Plasma Facing Materials, 313–318.

1. Introduction

Ion and plasma interactions with multi-component surface at the high temperature occur much difficult than with mono-component surfaces of metals, graphite, etc. The ion bombardment of the multi-component surfaces causes the surface stresses, acceleration of the diffusion, and the dislocation motion, recrystallization. As a result the development of the surface relief, structure and content modifications up to the microscopic depth are developed.

In such conditions the yield of the cascade and chemical sputtering are sufficiently changed, the another sputtering mechanisms could be activated (for instance, ion-stimulated desorption), the intensity and features of the ion retention and gas release could be modified. The character of the above-mentioned transformations depends on the surface temperature and on parameters of the irradiated particle flow. One can conclude that investigation of the main features and parameters of the multi-component material modification is necessary for prediction of their behavior under various conditions.

2. Experiments.

2.1. EXPERIMENTAL CONDITIONS.

The experiments were performed in a gas discharge switched in a three-electrode system similar to that described by Wehner [1]. The residual gas pressure did not exceed 5×10^{-4} Pa. The mean free path of the sputtered particles exceeded this value. The electron temperature and plasma density determined by a double Langmuir probe was 10 eV and 10^{10} cm in the vicinity of the target. The discharge area was surrounded by a jacket with an average dimension of 30 cm under floating potential. Sputtering of the jacket was negligibly small and did not affect conditions on the sample.

Helium ions were directed onto the sample using a negative potential. Experiments have been carried out with fine grain graphites of RG-B type (4%B_4C by mass), RUM type (4÷20 %B_4C by mass), RGT (3.5 at% Ti), one-dimensional graphite fiber material VMN-4+TiC. On the cross section of the carbide doped material of VMN-4 type perpendicular to the graphite fibers, the grains of the titanium or boron carbides were measured near 5 μm and was distributed approximately uniformly. The main distance between neighboring

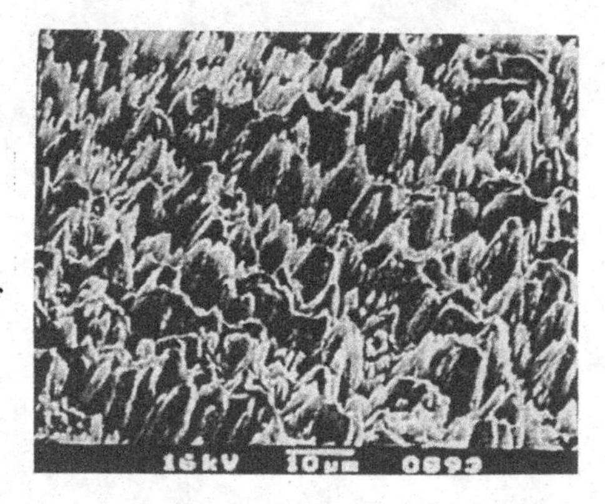

Fig.1. The irradiated surface of RGT type graphite.

grains did not exceed 100-200 μm. In the RGT practically all titanium was contained in the TiC precipitates measuring 1-5 μm.

The diameter of the samples was 1.3 cm and thickness – 0.02-0.03 cm. The irradiated spot area did not exceed 0.8 cm^2. Heating of the target was realized by radiation from a tungsten spiral incandescent lamp using direct current.

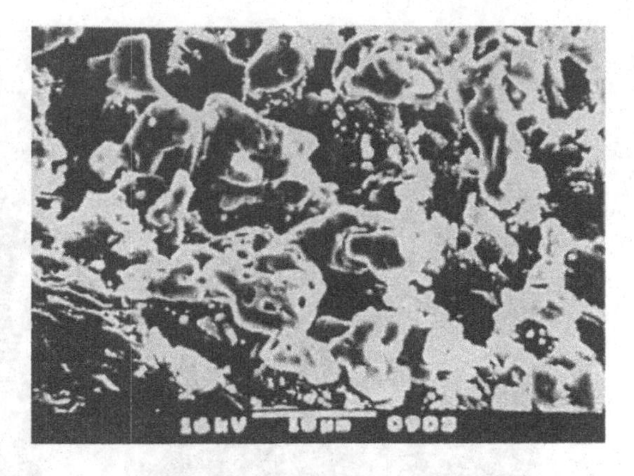

Fig.2. The irradiated surface of RGT type graphite, 1300C

2.2. RELIEF TRANSFORMATION.

Under ion bombardment the development of surface relief at all samples was observed. The relief structure significantly varied depending on graphite type and sample temperature.

Ion irradiation caused creation of highly developed relief at surface of graphite of RGT type. At temperature 500°C the intensive growth of cone type structure was observed (fig.1). The height of cones reached 10 μm and transverse dimension was 5 μm. It is compared with the thickness of removed layer calculated over measurement of sample mass loss. Some of cones had pointed top.

There were the formations of similar dimensions at surface of RGT irradiated at 1300°C (fig.2). In contrast to 500C case these formations was shapeless, the cone structure did not developed.

The relief formed under ion bombardment at surface of graphite of RGB type in contradistinction to RGT was rather smooth (fig.3). The round projections created at surface of graphite grains. The dimensions of these projections varied from 0.1 to 1 μm. The number of the projections with higher dimensions was larger at samples with larger concentration of boron. The samples were irradiated at 500 and 1300°C and the relief did not practically depend on the temperature.

On some grains of graphite with minimal concentration of boron (3%) the stratification was observed (fig.4). One can believe that similar stratification took place on samples with grater concentration of boron but was hidden by projections described above.

The development of the surface relief on the irradiated at the same conditions pure graphite MPG-8 differed from one on RGT as well as on RGB. On MPG-8 surface the pores appeared and grow under ion irradiation (fig.5). Neither projection nor stratification was observed.

Fig.3. The irradiated surface of RGB type graphite.

Fig.4. The exfoliation of RGB type graphite

Besides fine grain graphite the carbon-carbon composite of VMN type was investigated. The typical fibers of the composite of the VMN type are shown in fig.6. Irradiation was performed along carbon fibers.

Under ion bombardment the carbon fibers of composite dopped with titanium carbide grew together. The solid layer was formed at surface of sample (fig.7).

In the case of pure composite the fibers assumed smooth shape, its diameter grew and boundaries between fibers and graphite matrix smoothed over, but solid layer was not created. On composite with boron doping this process was weaker then on pure VMN.

3. Discussion.

The cones appeared on the surface of RGT under ion irradiation at 500°C have pointed tops and its dimensions are compared with total thickness of sputtered layer. It indicates that not only sputtering processes are responsible for relief development. The very similar cone structure was observed on the surface of the metal alloys sputtered under similar conditions. It was shown that cone growth on metal surface caused by stresses in near surface layers of material [2].

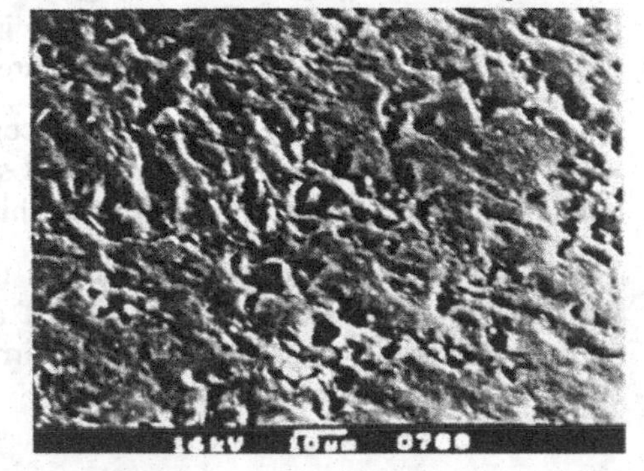

Fig.5. The irradiated surface of MPG-8 type graphite

Evidently, the cone growth on the surface of graphite with titanium carbide doping

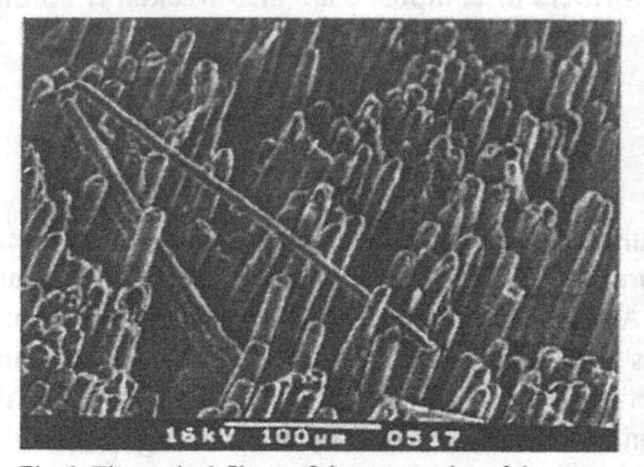

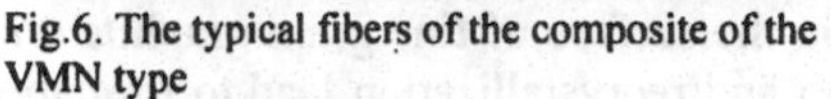
Fig.6. The typical fibers of the composite of the VMN type

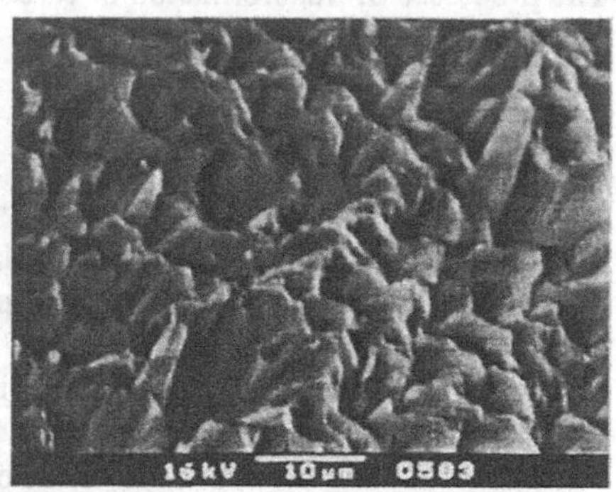

Fig.7. The surface of VMN type composite after ion irradiation

caused by the same reasons. The titanium plays the role of refractory component. One could mention that in the case of the metal samples the cones grow at temperatures $T>0.7\times T_m$, where T_m is melting temperature of material. Unlike this the cone growth on graphite is observed at comparably low temperature – 500°C.

At 1300°C the evaporation of titanium destroy the mechanism providing transport of material to the top of cone. As the result formless projection is created.

The C-C composite of VMN type doped with TiC gives one more example of strong influence of the dopant on the graphite modification under ion bombardment. In the C-C composite one can define at least three regions with different structure, namely: inner parts of fibers, outside of them, and graphite matrix. Under ion bombardment all of them were transferred in the uniform continuous layer. This transformation may be considered as result of the diffusion and recrystallization processes initiated by ion induced surface stresses. The presence of titanium carbide seems to provide the high intensity of the mentioned processes, what is needed for realization of observed transformations.

Another, more sophisticated processes at the bombarded surfaces of both the bulk graphite and the C-C composite we have observed in the case of the boron doping. The limited part (up to 1 at%) of the implanted boron is known to incorporated in the graphite structure, and the last one presents in the form of boron carbide One can believe, that both mentioned fractions of implanted boron stimulate the different transformation processes under ion bombardment. The desolved boron may be considered as a cause of stratification of the graphite structure. Stratification prevents development of the surface stresses, diffusion and crystallization, which stimulated by the presence of the refractory projections of boron carbide. That is why cone growth or uniform layer formation were not seen, respectively at the bulk graphite and at the C-C composite.

The small size projections appeared on the surface can be explained by redeposition of boron sputtered from carbide grains. This redeposited boron form islands of carbide on the surface of graphite grains. This conclusion is proved by secondary ion mass spectrometry and by measurements of sputtering yield.

The processes of transformation of carbon fibers of composite are also weaken if boron carbide is added to graphite.

4. Conclusion.

The doping brought into graphite materials can significantly influence to surface relief development and transformation. The presence of titanium increase stresses at near surface layers rose by ion bombardment. As the result the diffusion and recrystallization processes at surface are activated. This processes can produce different results in dependence of initial relief and bulk structure of material, but in all cases the significant relief modification is occur. The diffusion of material caused by stresses together with sputtering lead to formation of cone structure on the surface of fine grain graphite with titanium. In the case of C-C composite diffusion and recrystallization lead to formation of solid layer from carbon fibers grown together. In contrast with titanium, boron causes decrease of surface stresses and, as result, the diffusion and recrystallization place smoller role in processes at surface under ion irradiation.

5. References.

1. G.K. Wehner, J. Vac. Sci. Technol. A3(1985)182.
2. L.B. Begrambekov, A.M. Zakarov, A.A. Pustobaev and V.G. Telkovskij, Phys. Chim. Obr. Mater. 5(1989)26.

The simulation of the diagnostic mirror behavior under hydrogen isotope irradiation

V.V,Bandurko,V.A.Kurnaev[a],D.V.Levchuk,, N.N.Trifonov[a]

Moscow State Engineering and Physics Institute, Moscow 115409, Russia
e-mail: kurnaev@plasma.mephi.ru

1. Introduction

In this paper we present results of direct computer simulations based on binary collisions TRIM.SP like computer code as well as results of direct experimental measurements of reflection and deposition for possible materials of duct and mirror.

2. Computer simulations model.

The geometry of the duct was assumed to be of axial symmetry (see Fig 1). The angle α can be both positive and negative. We made calculations for the next parameters of the diagnostics duct: length of the duct is 2.0 m, diameter of the input orifice d is equal to 20 cm, diameter of the outlet orifice is equal to 35 cm. The mirror is inclined at 45° as respect to the duct axis.

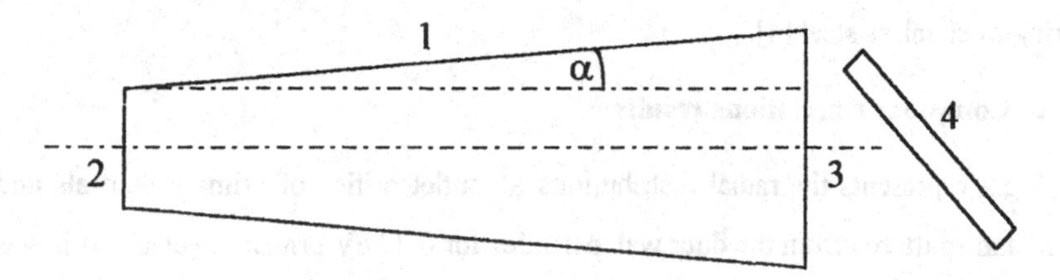

Fig.1. The diagnostic duct geometry for computer simulations.
1.Duct 2.Input orifice 3.Outlet orifice 4. Mirror

The code SCATTER [1] used in calculations physically is similar to TRIM.SP. Control calculations with both codes gave the same results for energy range of interest of the task.

C.H. Wu (ed.), Hydrogen Recycling at Plasma Facing Materials, 319–329.

The calculations were made in the following way. It was assumed that the free path length of the primary and reflected particles is mach more than the duct dimensions, so neutrals escaping plasma and reflected or sputtered from duct walls can collide only with solid surfaces (duct walls and mirror). The history of a particle after striking solid surface is calculated by SCATTER code. If this particle is reflected it can strike the duct wall in any other place (or return to plasma). After each impinging on the solid surface all collisions with target atoms and secondary collisions of knock out atoms of solid are directly calculated. Only particles in the energy range 5-10^4 eV are considered. Sputtered atoms are supposed to stick to the surface they meet. So secondary processes under influence of the energetic sputtered atoms are ignored. The duct surface is described by Eq. 1

$$-1.40625\cdot10^{-3}x^2 + y^2 + z^2 - 0.75x - 100 = 0 \quad (1)$$

For collisions with duct wall atoms target is considered to be flat.

The start points and direction of movement of primary particles escaping plasma are determined with conventional Monte-Carlo procedure with assumption of uniform distribution over inlet orifice and cosine angle distribution. Energy distribution of primary neutrals is drowned to follow expected one for ITER in accordance with Kukushkin calculations [2]. To get results on fluxes of sputtered atoms near mirror surface with reasonable accuracy about 10^8 trajectories of the primary particles were calculated.

The material of the duct is supposed to be a stainless steel but for simplification of simulations we used Fe in SCATTER calculations with binding energy E_s= 0.1 eV and displacement energy E_d= 7.5 eV to fit as close as possible the experimental data for D sputtering of stainless steel [4].

3. Computer simulations results

Fig.2 represents the radial distributions at outlet orifice of primary neutrals and both reflected and sputtered from the duct wall particles for 0.2 keV primary neutrals. It is seen that two latter distributions are uniform over mirror surface, but number of primary neutrals striking mirror within 10 cm radius is increased in accordance with the geometry of the duct with input orifice of the same dimensions. The energy distribution of primary neutrals as well as probability of primary and reflected deuterons with different energies to strike the mirror are shown in fig.3 (dashed line). The probability of Fe atoms sputtered by deuterons with different energies to get the mirror surface are also shown. It is seen that that maximum contribution to sputtering is due to particles with energies in keV range though high energy neutrals contribution to the primary energy spectra decreases. Fig.4 where contributions of primary and secondary particles fluxes from different parts of energy spectra as well as energy dependence of sputtering coefficient makes clear this result. Overlapping of primary flux of neutral decrease and sputtering coefficient as function of energy increase (see line in Fig.4) results in maximum value of the mirror surface sputtering at approximately 0.2 keV. Bold circles correspond to the sputtered Fe atoms deposition of the mirror, but empty circles correspond to the sputtering of stainless steel mirror (Fe) under bombardment with primary neutrals. Crosses represent the primary flux, bold squares - flux of particles reflected from the duct wall. It is seen that erosion of the mirror under flux of primary particles from plasma an order of magnitude greater deposition of material sputtered from the walls. This is in good qualitative agreement with results obtained in previous report for rectilinear geometry of the duct using analytical approach for sputtering.

It is of interest to compare sputtering and deposition fluxes for different mirror materials. Fig.5 represents the contribution of neutrals with different energies in sputtering of mirror made of different materials as compared with the flux of Fe atoms deposition (bold circles). The total values of erosion (the integral over energy) are presented in the Table 1. It is seen that only for Tungsten mirror erosion an order of magnitude less than deposition of Fe. In the case of Mo the deposition of Fe and erosion of Mo are approximately equal.

Table 1.

Material	Erosion, atoms·sec^{-1}·cm^{-2}	Erosion to deposition ratio
Al	$4{,}9\cdot10^{12}$	6,2
Fe	$2{,}8\cdot10^{12}$	3,5
Mo	$8{,}1\cdot10^{11}$	1,0
W	$8{,}1\cdot10^{10}$	0,11

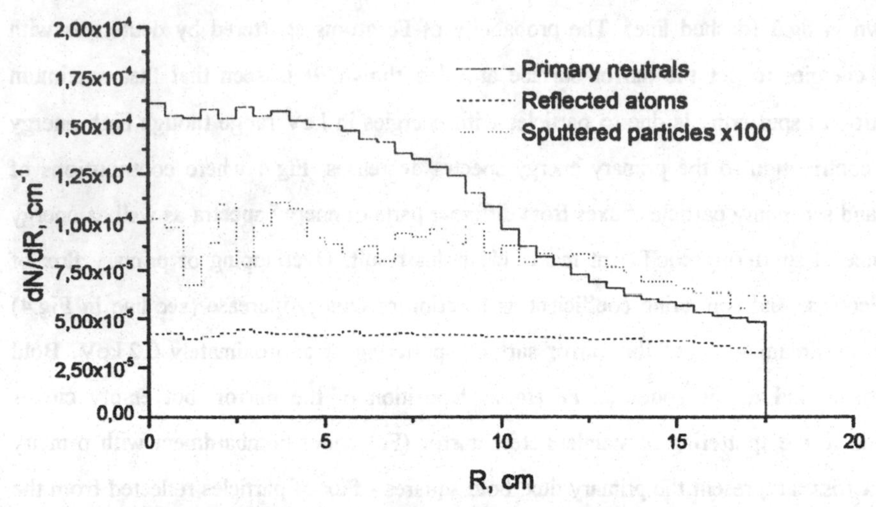

Fig. 2. The radial distributions of partiles at outlet orifice at launching of 0.2keV D atoms at inlet orifice.

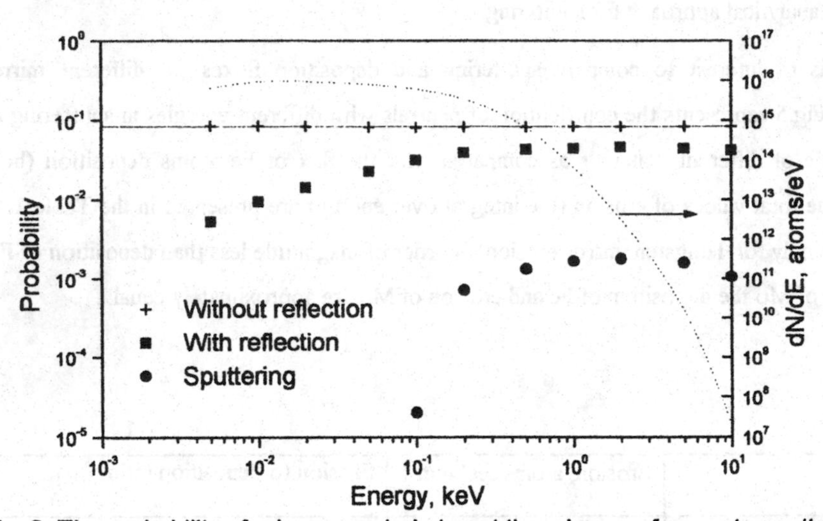

Fig. 3. The probability of primary neutrals to get the mirror surface or to sputter wall atoms. Energy distribution of primary neutrals is also shown (right axis).

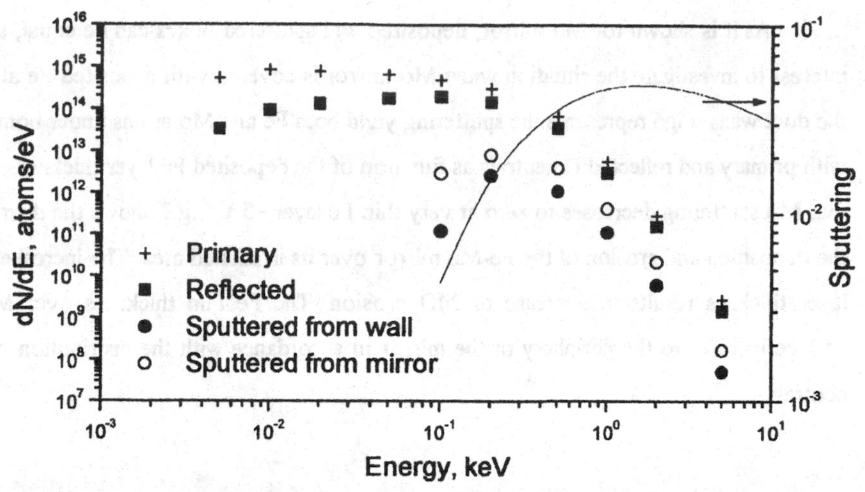

Fig. 4. Fluxes of primary and reflected neutrals, sputtered from wall and sputtered from mirror for different energies of primary neutrals. Sputtering coefficient of stainless steel for incidence angle 45°.

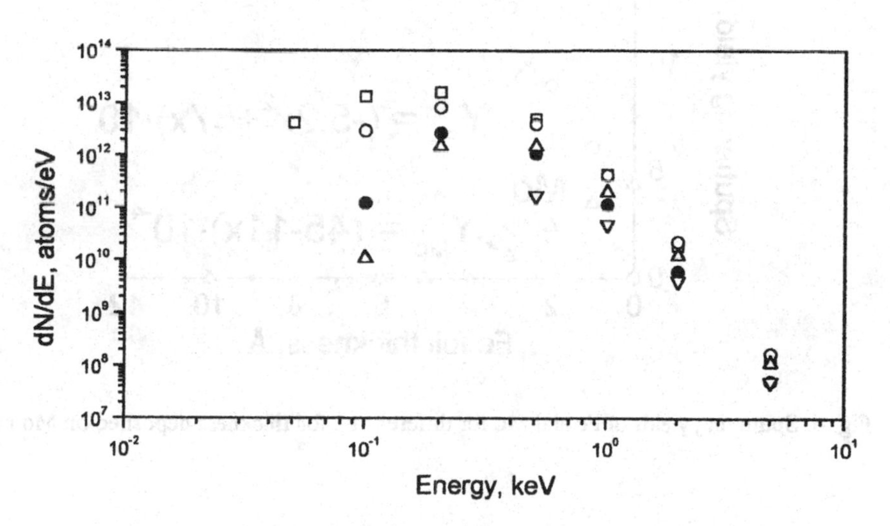

Fig. 5. The flux of sputtered atoms for different mirror's materials: □ - Al, ○ - Fe, △ - Mo, ▽ - W, ● - the flux of depositing atoms.

As it is shown for Mo mirror, deposited and sputtered fluxes can be equal, so, it is of interest to investigate the situation when Mo mirror is covered with deposited Fe atoms from the duct walls. Fig6 represents the sputtering yield both Fe and Mo atoms under bombardment with primary and reflected D neutrals as function of the deposited Fe layer thickness. It is seen that Mo sputtering decreases to zero at very thin Fe layer ~5A. Fig.7 shows the distribution of the deposition and erosion of the Fe-Mo mirror over its irradiated area. The increase of the Fe layer thicknes results in decrease or MO erosion. The Fe film thickness over Mo mirror surface increses to the periphery of the mirror in accordance with the distribution of primary neutrals.

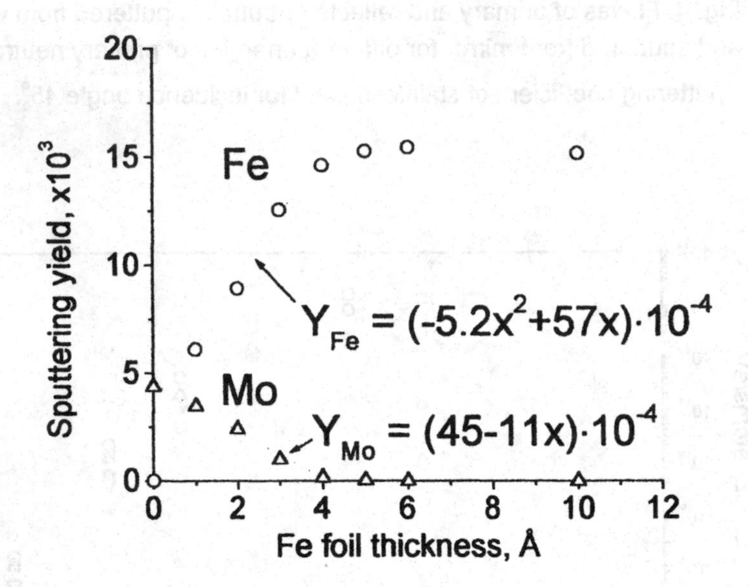

Fig. 6. Sputtering yields of Fe and Mo for different Fe foil thickness deposited on Mo mirror.

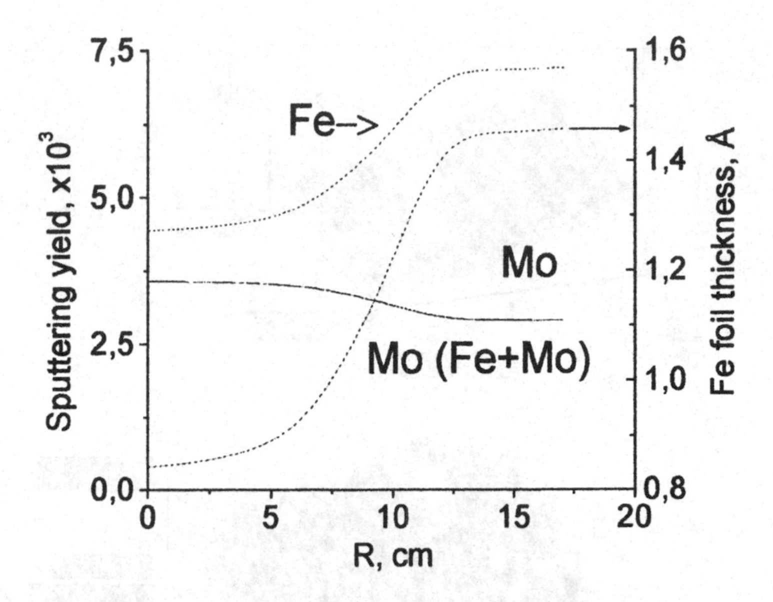

Fig. 7. Radial distribution of Mo sputtering yield and Fe foil thickness for Mo mirror.

4. Experimental

To compare computer simulations with experimental data on reflection and sputtering of possible duct material onto the mirror surface, we used deuterium beam with energy 200eV (0.83D3 +0.14D +0.02O). This composition of the beam corresponds to 0.94D of 67eV + 0.05D of 200 eV + 0.02 O 200eV. The Cu mirror was irradiated through the Mo diaphragm of 8mm in diameter situated at 1mm near Cu target. Fluence of irradiation was equal to 10E25m (-3). Investigation of the reflectivity of the mirror after irradiation showed that at wavelength 0.65-0.85mkm reflectivity decrease is ~ 0.08. But for wavelengths 0.5 –0.55 mkm reflectivity increased after irradiation. In the vicinity of the irradiated area the increase of reflectivity within 2-3 mm ring was found during radial profiling with the help of 0.63 mkm laser beam. The reflectivity in this ring is greater than the latter of nonirradiated target within 3-4 %.

The STM investigation of the target showed that the surface relief of the different areas differs (Fig.8a,b). The most developed relief corresponds to the irradiated zone, but the relief of the "aureole" zone is more smooth that that for unirradiated area.

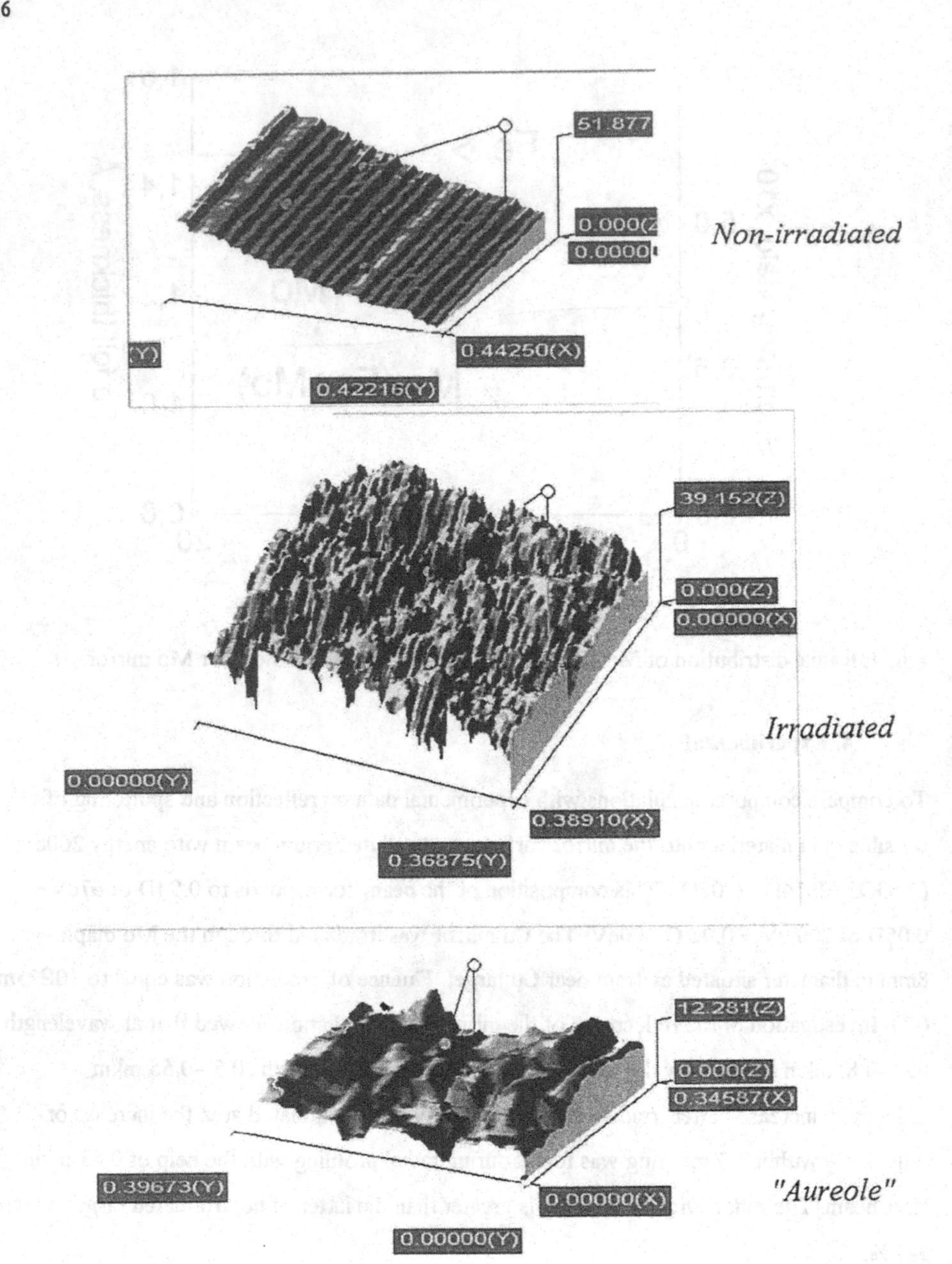

Fig. 8a. STM image of Cu mirror – X, Y - in μm, Z in nm

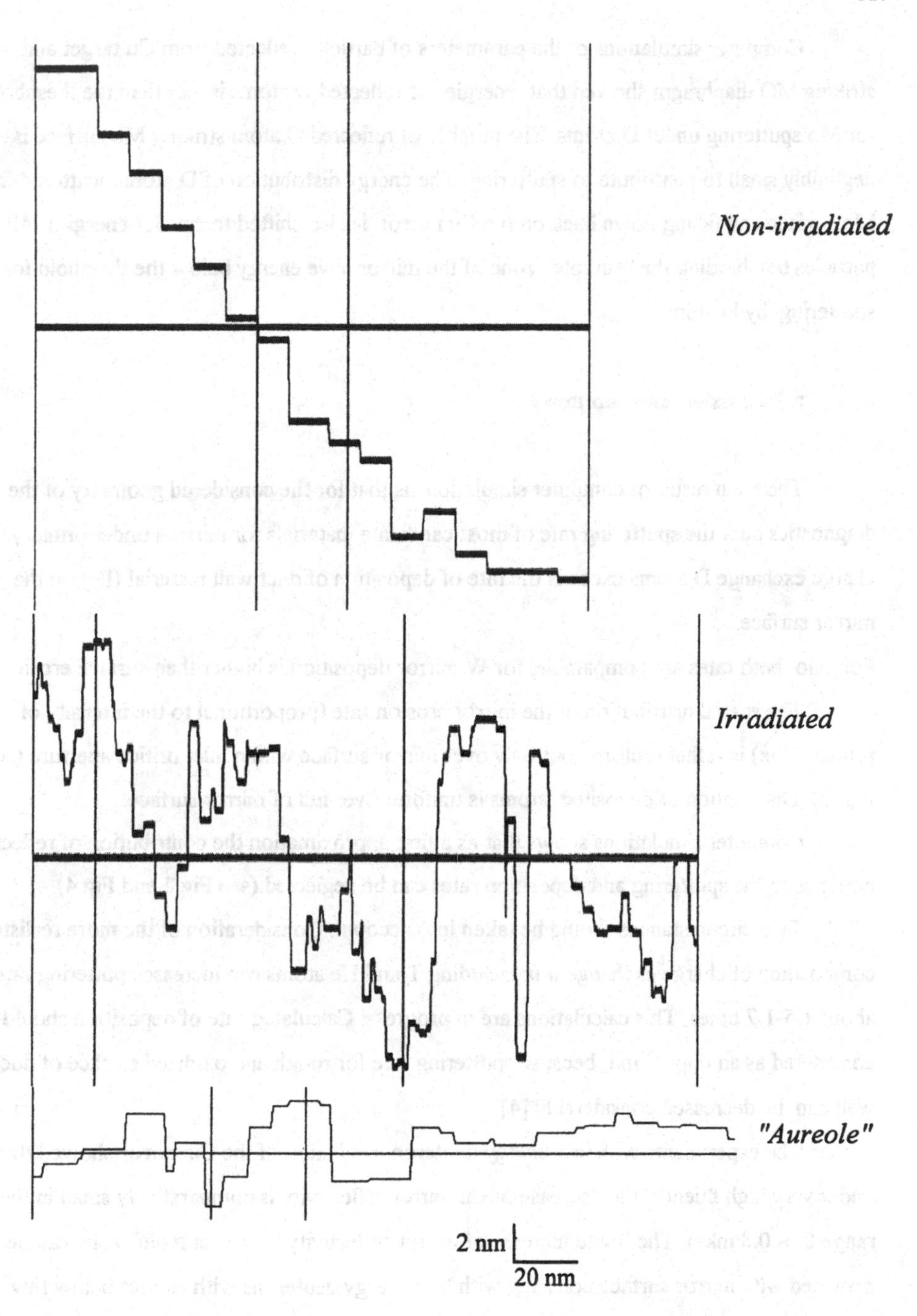

Fig. 8b. STM relief of Cu mirror

Computer simulations of the parameters of particles reflected from Cu target and striking MO diaphragm showed that energies of reflected D atoms is less than the theshhold for Mo sputtering under D atoms. The number of reflected O atom striking Mo surface is negligibly small to contribute to sputtering. The energy distribution of D atoms scattered from Mo surface and falling down back on the Cu mirror is also shifted to smaller energies. All particles bombarding the "aureole" zone of the mirror have energy below the threshold for Cu sputtering by D atoms.

5.Discussion and summary

The main result of computer simulations is that for the considered geometry of the diagnostics duct the sputtering rate of most candidate materials for mirrors under primary charge exchange D atoms exceeds the rate of deposition of duct wall material (Fe) on the mirror surface.

For Mo both rates are comparable, for W mirror deposition is higher than surface erosion.

The spatial distribution of the mirror erosion rate (proportional to the intensity of primary flux) is rather uniform but only over mirror surface within inlet orifice aperture (see Fig. 2). Distribution of deposited atoms is uniform over full of mirror surface.

Computer simulations shows that as a first approximation the contribution of reflected particles to the sputtering and deposition rates can be neglected.(see Fig.2 and Fig.4).

Two circumstances should be taken into account. Consideration of the more realistic composition of charge exchange flux including T and He atoms can increase sputtering rate about 1.5-1.7 times. This calculations are in progress. Calculated rate of deposition should be considered as an upper limit, because sputtering rate for rough and oxidized surface of duct wall can be decreased considerably [4].

The experiments with low energy deuteron irradiation of the Cu mirror showed that under very high fluence the decrease of Cu mirror reflectivity is comparatively small in the range 0.6- 0.8 mkm. The found increase of mirror reflectivity in the "aureole" zone can be explaned with mirror surface polishing with low energy deuterons with eneries below thw threshold for sputtering:

the intensity of reflection from titled real surface as much as several times is lower than predict computer simulations for flat clean surface, mean energy of reflected particles is less than in calculations.

Literature

1. N.N.Koborov et al, Nucl.Instr.Meth.B 129(1997)5
2. A.Kukushkin et al ITER report N (1996),
3. W.Eckstein, J.Bohdansky, J,Roth Physical sputtering // Atomic and Plasma-material interaction data for fusion V.1 (1991)51-61
4. V.A.Kurnaev, E.S.Mashkova, V.A.Molchanov Light ions backscattering from solids. Energoatonizdat. Moscow(1985)p.192(In Russian),
5. V.V.Bandurko et al Journ.Nucl Mater176/177(1990) 630

SURFACE LOSS PROBABILITIES OF NEUTRAL HYDROCARBON RADICALS ON AMORPHOUS HYDROGENATED CARBON FILM SURFACES: CONSEQUENCES FOR THE FORMATION OF RE-DEPOSITED LAYERS IN FUSION EXPERIMENTS

W. JACOB, C. HOPF, A. VON KEUDELL, T. SCHWARZ-SELINGER
Max-Planck-Institut für Plasmaphysik, EURATOM Association,
Boltzmannstr. 2, D-85748 Garching, Germany

1. Introduction

The success of present-day fusion experiments relies on the use of a divertor which efficiently pumps impurities generated by erosion of the first wall. In most fusion experiments the divertor surface consists of graphite tiles or carbon fiber composites. They are bombarded by ions from the scrape-off layer which are guided into the divertor by the magnetic field. This impinging ion flux leads to sputtering of the divertor tiles releasing carbon and hydrocarbon compounds into the gas phase. This emitted carbon flux is excited in the divertor plasma and dissipates the plasma power via radiation, leading to a reduction of the heat flux onto the divertor surface. Most emitted carbon and hydrocarbon species re-deposit promptly in or in proximity to the divertor. This balance between deposition and erosion is crucial for the performance of a divertor in a next-step device, since the total lifetime before replacement strongly depends on the ability to control this re-deposition.

However, thick re-deposited layers were observed in JET on surfaces not in direct contact with the plasma [1]. On these surfaces only neutral growth precursors can contribute to film deposition, because only they are able to cross the magnetic field. A major concern is the large amount of hydrogen which is trapped in these re-deposited films [2]. In a future fusion reactor this trapped hydrogen will be partly tritium, which will be lost for fusion reactions and represent a potential safety hazard. One aim of a future design is, therefore, to reduce this tritium retention for economy and safety reasons.

An attempt was made to model the occurrence of these films in the MARK IIa divertor of the JET experiment [1]. These re-deposited films in JET were predominantly found on the inner louvers, which are cold baffles at the exit of the inner leg of this divertor. This modeling showed, however, that the total amount of carbon transported from the divertor to the inner louvers is much larger than could be explained on the present basis of knowledge assuming physical sputtering of carbon by hydrogen isotopes as the main source for carbon [1]. Other mechanisms have, therefore, to be considered to explain the formation of these films on surfaces not in direct line-of-sight to the plasma.

The objective of this article is to present a plausible scenario to explain the phenomenon of re-deposition in remote parts of the vessel of fusion devices. It is based on a large number of very detailed investigations of deposition and erosion processes

C.H. Wu (ed.), Hydrogen Recycling at Plasma Facing Materials, 331–337.

and physical properties of amorphous, hydrogenated carbon (a-C:H) layers employing low-temperature laboratory plasmas and thin film analysis techniques, respectively [3-10]. The sticking coefficients for these species on the surface of amorphous hydrogenated carbon films are measured applying the cavity technique. On the basis of these data it is possible to identify the individual growth precursors and derive a scenario for formation of re-deposited films in existing fusion experiments.

2. Cavity measurements

2.1. EXPERIMENTAL

The surface loss probabilities of different hydrocarbon radicals can be measured by the cavity technique. Details of this technique, the concurrent modeling, and its application to C:H film growth were described in preceding publications [11,12]. They are shortly summarized as follows: reactive species impinging on a surface can adsorb with a sticking coefficient s, can react at the surface with a probability γ to form a non-reactive volatile molecule, or can be reflected with a probability r. The surface loss probability β, which describes the probability of losing a reactive particle in a surface collision corresponds to $\beta = s + \gamma$ with $r + s + \gamma = 1$. The surface loss probability β can be determined measuring the deposition profile inside a cavity. This cavity consists of a closed volume with a small entrance through which reactive species can enter. In our case the top and bottom of the cavity is made from two silicon substrates (15 mm by 20 mm) with a distance of 2.6 mm. A 1 mm wide slit, through which particles can enter, divides the top substrate in half. The deposition profiles are measured on the bottom and on the inner side of the top substrate using ex situ ellipsometry. A hydrocarbon discharge is used as particle source for measurement of the surface loss probabilities of C_xH_y radicals. The cavity is placed at a remote position with respect to this hydrocarbon plasma. Radicals emanating from the discharge are transported through the vessel from the plasma to the cavity and after entering the cavity they are reflected at the walls and build up a layer inside. From the spatial variation of the film thickness inside the cavity (= deposition profile) the surface loss probability can be deduced by comparing the experimental results to model calculations. Details of the model calculations are described in [11,13]. As mentioned above, β is not directly the required sticking coefficient s, but since in most cases γ is small compared to β, β can be considered a reasonable estimate for the sticking coefficient. If several reactive species with different values of β are present, which is in general the case, the variation of the film thickness inside the cavity at a given position x has to be modeled by superposing thickness profiles for different values of $\beta_{species}$ with weight factors $g_{species}$.

The experimental setup has also been described recently [11]. The deposition chamber is depicted in Fig. 1. An electron cyclotron resonance plasma was used as particle source. The following hydrocarbon source gases were used as precursor gases: CH_4, C_2H_6, C_2H_4, C_2H_2, C_3H_8, and C_4H_{10}. In each discharge, 3 cavity substrates were simultaneously exposed to the plasma in 3 different positions (see Fig. 1). Due to the remote position of the cavity substrates relative to the plasma ions do not contribute to deposition. All experiments were performed at room temperature.

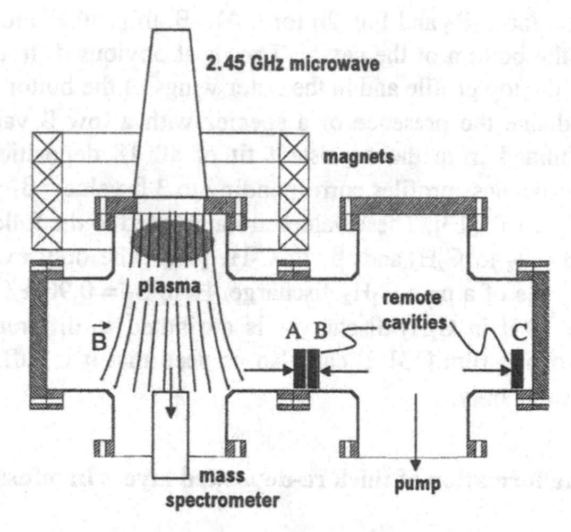

Figure 1: Sketch of the experimental setup. Cavity substrates were placed simultaneously in positions A, B, and C. For cavities A and C the slit is facing the plasma, for B it is facing away from the plasma. The orientation of the slit is indicated by the arrows.

2.2. RESULTS

The 6 used precursor gases yield together with the 3 different substrate positions 18 individual deposition profiles. A detailed study of these 18 profiles [13] has shown that they all together can be consistently modeled using 3 different reactive species, corresponding to 3 different β values, only [13]. Examples of measured deposition profiles in position A and the corresponding model results are represented in Fig. 2. Fig.

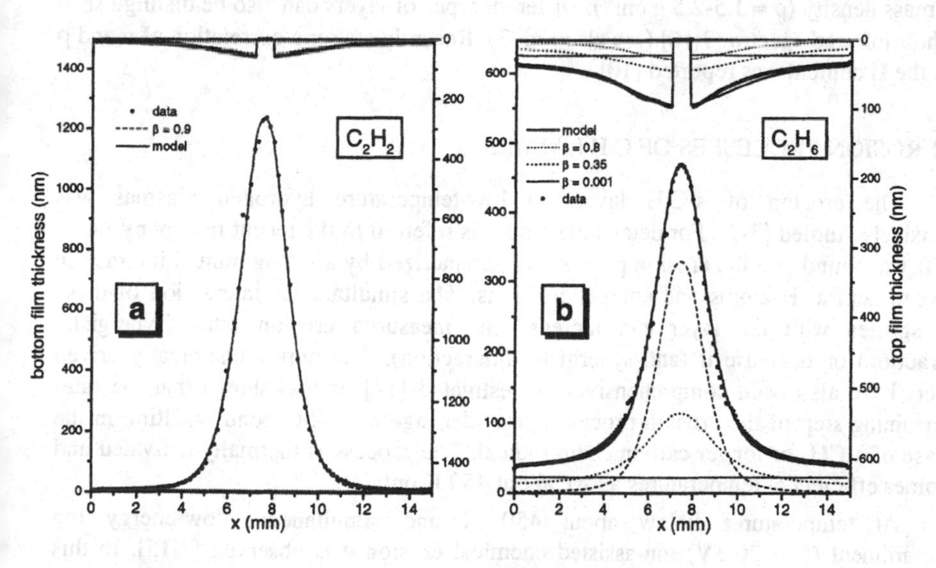

Figure 2: Deposition profiles inside cavity substrates exposed to C_2H_2 (a) and C_2H_6 (b) plasmas. Shown are the experimental (dots) and modeling (lines) results. The model comprises a superposition of 3 individual profiles, which are also shown.

2a shows the results for C_2H_2 and Fig. 2b for C_2H_6. Both profiles are characterized by a dominant peak at the bottom of the cavity. The most obvious differences are the much larger thickness of the top profile and in the outer wings of the bottom profile in the case of C_2H_6 which indicate the presence of a species with a low β value. The following values were determined from the consistent fit of all 18 deposition profiles using a superposition of 3 thickness profiles corresponding to 3 β values: $\beta_1 = 0.80 \pm 0.05$, $\beta_2 = 0.35 \pm 0.15$, and $\beta_3 < 10^{-2}$ [13]. These values are attributed to the following hydrocarbon radicals: β_1 to C_2H, β_2 to C_2H_3,and β_3 to C_2H_5 [13]. The only exception from these values is β_1 in the case of a pure C_2H_2 discharge. Here $\beta_1 = 0.90 \pm 0.05$ was used. This higher β value for C_2H in C_2H_2 discharges is attributed to differences in the surface structure of the growing film [13]. It can also be seen that it is sufficient to model the C_2H_2 using one β value only.

3. Scenario for the formation of thick re-deposited layers in a fusion device

3.1 TYPES OF C:H LAYERS

The structure, stoichiometry and physical properties of plasma-deposited amorphous hydrogenated carbon (a-C:H) films has been covered in a large number of publications (see Ref. 3 and references therein). The properties can vary over a wide range from soft polymerlike to very hard, scratch-resistant layers depending on deposition conditions. The most important parameter is the energy of the ions bombarding the growing film during deposition. Low ion energies (< 30 eV) yield soft layers, higher ion energies (> 50 eV) hard layers, with a smooth, continuos transition between both regimes. Soft film are characterized by a high H content (H/C > 0.8, up to 1.2 was measured) and low refractive index ($n \approx 1.5$) and mass density ($\rho \approx 1.0$ g cm^{-3}). Hard films posses a low H content (H/C ≈ 0.4) and high refractive index ($n \approx 1.8$-2.2) and mass density ($\rho \approx 1.5$-2.5 g cm^{-3}). Different types of layers can also be distinguished by their infrared spectra [7,10] (see also Fig. 3). Recently a strong correlation of n and ρ with the H content was reported [10]

3.2 EROSION PROCESSES OF C:H LAYERS

The erosion of a-C:H layers in low-temperature hydrogen plasmas was extensively studied [3-7]. For details the reader is referred to the recent review by Jacob [3]. It was found that the erosion process is characterized by a strong mutual interaction between neutral H atoms and energetic H ions. The simultaneous interaction of these two species with the layer can increase the measured erosion rates (synergistic interaction) or decrease it (anti-synergistic interaction). The purely chemically driven effects have also been comprehensively investigated [14]. It was shown that the rate-determining step of the erosion process is the cleavage of a C-C bond resulting in the release of a CH_3 or longer carbon chain radical. The process is thermally-activated and becomes efficient at temperatures above about 450 K only.

At temperatures below about 450 K and simultaneous low-energy ion bombardment (E < 20 eV) ion-assisted chemical erosion was observed [3,15]. In this regime, atomic H forms at the surface the precursor groups (e.g. terminal CH_3 groups), but the temperature is too low to allow thermal release. On the other hand, the ion energy is too low for physical sputtering [3], but the ions can transfer enough energy to directly break C-CH_3 bonds, thus releasing the volatile CH_3 groups. One could say that the ions take over the role of thermal activation in chemical erosion.

High-energy ion bombardment leads to the well-known physical sputtering. At intermediate energies (50 to about 200 eV) the erosion rates depend critically on the actual layer structure, the surface temperature, and the ion energy. However, important is not only the bulk composition of the material, but above all the actual structure at the surface (1-5 nm) that forms during interaction with the plasma.

3.3 THERMAL DESORPTION

In order to study the influence of substrate temperature, the thermally-released products from a-C:H layers have been investigated by thermal desorption spectroscopy (TDS) [8,9,14]. In these experiments, substrates covered with a-C:H layers are heated in vacuum applying a linear temperature ramp. The released products are measured using a mass spectrometer [9].

A typical result for a set of different a-C:H layers is shown in Fig. 3. The films for this study were deposited in an rf discharge in methane. The ion energy was varied

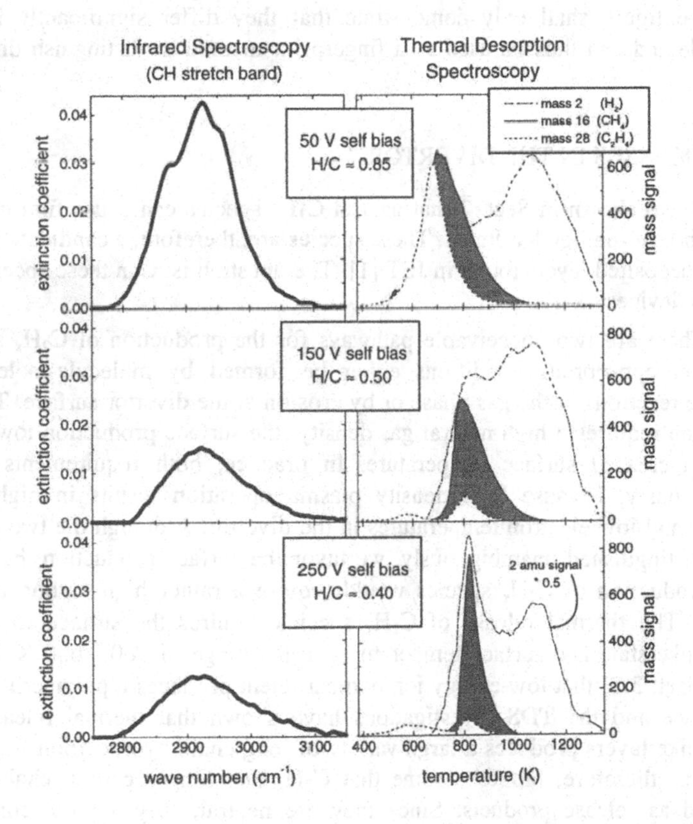

Figure 3: Infrared (left-hand side) and thermal desorption (right-hand side) spectra of rf plasma-deposited a-C:H layers. The negative dc self-bias during deposition and the resulting H/C ratio of the layers is indicated in the figure. The mass 2 signal of the 250 V layer (bottom, right) is scaled by a factor 0.5.

through the change of the applied rf power. The rf generated negative dc self-bias during deposition is a measure of the ion energy. It is indicated in the figure. Deposition at a bias of 50 V yields soft (H/C = 0.83), at 150 V intermediate (H/C = 0.50) and deposition at 250 V very hard layers (H/C = 0.40). Shown on the right-hand side of Fig.3 are thermal desorption spectra for H_2, CH_4, and C_2H_4 after background subtraction [7]. With increasing initial H content the onset temperature for thermal release shifts to lower temperatures and the relative composition of the released products changes dramatically. While for hard films the dominant product is H_2 (the 2 amu spectrum for the 250 V film in Fig. 3 is divided by 2), CH_4 and C_2H_4 become increasingly important with increasing initial H content. This trend is continued going to even softer films than in the top frame of Fig. 3. For soft, deuterated hydrocarbon films (a-C:D) with a D/C ratio larger than 1 the by far dominant release product was C_2D_2 and a large variety of C_xD_y species up to C_7D_7 were measured [9]. Although TDS is not a steady-state process, we assume that a similar difference of the erosion products occurs during plasma erosion of soft and hard a-C:H layers.

Also shown on the left-hand side of Fig. 3 are the corresponding infrared spectra of the layers. Plotted is the extinction coefficient as a function of wave number for the C-H stretching region. The algorithm to extract this value from infrared absorption spectra has been described recently [7]. The infrared spectra are not further discussed here. The figure shall only demonstrate that they differ significantly in shape and amplitude and can thus be used as a fingerprint method to distinguish different a-C:H layers.

3.4. PROCESSES IN THE DIVERTOR

It was shown in Sect. 2 that neutral C_2H_x species can cause film growth also in the absence of ion bombardment. These species are, therefore, a candidate to explain the thick re-deposited layers found in JET [1]. The question is: Can these species be formed in fusion devices?

There are two conceivable pathways for the production of C_2H_x species. They can under appropriate conditions either be formed by molecule-molecule or ion-molecule reactions in the gas phase or by erosion at the divertor surface. The gas phase production requires a high neutral gas density, the surface production low ion energies and an increased surface temperature. In practice, both requirements are fulfilled simultaneously, because high density plasma operation results in high neutral gas densities and low electron temperatures in the divertor. Although the two pathways can not be distinguished unambiguously, we favor the surface production, because the gas phase production of C_2H_x species would require a rather high carbon density in the divertor. The thermal release of C_2H_x species requires the surface to be in a soft, polymerlike state and surface temperatures in the range of 500 - 600 K. It was shown above (Sect. 3.2) that low-energy ion bombardment produces a polymerlike structure at the surface and the TDS investigations have shown that thermal release from such polymerlike layers produces a large variety of long chain hydrocarbon fragments (Sect. 3.3). It is, therefore, fair to assume that C_2H_x and longer carbon chain radicals are produced as release products. Since they are neutral, they are not confined by the magnetic field and contribute to deposition. If the temperature of the deposition area is above the threshold for thermal release, these species may be re-emitted and finally be deposited on the next accessible 'cold' surface. Depending on the individual sticking coefficient or surface loss probability these species may also be transported very far, so that deposition can occur on practically any surface in the vacuum system. This has, e.g., been observed in our low-temperature plasma deposition systems.

4. Conclusions

The surface loss probabilities of different hydrocarbon radicals on the surface of amorphous hydrogenated carbon (C:H) films were determined by depositing C:H films inside a cavity. This surface loss probability corresponds to the sum of the probabilities of effective sticking on the surface and of formation of non-reactive volatile products via surface reactions. By comparing a large number of deposition profiles we obtained for C_2H radicals $\beta = 0.80 \pm 0.05$, for C_2H_3 radicals $\beta = 0.35 \pm 0.15$, and for C_2H_5 radicals β below 10^{-2}. The growth rate of C:H films is, therefore, very sensitive to any contribution of C_2H_x species in the impinging flux.Similar growth precursors may be formed in the divertor plasma of fusion devices, in particular at high plasma density, thus leading to the formation of thick deposits also in regions not in direct line-of-sight to the divertor plasma.

References

1. P. Andrew, D. Brennan, J.P. Coad, J. Ehrenberg, M. Gadeberg, A. Gibson, M. Groth, J. How, O.N. Jarvis, H. Jensen, R. Lässer, F. Marcus, R. Monk, P. Morgan, J. Orchard, A. Peacock, R. Pearce, M. Pick, A. Rossi, B. Schunke, M. Stamp, M. von Hellermann, D.L. Hillis, J. Horgan : Tritium Recycling and recycling in JET, *J. Nuclear Materials* **266-269** (1999), 153-159.
2. G. Federici, R. Anderl, J.N. Brooks, R. Causey, J.P. Coad, D. Cowgill, R. Doerner, A. Haasz, G. Longhurst, S. Luckhardt, D. Mueller, A. Peacock, M. Pick, C.H. Skinner, W. Wampler, K. Wilson, C. Wong, C. Wu, D. Youchison: Tritium inventory in the ITER PFC's: Predictions, uncertainties, R&D status and priority needs, *Fusion Engineering and Design* **39-40** (1998), 445-464.
3. W. Jacob: Surface reactions during growth and erosion of hydrocarbon films, *Thin Solid Films* **326** (1998), 1-42.
4. A. von Keudell, W. Möller: R. Hytry, Deposition of dense hydrocarbon films from a nonbiased microwave plasma, *Applied Physics Letters* **62** (1993), 937-939.
5. A. von Keudell, W. Möller: A combined plasma-surface model for the deposition of C:H films from a methane plasma, *J. Applied Physics* **75** (1994), 7718-7727.
6. A. von Keudell, W. Jacob: Growth and erosion of hydrocarbon films, investigated by in situ ellipsometry; *J. Applied Physics* **79** (1996), 1092-1098.
7. A. von Keudell and W. Jacob: Interaction of hydrogen plasmas with hydrocarbon films, investigated by infrared spectroscopy using an optical cavity substrate, *J. Vacuum Sci. Technol. A* **15** (1997), 402-407.
8. F.P. Bach: Diploma thesis (Univ. of Bayreuth 1996, in German); F.P. Bach, W. Jacob, *to be published.*
9. K. Maruyama, W. Jacob, J. Roth: Erosion behavior of soft amorphous, deuterated carbon films by heat treatment in air and under vacuum, *J. Nuclear Materials* **264** (1999), 56-70.
10. T. Schwarz-Selinger, A. von Keudell, W. Jacob: Plasma chemical vapor deposition of hydrocarbon films: The influence of hydrocarbon source gas on the film properties, *J. Applied Physics* (1999), in print.
11. A. von Keudell, C. Hopf, T. Schwarz-Selinger, W. Jacob: Surface loss probabilities of hydrocarbon radicals on amorphous hydrogenated carbon films: consequences for the formation of re-deposited layers in fusion experiments, *Nuclear Fusion* (1999), in print.
12. C. Hopf, K. Letourneur, W. Jacob, T. Schwarz-Selinger, A. von Keudell: Surface loss probabilities of the dominant neutral precursors for film growth in methane and acetylene discharges, *Applied Physics Letters* **74**, (1999), 3800-3802.
13. C. Hopf, T. Schwarz-Selinger, W. Jacob, A. von Keudell: Surface loss probabilities of hydrocarbon radicals on amorphous hydrogenated carbon film surfaces, to be published.
14. J. Küppers: The hydrogen surface chemistry of carbon as plasma facing material, *Surface Science Reports* **22** (1995), 249-322.
15. A. Annen, W. Jacob: Chemical erosion of amorphous hydrogenated boron films, *Applied Physics Letters* **71** (1997), 1326-1328.

DEVICE FOR INVESTIGATIONS OF TRITIUM RETENTION IN AND PERMEATION THROUGH METALS AND STRUCTURAL MATERIALS

S.K. GRISHECHKIN[1], A.N. GOLUBKOV[1], E.V. GORNOSTAEV[1],
B.S. LEBEDEV[1], R.K. MUSYAEV[1], V.I. PUSTOVOI[1],
YU.I. VINOGRADOV[1], A.A. YUKHIMCHUK[1], A.I. LIVSHITS[2],
A.A. SAMARTSEV[2], M.E. NOTKIN[2], A.A. KURDYUMOV[3],
I.E. GABIS[3]

[1] *Russian Federal Nuclear Center- VNIIEF, Sarov, Nizhny Novgorod Region, Russia*

[2] *Bonch-Bruyevich University of Telecommunications, St. Petersburg, Russia*

[3] *Saint Petersburg State University, Institute of Physics, Peterhof, St. Petersburg, Russia*

Abstract

Apparatus being developed at the Russian Federal Nuclear Center-VNIIEF for research of tritium retention in and permeation through metals and structural materials is described. The apparatus is composed of two combined research complexes involved in the solution of related scientific problems. The first problem is to study tritium superpermeability through metal membranes. The second – is to investigate tritium permeation through and retention by structural materials, to search and analyze protective coatings which increase safety of tritium-bearing gas media application.

Presented are the units of the apparatus, schemes of their interaction and conjugation, technical and physical characteristics of research complexes, as well as possible test conditions.

In the apparatus design, particular attention was given to the radiation safety issues.

1. Introduction

It is common knowledge that the rates of surface reactions are strongly dependent on chemical composition of the metal surface at the level of the monolayer units. Nonmetal impurities result in the occurrence of potential barrier at the phase

C.H. Wu (ed.), Hydrogen Recycling at Plasma Facing Materials, 339–348.

boundary gas-metal. In this case, velocities of interphase processes are suppressed as compared with clean surface, and the penetration rate decreases, if we are dealing with molecular hydrogen in a gas phase [1-5]. These very barriers have a profound impact on the interaction between metal and above-thermal particles of hydrogen isotopes [1,5,6]. In this case the surface films may affect hydrogen penetration and absorption in absolutely different manner [2,3,5,7-15], and superpermeability can be achieved [1,5,6].

The phenomenon of superpermeability was offered to be used for active control over the fuel mix flows in thermonuclear reactions, for fuel removal from helium, the product of a thermonuclear reaction and from other impurities [5,13,16]. For example, it is anticipated that effective pumping of fuel particles out of the divertor area can provide significant improvement in the basic plasma parameters [17-19].

There is one more important aspect in the arrangement of fuel recycling in fusion reactors, namely, vacuum pumping of a spent fuel mix by cryogenic pumps and, thus, tritium accumulation on cryopanels. Removal of the most of tritium ahead of cryopanels might significantly reduce tritium amount in a fuel cycle, increase the service life of cryopanels and result in the gain both in safety and cost [13,18].

On the other hand, there is the risk of radioactivity leak caused by tritium penetration through the walls of tritium-bearing equipment and facilities, and unwanted accumulation of high tritium doses in materials of the facility's parts. In the first case this results in the deterioration of radiation situation, and in the second – causes "necrosis" of large tritium amounts and degradation of structural materials. Development of protective coatings, to ensure safety operation of this equipment, becomes extremely urgent due to the development of controlled – thermonuclear fusion programs where a great quantity of tritium must circulate in the reactor loop. It is also important to prevent significant tritium accumulation in the internal nodes of thermonuclear reactors exposed to the fluxes of high-energy deuterium and tritium particles [15,20].

For practical creation of diffusion membrane pumps, which use the phenomenon of superpermeability, one should have reliable experimental data on behavior of the system "hydrogen isotopes – coating – metal". Most of experimental data, gained by now, is related to hydrogen and, partially, to deuterium. It is known that tritium behavior in chemical reactions and as a solid solution, can not be always reliably predicted by simple multiplying thermodynamic and kinetic parameters by isotopic coefficients. Radioactive tritium decay leading to progressing defectiveness of the lattice and retention of helium may have even stronger effect.

To get direct experimental data on tritium, appropriate techniques and equipment should be developed. Currently this work is being performed at RFNC-VNIIEF with direct involvement of experts from Bonch-Bruyevich University and Saint Petersburg State University.

2. Approach to the device development

When developing the device, the following circumstances were taken into account:

- device must provide pursuance of research in two directions: investigation of superpermeability on cylindrical membranes with relatively large area, and the study and selection of protective coatings of metals and structural materials on flat membranes;
- investigations of superpermeability (Livshits's cell, LC) must allow for the exhaust pump operation in a demo mode with the capacity up to 100 l/s;
- cell for research in protective coating of metals and structural materials (Kurdyumov's cell, KC) must provide for the creation of concentration meander and for the investigation of protective coating under the impact of above-thermal particles;
- the following systems must be common: the leak-in system, vacuum pumping, analysis, tritium recovery, radiation safety, automatic monitoring and control and some other auxiliary systems;
- device design must ensure the safety of personnel and environment.

The list of these requirements determined block diagram of the device (Fig. 1).

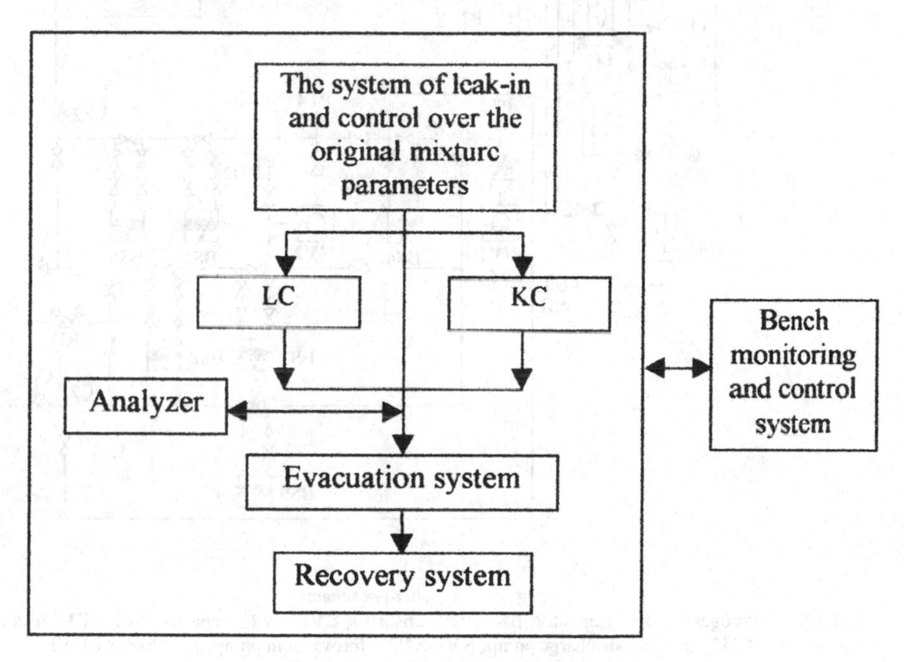

Fig. 1. Device block diagram.

The device is composed of:

- the system of leak-in and original mixture control intended to supply a given amount or a flow of a working gas (mix) into research cells;
- KC for investigation of hydrogen isotopes permeability through structural materials;
- LC for research in hydrogen isotopes superpermeability through metals;

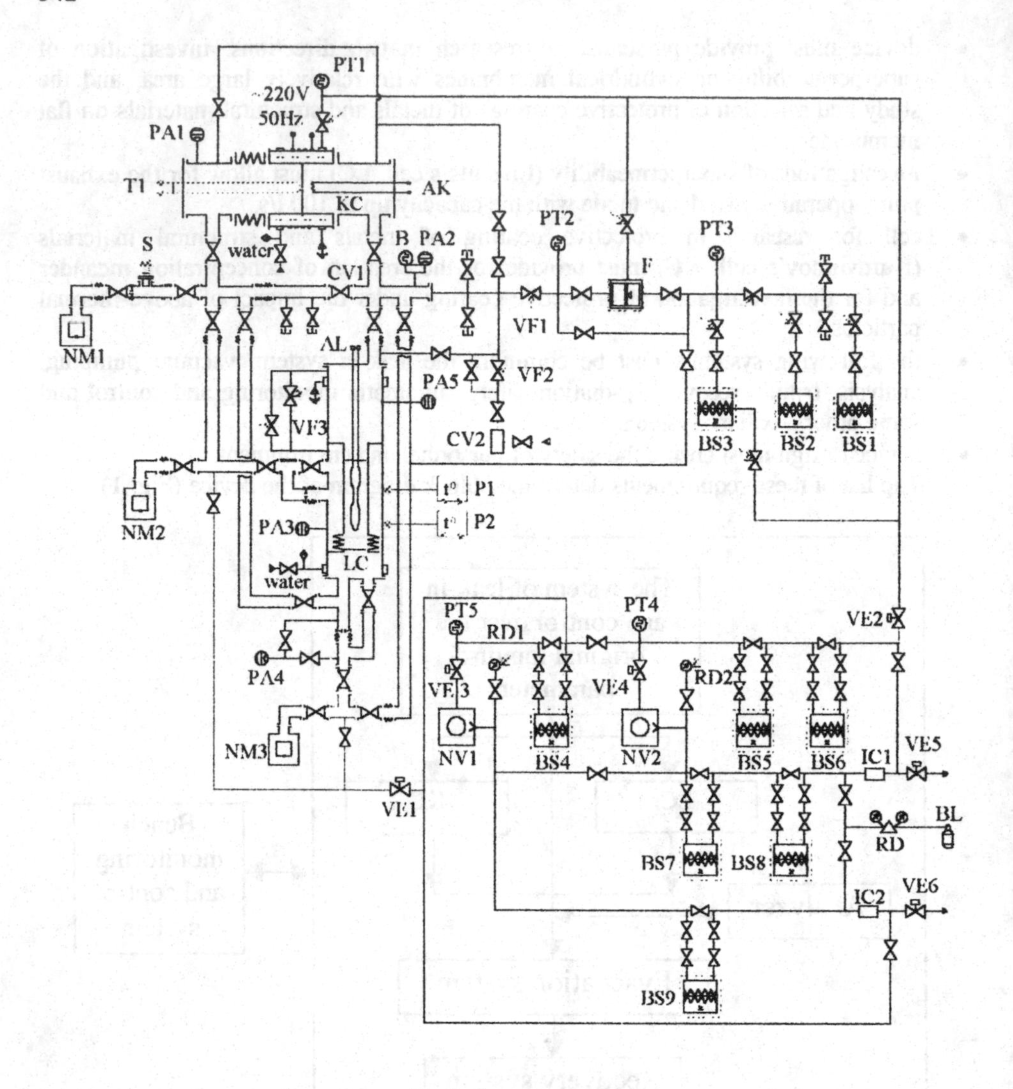

Fig. 2. Gas vacuum scheme:

BS1-BS3 – hydrogen isotopes generator; BS4-BS9 – absorber; CV1, CV2 – capacity; VF1-VF3 – leak valve; NM1-NM3 – magnetodischarge pump; NV1-NV2 – forevacuum pump; F – filter; VE1-VE6 – electromagnetic valve; IC1, IC2 – ionization chamber; AK, AL – atomizers of LC and KC; T1 – thermotransducer; S – mass-analyzer; RD – pressure reducer; BL – container; PA1-PA5, PT1-PT5, RD1-RD2 – vacuummeters; P1, P2 – pyrometers.

- mixture composition analyzer, availability of impurities, so on. When working with KC, the analyzer is used as a diffusing flow recorder;
- vacuum system providing preparation and performance of experiments with research cells (LC, KC);
- recovery system responsible for tritium recovery during experiments;

- the system of the apparatus monitoring and control intended to acquire information on the state of the apparatus nodes and assemblies (temperature, pressure, state of valves), to record and process experimental data, to operate apparatus in automatic mode in an emergency and to block nonstandard actions of the operator.

3. Basic functional elements of the device

The device is a compact, general-purpose apparatus with the elements interfaced by means of gas lines. Fig.2 shows a key scheme of the device gas lines. Design of the apparatus will present a double-level hinged construction with easy access to manual controls and the cells (LC, KC).

3.1. THE CELL FOR RESEARCH OF TRITIUM SUPERPERMEABILITY THROUGH METAL MEMBRANES

The LC, intended for research into hydrogen isotopes superpermeability through metals, shown in fig.3, includes a cylindrical membrane under investigation, atomizer, recirculation system, pressure and temperature meters.

Vertical cylindrical membrane of Ø100×180 mm devides the cell into input and output sides. A flow of hydrogen isotopes is continuously pumped by the pump through the input side (inner cylinder volume).

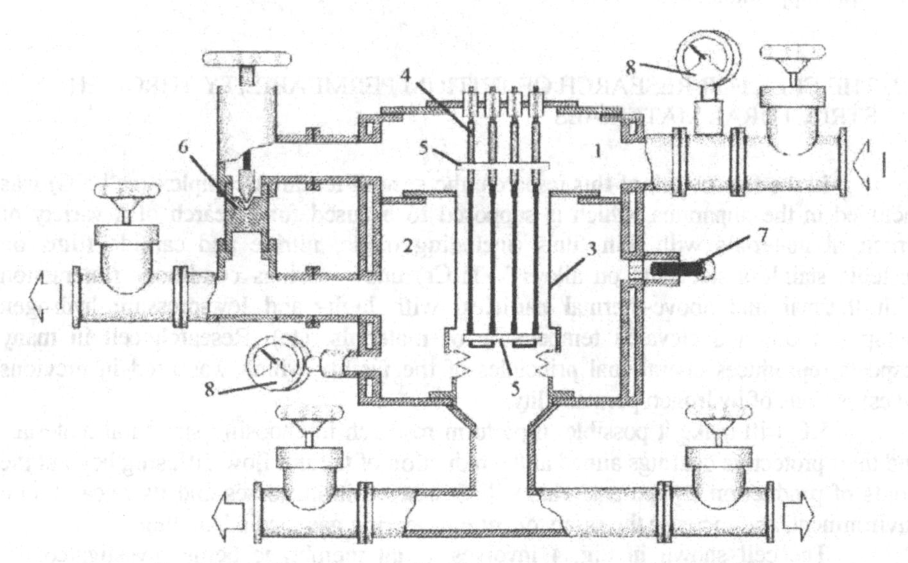

Fig. 3. The cell for research of hydrogen isotopes superpermeability (LC):
1 – input volume; 2 – output volume; 3 – cylindrical membrane; 4 – atomizer; 5 – quartz disks; 6 – leak valve; 7 – pyrometer; 8 – pressure gauge.

Inside the membrane an atomizer is placed which presents a set of tantalum plates arranged through the whole length of membrane. Surface area of the atomizer makes ~20% of the membrane area. The atomizer flange has six lead-in wires rated at the current up to 70 A in stationary operating conditions, and up to 100 A in a short-term technological heating (power supply to 1 and 2 kW, respectively). Operating temperature of atomizer reaches 1800 K, therefore the cell housing has a water cooling jacket.

The ends of cylindrical membranes are covered by quartz disks for shielding structural elements of the cell from the impact of atomic particles. The clearance was chosen so that the pumping rate was not restricted. The output side of the cell (external cylinder side) is continuously evacuated by the pump. It is anticipated that during the experiment at least 70% of the gas pumped through the membrane will arrive at the membrane output side.

Operating pressure at the input side of the cell may be from 2×10^{-7} Pa to 10 Pa, and at the output side it may be between 5×10^{-7} Pa and 10^{5} Pa. Gas pressure is recorded by sensors with appropriate measurement ranges. Membrane temperature is measured at least at two points using pyrometric sensors installed in special windows.

To demonstrate the effect of the pressure growth, on the output side of the membrane a provision is made for parallel pipelines which ensure the constant pumping rate. Gas recirculating system includes the leak valve, closing the input and output sides of the cell. During the demonstration of recirculation, operating gas pumping through the cell and evacuation of the membrane output side are stopped.

The cell LC is similar to the apparatus earlier used in activities on superpermeability. Design of the cell allows rather effective replacement of membrane and working elements of the atomizer without disassembling of the cell pipelines mated with other apparatus units.

3.2. THE CELL FOR RESEARCH OF TRITIUM PERMEABILITY THROUGH STRUCTURAL MATERIALS

In the framework of this research, the second research complex (cell KC) was included in the apparatus, which is supposed to be used for research of a variety of structural materials with thin films (including oxide, nitride and carbide films on austenic stainless steel and on alloys V-Ti-Cr) under various conditions (interaction with thermal and above-thermal particles, with high- and low-pressure hydrogen isotopes, room and elevated temperature of materials, etc). Research cell in many respects reproduces operational principles of the facility which was used in previous investigations of hydrogen permeability.

KC will make it possible to perform research in choosing structural materials and their protective coatings aimed at the reduction of tritium flow diffusing beyond the limits of production equipment. This will decrease tritium losses and its releases into environment, and increase the safety of tritium-bearing gas media handling.

The cell shown in Fig. 4 involves a flat membrane being investigated, its heater, atomizer, vacuum and temperature meters.

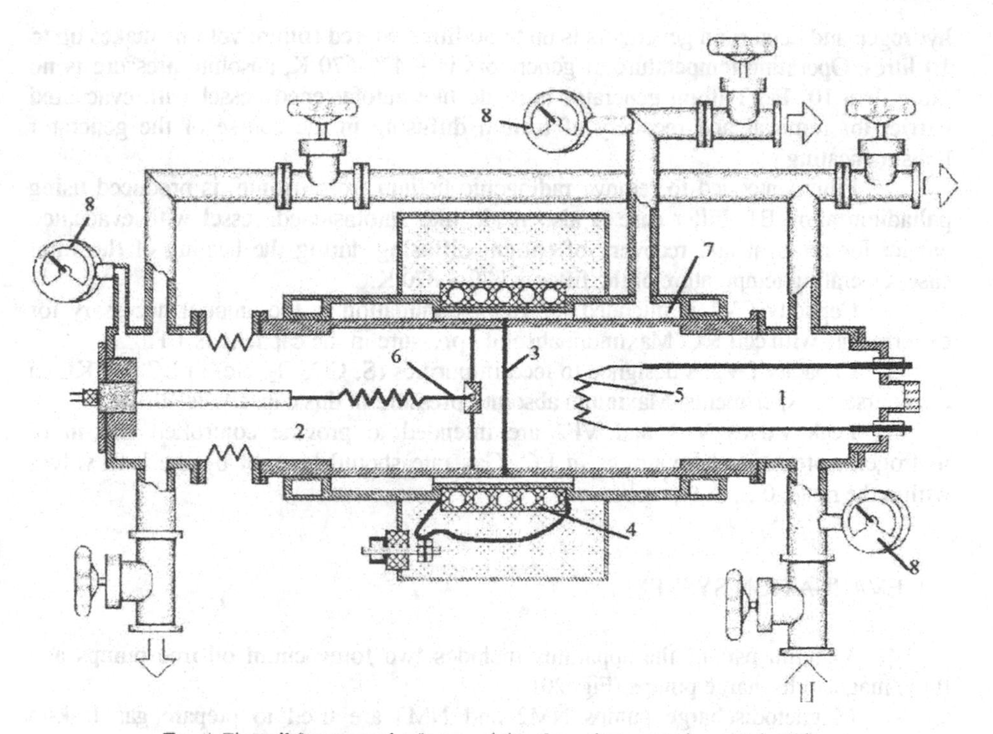

Fig. 4. The cell for research of permeability through structural materials (KC):
1 – input volume; 2 – output volume; 3 – flat membrane; 4 – membrane heater; 5 – atomizer; 7 – cavity for diffusing gas collection; 8 pressure gauge.

The cell is devided by the membrane into input and output sides. Diameter of the investigated membrane reaches 50 mm, electrical heater provides its warm-up to 1000 K. The warm-up area of the cell housing has a cavity for diffusing tritium collection that is continuously evacuated during the experiment. Operating pressure of hydrogen isotopes within the range 2×10^{-7}...1 Pa is supplied at the input side. Provision is made for the meters of gas pressure and the temperature of the membrane under investigation. Output side of the membrane is continuously evacuated by the pump. Penetrating flow of hydrogen isotopes is measured with a mass-spectrometer calibrated using pressure gauges. The possibility is envisaged for hydrogen isotope molecules transfer in atomic state, for this purpose an atomizer (filament) of at least 1 W power is placed inside the input side of the cell.

3.3. ORIGINAL MIXTURE LEAK-IN AND CONTROL SYSTEM

The system of original mixture leak-in and control includes three hydrogen isotopes generators (protium BS1, deuterium BS2 and tritium BS3), filter F, two cavities CV1, CV2, leak valves VF1, VF2 and pressure gauges (Fig.2). In these generators U^{238} hydride is used as a generating composition. Gas volume stored in hydrogen and deuterium generators is up to 50 litres. Stored tritium volume makes up to 10 litres. Operating temperature of generators is ~ 470-670 K, absolute pressure is no more than 10^4 Pa. Tritium generator is made like autofastened vessel with evacuated barrier for removal and recovery of tritium diffusing in the course of the generator housing heating.

Filter, intended to remove radiogenic helium from tritium, is produced using palladium alloy B1. Filter case is also made like autofastened vessel with evacuated barrier for removal and recovery of tritium diffusing during the heating of the filter case. Operating temperature of the filter is 770 – 870 K.

Capacity CV1 is intended for gas accumulation in the amount necessary for experiments with cell KC. Maximum absolute pressure in the capacity is 1 Pa.

Capacity CV2 is designed to feed impurities (S, CO, O_2, etc) in LC and KC in the course of experiments. Maximum absolute pressure in this capacity is 10^5 Pa.

Leak valves VF1 and VF2 are intended to provide controlled leak-in of hydrogen isotopes and impurities in LC. Gas rate should be kept by the leak valves within the range 0 ...1×10^{-2} m^3/s.

3.4. EVACUATION SYSTEM

Vacuum part of the apparatus includes two forevacuum oil-free pumps and three magnetodischarge pumps (Fig. 20).

Magnetodischarge pumps NM2 and NM3 are used to prepare gas leak-in system, the cells LC and KC, for pumping through the cell LC and for pumping out of the membrane output side. Productivity of pumps must be at least 200 l/s, operating pressure - 2×10^{-5} Pa.

Forevacuum pumps NV1 and NV2 should ensure normal activation of magnetodischarge pumps. Productivity of the pump NV1 must be no less than 0,8 l/s, and that of the pump NV2 – at least 5 l/s.

Magnetodischarge pump NM1 is employed to evacuate the output side of the KC membrane during experiment. Its output should be no less than 100 l/s, operating pressure - 2×10^{-5} Pa.

Vacuum at the input of forevacuum pumps is measured by pressure gauges with a working range 0,13 – 13 Pa.

3.5. AUTOMATED MONITORING AND CONTROL SYSTEM

Monitoring and control system of the apparatus is based on module equipment with standard PC LAB-cards of analog output and a 6-channel controller MaxiGauge. Personal computer of IBM Pentium type is used as a control computer.

The system will provide temperature and pressure measurements, recording and processing of the mass-spectrometer measuring channel, monitoring of electric furnaces, electromagnetic valves, control over ionization chambers, graphic image of the apparatus state on mimic diagram, storage of recorder parameters in the data base, the provision of work with data by using graphic interface.

The system will also provide for automatic switching of the gas flow being pumped through from the main to a backup purification line (according to the readings of the ionization chamber IC1), operation of blockings envisaged for increasing operational safety of hydrogen isotopes generators, filter, LC and KC cells.

4. Tritium safety ensuring

Recovery system is designed to remove tritium from the gas released in the atmosphere up to the level of 1×10^{-6} Ci/l.

The system involves two stages of tritium removal from the gas and is composed of the main and backup lines (Fig.2). The first stage of the main line includes two Ti-base absorbers BS5 and BS6 and is located in front of the forevacuum pump NV2 input. The second stage of the main line incorporates two absorbers BS7 and BS8 with intermetal compound Zr $(V_{0,8}Cr_{0,2})_2$, it is placed behind the exhaust of the forevacuum pump NV2. During the operation both absorbers are continuously functioning at ~200°C.

Gas is injected in the atmosphere through the ionization chamber IC1. If a signal in the chamber IC1 exceeds a specified level (1×10^{-6} Ci/l), electromagnetic valve VE5 shuts off gas release in the atmosphere by means of electromagnetic valves VE1...VE4, gas purification is switched to a backup line of absorbers BS4 and BS9 with gas leak-in in the atmosphere through the ionizing chamber IC2. As this takes place, mixed gas in the main line is manually subjected to recycling purification. Absorber BS4 is made on the basis of Ti and is located ahead of the input of forevacuum pump NV1, absorber BS9 is made with an intermetal compound based on $Zr(V_{0,8}Cr_{0,2})_2$ and is placed behind the pump NV1 exhaust. Bottle post with the container BL of an inert gas is designed to blow through ionization chambers.

In the operation of tritium generator, permeator and the cell KC, provision is made for continuous evacuation of the housing cavities with collection of diffused tritium. Evacuation degree is measured by a pressure gauge with an operating range between 0,13 – 13 Pa.

5. Conclusion

These are the tasks to be accomplished during the work performance:

- Research in thin nonmetal films impact on hydrogen and tritium retention and permeation in film-metal systems;
- Creation of a data base on parameters of hydrogen isotopes interaction with thin-film coatings and structural materials;
- Development and creation of mathematical models and computer programs to calculate hydrogen isotopes retention and permeation in multilayer systems metal-coating;

- Elaboration of recommendations on the use of superpermeable membranes in nuclear fusion and hydrogen power programs;
- Elaboration of recommendations on the use of protective coatings against hydrogen isotopes retention and leaking in structural materials.

This work is supporting by ISTC (project #1110).

References

1. A. I. Livshits, M. E. Notkin, Yu. M. Pustovoit and A. A. Samartsev, *Vacuum* 29 (1979) 103; 113.
2. T. N. Kompaniets and A. A. Kurdyumov, *Progr.Surf.Sci.* 17 (1984) 75.
3. W. R. Wampler, *Appl.Phys.Lett.* 18 (1986) 405.
4. M. Yamawaki, T. Namba, K. Yamaguchi and T. Kiyoshi, *Nucl.Instr.Meth.* B23 (1987) 498.
5. A. I. Livshits, M. E. Notkin, and A. A. Samartsev, *J.Nucl.Mater.* 170 (1990) 74.
6. A. I. Livshits, *Sov.Phys.Tech.Phys.* 20 (1975) 1207; 21 (1976) 187.
7. B. L. Doyle and D. K. Brice, *J.Nucl.Mater.* 145 147 (1982) 288.
8. S. M. Myers, P. M. Richards, W. R. Wampler and F. Besenbacher, *J.Nucl.Mater.* 165 (1989) 9.
9. M. A. Pick and K. Sonnenberg, *J.Nucl.Mater.* 131 (1985) 208.
10. T. Tanabe, *Fus.Eng.Design* 10 (1989) 325.
11. A. I. Livshits, M. E. Notkin, A. A. Samartsev and I. P. Grigoriadi, *J.Nucl.Mater.* 178 (1991) 1.
12. P. L. Andrew and A. Haasz, *J.Less-Comm.Met.* 172–174 (1991) 732.
13. A. I. Livshits, M. E. Notkin, A. A. Samartsev, A. O Busnyuk, A. Yu. Doroshin and V. I. Pistunovich, *J.Nucl.Mater.* 196-198 (1992) 159.
14. M. Yamawaki, N. Chitose, V. Bandurko and K. Yamaguchi, *Fus.Eng.Design* 28 (1995) 125.
15. W. M. Shu, Y. Hayashi and K. Okuno, *J.Nucl.Mater.* 220–222 (1995) 497.
16. M. Sugihara and T. Abe, *Nucl.Fusion* 21 (1981) 1024.
17. N. Ohyabu, A. Komori, K. Akaishi, N. Inoue, Y. Kubota, A. I. Livshits, N. Noda, A. Sagara, H. Suzuki, T. Watanabe, O. Motojima, M. Fujiwara and A. Iiyoshi, *J.Nucl.Mater.* 220-222 (1995) 298.
18. A. I. Livshits, M. E. Notkin, V. I. Pistunovich, M. Bacal and A. O. Busnyuk, *J.Nucl.Mater.* 220-222 (1995) 259.
19. G. Janeschitz, K. Borrass, G. Federici, Y. Igitkhanov, A. Kukushkin, H. D. Pacher, G. W. Pacher and M. Sugihara, *J.Nucl.Mater.* 220–222 (1995) 73.
20. Fusion Reactor Materials VII (Part I and II), *Proceedings of the Seventh International Conference on Fusion Reactor Materials (ICFRM-7), Obninsk, Russia, 25 to 29 September 1995, J.Nucl.Mater.* (1996) v. 233-237.

THERMOCYCLING IN GASEOUS HYDROGEN AS A WAY OF THE HYDROGEN DEGRADATION ACCELERATION

B. P. LONIUK
Karpenko Physico-Mechanical Institute of the NAS of Ukraine
5 Naukova St, 290601 Lviv, Ukraine
E-mail: nykyfor@ah.ipm.lviv.ua

1. Introduction

In last decades, much attention has been paid to the material damage without significant residual deformation. The expert conclusions for some strainless fracture of power plant components regard them as negative effect of hydrogen [1]. The known methods of workability evaluation of the power plant materials allow to predict with comparatively high accuracy the life time of steam pipe line under the creep conditions [2]. However, it does not enable to predict the brittle fractures, which are caused by the metal hydrogenating in the service conditions. At the same time hydrogen formation during the contact of stressed metal with corrosive environment and its penetration into bulk metal causes number undesirable effects. At high temperatures typical for power plant operation, first of all, there is a problem of degradation of the microstructure and properties of the virgin material.

This paper continues the previous investigation [3, 4]. The aim of this work is to validate the use of thermocycling in hydrogen environment as an efficient way of in-laboratory simulation of steel microstructural degradation, occurring in long-term operation conditions. Based on this the approaches of fatigue fracture mechanics should be applied for evaluation of the level of material damage. As a result, the approach, for the life time prediction for steam pipe line metal should be developed.

2. Experimental procedure

The researches are carried out on steam pipeline 12Kh1MF steel (0.1C, 1.1 Cr, 0.26 Mo, 0.17V, 0.26 Si, 0.54 Mn) widely used in power plant industries. The tensile properties of the steel in virgin state are: σ_{UTS} = 470 MPa, σ_{YS} = 280 MPa, ε = 29%, ψ = 75%.

Material workability was evaluated by using the linear fracture mechanics approach. Fatigue crack growth tests were carried out in air at the ambient temperature. Beam specimens with the cross-section of 6 mm x 22 mm with an edge crack were subjected to a cyclic bend loading with a frequency of 10 Hz. The threshold parameters of the fatigue crack growth were analysed.

C.H. Wu (ed.), Hydrogen Recycling at Plasma Facing Materials, 349–354.

The method of the accelerated ageing of steels was developed. Its principal point is in thermocycling of specimens in a hydrogen environment. The method is based on the facts that equilibrium concentrations of hydrogen, which is dissolved in metal are different for different temperatures (i.e. higher for solubility corresponds to higher temperature) and the reduction of the hydrogen mobility because of the temperature decrease.

On Fig. 1 the scheme of implementation of this ageing method is shown. The 280x22x15 mm beam specimen was placed in a hermetic chamber, which was filled with hydrogen of pressure 0.3 MPa. The sample was heated up by electric current to the operation temperature 570°C at a rate of 2°C/s. The specimen was being kept under such conditions during 1 hour. This time is enough to achieve the equilibrium hydrogen concentration at this temperature inside the specimen. After that the specimen was exposed to the sharp decrease of temperature down to 100°C at a rate of 3°C/s, that was provided by cooling of the current lead blocks with passage water. Since in proposed ageing method the specimen was not restrain in grabs, the change of its dimensions during the heating and cooling did not cause any thermal stresses. Moreover, the distance between the current lead blocks few times exceeded the specimen width. Thus, the thermocycling itself does not cause any essential internal stresses in the unrestrained specimen. However, the sharp temperature drop causes the retardation of diffusion processes (hydrogen diffusion factors at 100 and 570°C are equal $4.4 \cdot 10^{-5}$ cm^2/s and $2.0 \cdot 10^{-4}$ cm^2/s respectively) and essential decrease of the hydrogen equilibrium solubility in the metal (0.1g/cm^3 and 1.0 g/cm^3 respectively). The proposed treatment enables to keep at 100°C the equilibrium hydrogen concentration in metal body, which corresponds to the temperature of 570°C. Thus the sharp cooling of hydrogenated metal may cause an increase of internal stresses in the metal because of excessive equilibrium fraction of hydrogen. When such a cycle is repeated many times, it may cause initiation and growth of microdefects, that essentially reduces its serviceability. It was confirmed concerning to the specimen with a stress concentrator. After 100 thermocycles in hydrogen exposed a net of cracks appeared from the notch top towards the specimen, which were maximum 0.5 mm long. This indicates that a critical stress-strain state was achieved at a notch tip despite of absence of external mechanical load.

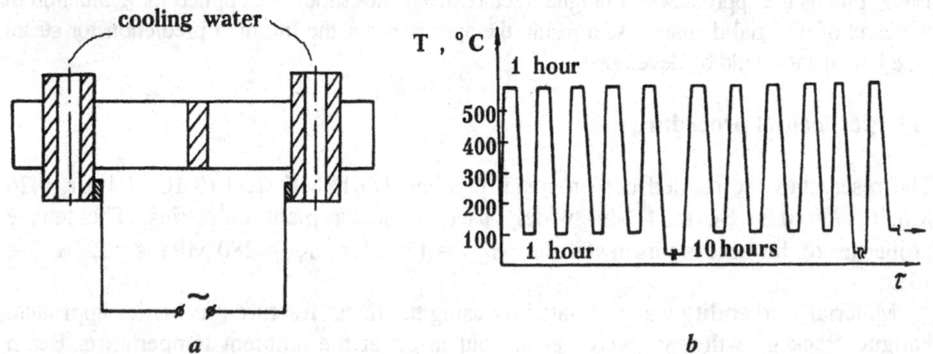

Figure 1. Schemes of heating and cooling of specimens (a) and its temperature changes (b).

After the thermocycling in air the cracks were not observe. To initiate a crack by thermocycling nearly 10^3 cycles would be needed [5]. This allows to assume, that the fixed cracking is a not result of normal thermocycling but is caused by the specimen self-loading due to the cyclic tension of the crystalline metal lattice by the excessive hydrogen or/and fluctuation of hydrogen pressure in the defects.

3. Result and discussion

3.1. MICROSTRUCTURE

The microstructure of the virgin 12Kh1MF steel is a mixture of the ferrite and pearlite grains with average-size of a ferrite grains did not exceed 40 μm (Fig. 2a). The thermocycling in the hydrogen environment allows transform it very quickly. These microstructural changes are follows: decomposition of pearlite, growth of ferrite grains and forming of some carbides at grain boundaries (their quantity and sizes are growing during the ageing). After 100 thermocycles in hydrogen environment the microstructure consists of almost equiaxial ferrite grains with a lot of fine globular carbide particles inside and long penny-shaped carbides at the grain boundaries (Fig. 2b). Similar changes of the microstructure were observed in the service matcrial after 20 years [3, 4].

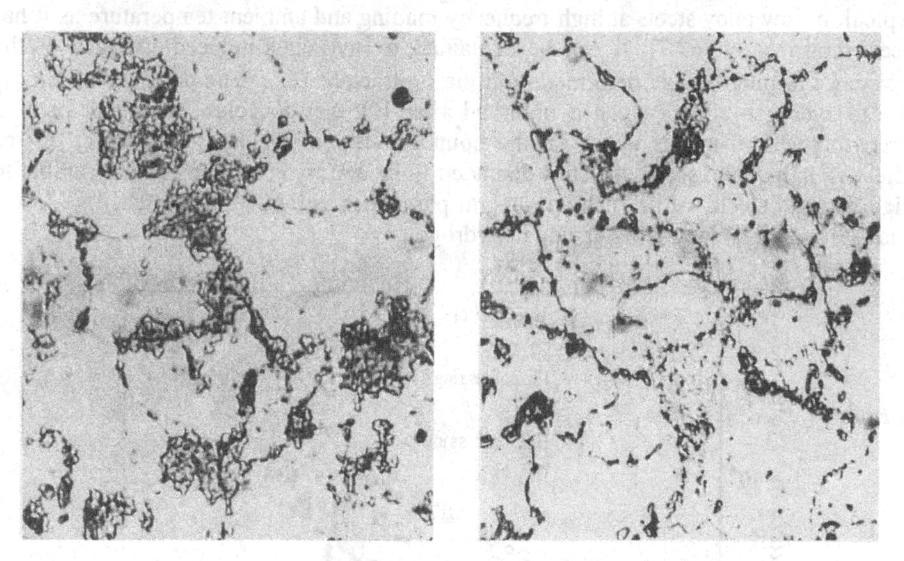

Figure 2. The microstructure of the 12Kh1MF steel in virgin state (a) and after 100 thermocycling in hydrogen (b).

3.2. CRACK GROWTH RESISTANCE

In Fig. 3 the fatigue crack growth rates for this steel after different numbers of thermocycles are presented. The kinetic diagrams da/dN - ΔK and da/dN - ΔK_{eff} for the specimens tested just after the thermocycling and outgassing are showed. da/dN - ΔK curves do not enable to relate the degradation rate during the ageing, with the number cycles n, because there is no direct correlation between ΔK_{th} and n. Only the da/dN - ΔK_{eff} dependencies, which take into account the crack closure effect, indicate the decrease of effective fatigue thresholds $\Delta K_{th\ eff}$ with the increase of number of thermocycles. Comparison of these results and results presented in works [3, 4] for materials after different time of operation indicates the similar changes of the threshold characteristics of crack growth. Only $\Delta K_{th\ eff}$ decreases with the increase of operation time.

The analysis of experimental results confirms conclusions made by [3, 4] about the prospects for using the fatigue fracture mechanics approaches in evaluation of the degree of high-temperature degradation of structural alloys. Obviously, it is that parameter to be the most sensitive to material degradation, which provides the most local fracture. In this connection the threshold characteristics of fatigue crack growth are the most preferable. Thus only $\Delta K_{th\ eff}$ can unambiguously characterise the degree of material degradation since ΔK_{th} is the integral characteristic of the fatigue crack growth resistance and the crack closure.

The data presented in Fig. 3 allow to compare $\Delta K_{th\ eff}$ of the hydrogenated and outgassed specimens at different values of n. The results show an ambiguous effect of dissolved hydrogen depending on n: it increases $\Delta K_{th\ eff}$ level for 13 thermocycles and decreases one for 100. The increase of $\Delta K_{th\ eff}$ due to hydrogen after 13 thermocycles is typical for low alloy steels at high frequency loading and ambient temperature as it has been shown early [6, 7]. It can be explained by hydrogen-induced increase of the resistance to microplastic deformation in the prefracture zone. The decrease of $\Delta K_{th\ eff}$ due to hydrogen effect, which is observed after 100 thermocycles, is typical for high strength steels. It means that from the point of view of the hydrogen effect on the effective fatigue threshold, the steel degraded in laboratory conditions is comparable to high strength steels, while such an integral parameter as hardness is insensitive to the changes caused by laboratory ageing in hydrogen.

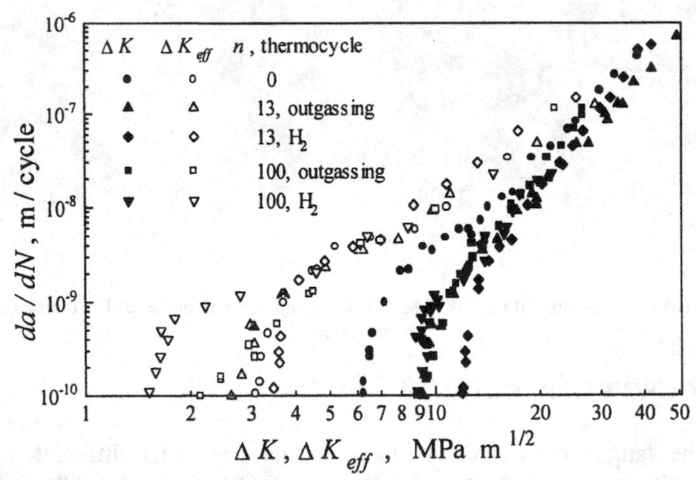

Figure 3. Fatigue crack growth rates for the 12Kh1MF steel after the thermocycling in hydrogen.

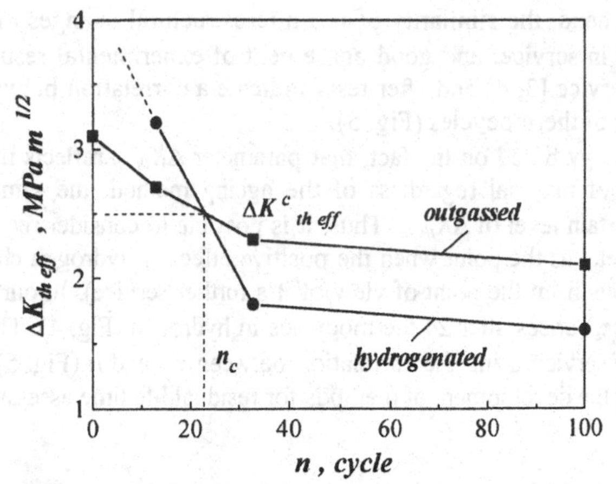

Figure 4. Dependencies $\Delta K_{th\,eff}$ - n for steel tested directly after thermocycling in hydrogen and after additional outgassing.

Thermocycling in hydrogen intensify the transformations of the microstructure of the material and promote the negative influence of internal hydrogen on the resistance of subthreshold fatigue crack growth. This means that, from the point of view of the hydrogen effect on the effective fatigue threshold, the steel degraded in laboratory conditions is comparable to high strength steels. This observation should be considered as a warning because even integrally insignificant concentrations of hydrogen in metal after long-term operation may provoke the formation of local regions of supersaturation in the stressed metal and, hence, led to a considerable decreasc in the actual fatigue threshold.

The analysis of obtained results enable to propose the $\Delta K_{th\,eff}$ parameter as indicator of the material state. The found inversion of hydrogen effect on $\Delta K_{th\,eff}$ in dependence on cycles number (Fig. 4) can be useful for determining the limiting state of metal, when it becomes sensitive to hydrogen embrittlement on the local scale.

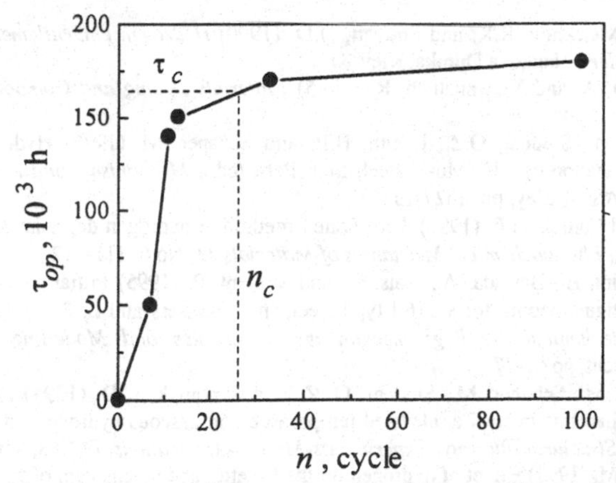

Figure 5. Correlation between time operation τ_{op} and number of thermocycling n in hydrogen environment.

On the other hand, the similarity of the microstructural changes after thermocycles in hydrogen and in service, and good agreement of experimental results on $\Delta K_{th\ eff}$ for materials after service [3, 4] and after tests indicate a correlation between the operation time and number of thermocycles (Fig. 5).

This methodology based on the fact, that parameter $\Delta K_{th\ eff}$ reflects unambiguously the mechanical state of material regardless of the ageing method: the same level of ageing corresponds to certain level of $\Delta K_{th\ eff}$. Thus, it is possible to consider reaching of structural degradation of metal as the point when the positive effect of hydrogen changes to negative one (it is dangerous from the point of view of it's further service). In our case such critical state of material (τ_c) arises after 24 thermocycles in hydrogen (Fig. 4). That corresponds to $16.5 \cdot 10^4$ hours of service using the correlation between τ_{op} and n (Fig. 5). This correlation can be helpful for the development of methods for residual life time assessment of structures.

4. Conclusions

The themocycling of the material containing dissolved hydrogen is proposed as express method of high temperature degradation of pipe line steel.

The effective fatigue threshold of $\Delta K_{th\ eff}$, which accounts crack closure phenomenon, can be considered as the mechanical parameter sensitived to metal degradation.

Inversion of the effect of hydrogen dissolved in the metal on the $\Delta K_{th\ eff}$ depending on number of thermocycles in hydrogen, which determines the metal degradation degree, has been found.

Criterion of the limiting state of metal due its high temperature degradation is formulated. The critical state of metal is reached, when the absorbed by metal hydrogen starts to affect negatively the effective fatigue threshold $\Delta K_{th\ eff}$.

An approach to evaluation of life time of steam pipe line metal is proposed. It enables to take into account the negative effect of hydrogen.

References

1. Vainman, A.B., Melekhov, R.K. and Smiyan, O.D. (1990) *Hydrogen embrittlement of components of high pressure boilers,* Naukova Dumka, Kiev.
2. Bicego, V., Nitta, A. and Viswanathan, R. (1995) *Materials Ageing and Component Life Extencion,* EMAS, Warley.
3. Nykyforchyn, H.M., Student, O.Z., Loniuk B.P. and Schaper, M. (1996) Hydrogen effect on high temperature degradation of a Cr-Mo-V steel, in J. Petit (ed.), *Mechanisms and Mechanics of Damage and Failure,* EMAS, Warley, pp. 1527-1532.
4. Student, O.Z. and Loniuk, B.P. (1997) Accelerated method of hydrogen degradation of structural steels by thermocycling, *Physicochemical Mechanics of Materials* **33**, No 6, 111-112.
5. Fissolo, A., Marini, B., Berrada, A., Nais, G. and Wident, P. (1995) Initiation and growth of cracks under thermal fatigue loading for a 316 l type steel, in J. Bressers and L. Rémy (eds.), *Fatigue under Thermal and Mechanical Loading: mechanisms, Mechanics and Modelling,* Kluwer Academic Publishers, Dordreht, pp.67-77.
6. Nykyforchyn, H. M., Schaper, M., Student, O. Z. and Skrypnyk, I. D. (1994) Fatigue crack growth kinetics and mechanisms in steel at elevated temperatures in gaseous hydrogen, in K-H. Schwable and C. Berger (eds.), *Structural Integrity: Experiments-Models-Applications,* EMAS, Warley, pp. 309-314.
7. Nykyforchyn, H.M. (1997) Effect of hydrogen on the kinetics and mechanism of fatigue crack growth in structural steels, *Physicochemical Mechanics of Materials,* **33**, No 4, 97-106.

INDEX

Participants

Alimov, V.Kh.	Bonch-Bruyevich University Surface Physics & Electronics Center 61 Moika 191186 St.Petersburg, Russia
Arkhipov, I.	Russian Academy of Science 31 Leninsky Prospect 117915 Moscow, Russia
Bacal, M.	Laboratoire de Physique des Milieux Ionisé, Ecole Polytechnique, 91128 Palaiseau, France
Baksht, F.	Ioffe Physico-Technical Institute 26 Politekhnicheskaja 194021 St. Petersburg, Russia
Begrambekov, L. B.	Moscow State Engineering & Physics Institute 31 Kashirskoe sh., 115409 Moscow, Russia
Bondarev, B.	Bonch-Bruyevich University, Surface and Electronics Center 31 Moika 191186 St. Petersburg, Russia
Braithwaite, N.St.J.	The Open University, Oxford Research Unit Boars Hill Oxford, 0X1 5HR, UK
Burtseva, T.	Efremov Scientific Research Institute of Electrophysical Apparatus 194021 St. Petersburg, Russia
Busnyuk, A.	Nuclear Engineeering Research Laboratory University of Tokyo, 2-22 Shirakata-Shirane City Tokai-mura, Ibaraki-ken, 319-1106, Japan
Denisov, E.A.	Scientific Research Institute of Physics St. Petersburg University Ulianivskaja 1 198904 St. Petersburg, Russia
Dobrotvorsky, A.	VNIINeftekhim, 196084, Zaozernaya 1 St. Petersburg, Russia
Doroshin, A.	Bonch-Bruyevich University Surface Physics and Electronics Center 61 Moika 191186 St. Petersburg, Russia

Emmoth, B. Alfven Laboratory, Royal Institute of Technology
10044 Stockholm, Sweden

Evard, E. Institute of Physics
State University of St. Petersburg
1 Ulianivskaja
198904 St. Petersburg, Russia

Gabis, I.E. Institute of Physics
St. Petersburg University
Ulianivskaja 1
198904 St. Petersburg, Russia

Glazanov, D. Bonch-Bruyevich University
Surface Physics and Electronics Center
61 Moika
191186 St. Petersburg, Russia

Grigoriadi, I. Bonch-Bruyevich University
Surface Physics and Electronics Center
61 Moika
191186 St. Petersburg, Russia

Grisolia, Ch. CEA Cadarache, DRFC/STEP bat 513
13108 St. Paul lez Durance, France

Guseva, M.I. RRC Kurchatov Institute
1 Kurchatov Square
123182 Moscow, Russia

Hassanein, A. Argonne National Laboratory
Bld. 2059700 South Cass Avenue
Argonne, IL 60439, USA

Hayashi, T. Tritium Engineering Laboratory, JAERI,
2-4 Shirakata-Shirane
Tokai-mura, Naka-gun, Ibaraki-ken,
319-1195, Japan

Jacob, W. Max-Planck-Institut für Plasmaphysik
Boltzmannstr. 2
85748 Garching, Germany

Klepikov, A.Kh. Science Research Institute of Experimental and Theoretical Physics
Kazakh State University
96a Tole bi Str.
480012 Almaty, Kazakhstan

Kompaniets, T.	St. Petersburg State University Institute of Physics 1 Ulianivskaja St. 198904 St. Petersburg, Russia
Kurdyumov, A.	St. Petersburg State University Institute of Physics 1 Ulianivskaja St. 198904 St. Petersburg, Russia
Kurnaev, V.A.	Moscow State Engineering and Physics Institute Kashirskoe sh. 31 115409 Moscow, Russia
Lapidus, O.	International Science and Technology Center Moscow, Russia
Livshits, A.I.	Bonch-Bruyevich University Surface Physics and Electronics Center 61 Moika 191186 St. Petersburg, Russia
Loniuk, B.P.	Karpenko Physico-Mechanical Institute of the NAS of Ukraine 5 Naukova St. 290601 Lviv, Ukraine
Mayer, M.	Forschungszentrum Jülich GmbH Institut für Plasmaphysik, EURATOM Ass., 52425 Jülich, Germany
Mioduszewski, P.	Oak Ridge National Laboratory Fusion Energy Division, P.O. Box 2009 Oak Ridge, Tennessee 37716, USA
Morita, K.	Graduate School of Engineering Department of Crystalline Materials Furo-cho, Chikusa-ku, Nagoya 464-8603, Japan
Nakahara, Y.	National Institute für Fusion Science 322-6 Oroshi-cho Toki, Gifu-ken 509-5292, Japan
Nakamura, Y.	National Institute for Fusion Science 322-6 Oroshi-cho Toki, Gifu-ken 509-5292, Japan
Noda, N.	National Institute for Fusion Science 322-6 Oroshi-cho Toki, Gifu-ken 509-5292, Japan

Notkin, M.E.	Bonch-Bruyevich University of Telecommunications 61 Moika, St. Petersburg 191186, Russia
Pereslavtsev, A.V.	RRC Kurchatov Institute 1 Kurchatov Square 123182 Moscow, Russia
Pisarev, A.A.	Moscow State Engineering and Physics Institute Kashirskoe sh. 31 Moscow 115409, Russia
Pistunovich, V.I.	RRC Kurchatov Institute 1 Kurchatov Square 123182 Moscow, Russia
Polosukhin, B.G.	Sverdlovsk branch of Research and Development Institute of Power Engineering Sverdlovsk Region Zarechny 624051, Russia
Samartsev, A.	Bonch-Bruyevich University, Surface Physics and Electronics Center 61 Moika 191186 St. Petersburg, Russia
Scarin, P.	Consorzio RFX 4 Corso Stati Uniti 35127 Padova, Italy
Shmal'ko, Yu.F.	Podgorny Institute of National Academy of Sciences of Ukraine 2/10 Pozharsky St. 310046 Kharkov, Ukraine
Skovoroda, A.A.	INF RRC Kurchatov Institute 1 Kurchatov Square 123182 Moscow, Russia
Sugisaki, M.	Department of Advanced Energy Engineering Science, Interdisciplinary Graduate School of Engineering Sciences Kyushu University Hakozaki, Fukuoka, 812-8581, Japan
Takamura, S.	Department of Energy Engineering and Science Graduate School of Engineering Nagoya University Nagoya 464-8603, Japan
Tanabe, T.	Center for Integrated Research in Science and Engineering, Nagoya University Nagoya 464-8603, Japan

Vainonen-Ahlgren, E.
Accelerator Laboratory
P.O. Box 43 University of Helsinki
00014 Helsinki, Finland

Venhaus, Th. J.
Los Alamos National Laboratory ESA-TSE
P.O.Box 1663
Los Alamos, NM 87545, USA

Vietzke, E.
Institut für Plasmaphysik, Forschungszentrum
Jülich GmbH, Euratom Association, Trilateral
Eurigio Cluster,
52425 Jülich, Germany

Voitsenya, V.S.
IPP NSC KIPT
310108 Kharkov, Ukraine

Wu, C.H.
NET Team Max-Planck-Institut für Plasmaphysik
Boltzmannstr. 2
85748 Garching, Germany

Xiao, B.
Department of Quantum Engineering and Systems
Sciences, Faculty of Engineering, Tokyo University
NERL, 7-3-1 Hongo,
Bunkyo-ku City, Tokyo, Japan

Yamaguchi, K.
Nuclear Engineering Research Laboratory
University of Tokyo, 2-22 Shirakata-Shirane
Tokai-mura, Ibaraki-ken, 319-1106, Japan

Yamawaki, M.
Graduate School of Quantum Engineering and
Systems Science, School of Engineering,
University of Tokyo 7-3-1 Hongo, Bunkyo-ku,
Tokyo 113-8656, Japan

Yukhimchuk, A.A.
Russian Federal Nuclear Center VNIIEF
37 Mira Ave. Novgorod Region,
Sarov 607190, Russia

Zakharov, A.
Russian Academy of Science, Institute of Physical
Chemistry, 31 Leninsky Prospekt
117915 Moscow, Russia